The ASQ Certified Pharmaceutical GMP Professional Handbook

The ASQ Certified Pharmaceutical GMP Professional Handbook

Third Edition

Mark Allen Durivage, Editor

ASQExcellence
Milwaukee, Wisconsin

Published by ASQExcellence, Milwaukee, WI
Produced and distributed by Quality Press, ASQ, Milwaukee, WI

Publisher's Cataloging-in-Publication data

Names: Durivage, Mark Allen, editor.
Title: The ASQ certified pharmaceutical GMP professional handbook , third edition / Mark Allen Durivage, Editor
Description: Includes bibliographical references and index. | Milwaukee, WI: ASQExcellence, 2024.
Identifiers: LCCN: 2024942831 | ISBN: 9781636941493 (hardcover) | 9781636941509 (PDF) | 9781636941516 (epub)
Subjects: LCSH Pharmaceutical industry—Quality control. | Drug development. | Drugs—Standards. | BISAC BUSINESS & ECONOMICS / Quality Control | BUSINESS & ECONOMICS / Industries / Pharmaceutical & Biotechnology | BUSINESS & ECONOMICS / Industries / Healthcare
Classification: LCC RM301.25 A77 2024 | DDC 338.4/76151—dc23

ASQ advances individual, organizational, and community excellence worldwide through learning, quality improvement, and knowledge exchange.

Bookstores, wholesalers, schools, libraries, businesses, and organizations: Quality Press books are available at quantity discounts for bulk purchases for business, trade, or educational uses. For more information, please contact Quality Press at 800-248-1946 or books@asq.org.

To place orders or browse the selection of all Quality Press titles, visit our website at: http://www.asq.org/quality-press.

28 27 26 25 24 GP 7 6 5 4 3 2 1

Quality Press
600 N. Plankinton Ave.
Milwaukee, WI 53203-2914
Email: books@asq.org

ASQ Excellence Through Quality™

ASQE™
Inspire Excellence

Table of Contents

List of Figures and Tables

Acronyms and Abbreviations

A

AAF—accelerated aging factor

AAT—accelerated aging time

ACC—air chemical cleanliness

ACCME—Accreditation Council for Medical Education

ADE Report—Adverse Drug Experience Report

ADME—absorption, distribution, metabolism, and excretion

AE—adverse event

AHU—air-handling unit

AIP—Application Integrity Policy

ANDA—Abbreviated New Drug Application

AOAC—Association of Official Analytical Chemists

APA—Administrative Procedure Act *or* antiseptic processing area

APHIS—Animal and Plant Health Inspection Service

API—active pharmaceutical ingredient

APR—annual product review

AQL—acceptable quality level

ASTM—American Society for Testing and Materials

B

BCMS—business continuity management system

BET—bacterial endotoxin test

BI—biological indicator

BIMO—Bioresearch Monitoring Program

BLA—Biological License Application

BOM—bill of materials

BP—British Pharmacopeia

BSE—bovine spongiform encephalopathy

C

CA—clinical affairs

CANDA—Computer-Assisted New Drug Application

CAPA—corrective and preventive action

CBE-0—Supplement—Changes Being Effected

CBE-30—Supplement—Changes Being Effected in 30 Days

CBER—Center for Biologics Evaluation and Research

CCP—critical control point

CDC—Centers for Disease Control and Prevention

CDER—Center for Drug Evaluation and Research

CDRH—Center for Devices and Radiologic Health

CFR—Code of Federal Regulations

CDMS—Clinical data management system

CFSAN—Center for Food Safety and Applied Nutrition

cfu—colony-forming unit

cGMP—current good manufacturing practices

CIOMS—Council for International Organizations of Medical Sciences

CIP—clean in place

CMC—chemistry, manufacturing, and controls

CME—continuing medical education

CMO—contract manufacturing organization

CMS—Centers for Medicare and Medicaid Services

COA—certificate of analysis

COC—certificate of compliance

COP—clean out of place

COTS—commercial off-the-shelf

CPG—Compliance Policy Guide

CPGP—certified pharmaceutical good manufacturing practices professional

C_{pk}—process capability index

CPP—critical process parameters

CPSC—Consumer Product Safety Commission

CQ—component qualification

CQA—critical quality attributes

CRA—clinical research associate

CRF—case report form

CRO—contract research organization

CSA—Controlled Substances Act

CTA—Clinical Trial Application

CTD—common technical document

CTMS—Clinical trial management system

CVM—Center for Veterinary Medicine

CVMP—cleaning validation master plan

D

DDMAC—Division of Drug Marketing, Advertising, and Communications

DEA—Drug Enforcement Agency

DESI—drug efficacy study implementation

DHR—device history record

DMF—drug master file

DOE—design of experiments

DOJ—Department of Justice

DOP—drop-off point

DQ—design qualification

DSHEA—Dietary Supplement Health and Education Act

DSUR—Development Safety Update

DTC—direct to consumer

E

ECO—engineering change order

EDMS—electronic document management system

EDQM—European Directorate for the Quality of Medicines & HealthCare

EDR—electronic document room

EEA—European Economic Area

EIR—Establishment Inspection Report

EMA—European Medicines Agency

EPA—Environmental Protection Agency

EPCD—external process challenge device

ERSR—electronic regulatory submission and review

ESM—electronic secure messaging

EU—European Union

F

FAERS—FDA Adverse Event Reporting System

FAP—food additive petition *or* filling assembly procedure

FAT—factory acceptance testing

FCR—facility change request

FDA—Food and Drug Administration

FDAAA—Food and Drug Administration Amendments Act

FDAMA—Food and Drug Administration Modernization Act

FDCA—Food, Drug, and Cosmetic Act

FEFO—first-expired/first-out

FIFO—first-in/first-out

FMEA—failure mode and effects analysis

FMECA—failure mode, effects, and criticality analysis

FOIA—Freedom of Information Act

FTA—fault tree analysis

FTC—Federal Trade Commission

G

GAMP—good automated manufacturing practices

GC—gas chromatography

GDP—good documentation practices

GEP—good engineering practices

GLP—good laboratory practices

GMP—good manufacturing practices

GRASE—generally recognized as safe and effective

H

HACCP—hazard analysis and critical control points

HAZOP—hazard operability analysis

HCFA—Health Care Financing Administration

HEPA—high-efficiency particulate air (filter)

HHS—Department of Health and Human Services

HIC—hydrophobic interaction chromatography

HPFB—Health Products and Food Branch inspectorate

HPKB—human pharmacokinetics and bioavailability

HPLC—high-performance (pressure) liquid chromatography

HVAC system—heating, ventilation, and air conditioning system

HW—hardware

I

IB—investigator's brochure

ICDRA—International Conference of Drug Regulatory Authorities

ICF—informed consent form

ICH—International Conference on Harmonization

ICSR—Individual Case Safety Report

IDE—Investigational Device Exemption

IEC—ion-exchange chromatography

IFAP—Investigational Food Additive Petition

IMP—investigational medicinal product

IMTE—inspection, measuring, and test equipment

INAD—investigational new animal drug

IND—investigational new drug

INDA—Investigational New Drug Application

IPC—in-process control

IPCD—internal process challenge

IPEC—International Pharmaceutical Excipients Council

IQ—installation qualification

IR—infrared

IRB—Institutional Review Board

ISE—integrated summary of effectiveness

ISO—International Organization for Standardization

ISS—integrated summary of safety

J

JP—Japanese Pharmacopeia

L

LAL—limulus amebocyte lysate

LIMS—laboratory information management system

LOA—letter of authorization

M

MAHs—marketing-authorization holders

MBR—master batch record

MDD—microbial data deviation

MDR—Medical Device Reporting

MDUFMA—Medical Device User Fee and Modernization Act

MHLW—Ministry of Health, Labor, and Welfare

MHRA—Medicines and Healthcare Products Regulatory Agency

MKT—mean kinetic temperature

MPN—most probable number

MRA—mutual recognition agreement

MSA—measurement system analysis

N

NAI—no action indicated

NCE—new chemical entity

NCR—nonconformance

NDA—New Drug Application

NDC—National Drug Code

NEPA—National Environmental Policy Act

NF—National Formulary

NIH—National Institutes of Health

NIST—National Institute of Standards and Technology

NMR—nuclear magnetic resonance

NMT—no more than

NOEL—no observable effect level

NOV—Notice of Violation

NSR—nonsignificant risk

O

OAI—official action indicated

ODA—Orphan Drug Act

ODE—Office of Device Evaluation

OECD—Organization for Economic Cooperation and Development

OFAT—one factor at a time

OOC—out of calibration

OOPD—Office of Orphan Products Development

OOS—out of specification

OOT—out of tolerance

OQ—operational qualification

QR—quick-response code

OSHA—Occupational Safety and Health Administration

OTC—over the counter

P

PAI—pre-approval inspection

PAS—Prior Approval Supplement

PAT—process analytical technology

PBR—production batch record

PDCA—plan–do–check–act

PCD—process challenge

PDMA—Prescription Drug Marketing Act

PDUFA—Prescription Drug User Fee Act

PFDA—Pure Food and Drug Act

PFSB—Pharmaceutical and Food Safety Bureau

Ph Eur—European Pharmacopeia

PHA—preliminary hazard analysis

PhRMA—Pharmaceutical Research and Manufacturers of America

PHS—Public Health Service

PI—package insert

PIC/S—Pharmaceutical Inspection Convention and Pharmaceutical Inspection Co-operation Scheme

PK—pharmacokinetics

PL—Public Law

PMA—premarket approval

PPQ—process performance qualification

PQ—performance qualification

PQR—product quality review

PRO—patient-reported outcome

PSURs—Periodic safety update

PTC—points to consider

PV—pharmacovigilance

Q

QA—quality assurance

QAC—quaternary ammonium compounds

QbD—quality by design

QC—quality control

QCU—quality control unit

QMS—quality management system

QOL—quality of life

QP—qualified person

QSIT—quality system inspection technique

QU—quality unit

R

RA—regulatory affairs

R&D—research and development

RAS—rapid alert system

RDP—regulatory development plan

REB—Research Ethics Board

RFID—radio-frequency identification

RH—relative humidity

RODAC—replication organism detection and counting (plate)

RTD—resistance temperature detector

RTU—ready to use

S

SAT—site acceptance testing

SBS—sterile barrier system

SCHIP—State Children's Health Insurance Plan

SDA—Sabouraud dextrose agar

SDS—material safety data sheet

SDV—source document/data verification

SEC—Securities and Exchange Commission

SGE—special government employees

SIP—sample item portion

SMF—site master file

SNDA—Supplemental New Drug Application

SOP—standard operating procedure

SPC—statistical process control *or* surface particle cleanliness

T

TAMC—total aerobic microbial count

TMV—test method validation

TGA—Therapeutic Goods Administration

TOC—total organic carbon

TIR—technical information report

TRO—temporary restraining order

TSA—trypticase soy agar

TSE—transmission spongiform encephalopathy

U

UK—United Kingdom

ULPA—ultra-low penetration air (filter)

US—United States

USDA—United States Department of Agriculture

USP—United States Pharmacopeia

UV—ultraviolet

V

VAI—voluntary actions indicated

VMP—validation master plan

VNR—video news release

VSTA—Virus-Serum-Toxin Act

v/v—volume solute per volume solvent

W

WFI—water for injection

WHO—World Health Organization

WIP—work in progress

w/v—weight by volume

Preface

The purpose of this handbook is to assist individuals in preparing for the Certified Pharmaceutical Good Manufacturing Practices Professional (CPGP) examination and to provide a reference for the practitioner. The third edition reflects the Body of Knowledge that was updated in 2024. This edition has also incorporated additional information, including updated references. The updates reflect the current trends and expectations of the evolving pharmaceutical industry, driven by consumer expectations and regulatory oversight. As the world becomes ever more paperless and electronically nimble and adaptive to changes, reference books such as the *The ASQ Certified Pharmaceutical GMP Professional Handbook* are obligated to provide as many means as possible of bridging, if not transcending, the predictable (and inevitable) changes to laws, regulations, guidelines/guidance documents, compendia, and consensus standards. This handbook inculcates the reader in where and how to tap appropriate websites and other sources of information on the changing laws, regulations, guidelines, compendia, and consensus standards. Throughout this handbook, several "typical" examples are provided based on the collective experience and knowledge of the authors and editor. However, these "typical" examples are not explicitly specified in regulations, leaving decisions and the burden of justifying practices using sound scientific principles—which provide the context of the rationale—up to the company.

Acknowledgments

The third edition of *The ASQ Certified Pharmaceutical GMP Professional Handbook* is dedicated to the hardworking individuals globally who design, develop, manufacture, store, handle, and distribute medicinal products that impact the identity, strength, quality, and purity of drug products.

The following individuals are to be recognized as contributing chapter authors for the previous edition handbook: Belinda Bagnal, Bob Seltzer, Jennifer Ish, Lance Harding, John Lyall, Leann Christman, Anthony DeMarinis, Gerle Peña, Mark Durivage, Adam Foresman, Ajit Basrur, Scott Kochendoerfer, Stephanie Gaulding, Kelli Turner, Bob Mehta, Elena Mack, Cathelene Compton, Mary Chris Easterly, John English, Jeremiah Genest, Kanti Thirumoorthy, Jim Arnold, Carolina Valoyes, José Rodríguez-Pérez, Nancy Van Gieson, Armand Niangara, Sofia Hernandez, Shawn Armstrong, James Eric Miller, Mike Villa, Vidhya Sivakumar, Sandra Storli, Robert Johnson, Mary Kearnes, Yoko Doucette, and Frank Settineri. By the inclusion of several individual subject matter experts, the overall quality and technical content of this handbook has been greatly enhanced. Melvin Alexander and Jane Keathley are to be acknowledged for reviewing the draft manuscript.

Additionally, the following individuals are recognized for their contributions: Roland Bizanek, Jennifer Asleson, Ted Hilliard, David Schultenover, Rosemarie Christopher, Janet Rea, Richard Dolejan, Mairead Goetz, and Raj Rajamani.

I would also like to recognize the Food, Drug, and Cosmetic Division leadership committee. Without their combined vision, passion, and support, this project would not have been completed.

Lastly, I would like to acknowledge my wife Dawn and my boys, Jack and Sam, for being patient and allowing me the time to organize, write, and edit this handbook.

Mark Allen Durivage

LIMIT OF LIABILITY/DISCLAIMER OF WARRANTY

Part I

Regulatory Agency Governance

Chapter 1
Global Regulatory Framework

> Identify the acts, statutes, and directives that apply to pharmaceuticals. (Understand)
>
> **Body of Knowledge I.A**

UNITED STATES FEDERAL STATUTES RELEVANT TO PHARMACEUTICALS

The Administrative Procedure Act

The Administrative Procedure Act (APA) of 1946 is the United States federal law governing how administrative agencies (for example, the Food and Drug Administration [FDA]) within the federal government of the United States of America propose and establish regulation. The APA sets up a process for the United States federal courts to review directly (that is, judicial review) any agency decisions or regulations. The difference between *federal statutory law* and *federal administrative law* (agency regulation) is explained as follows: Congress finalizes the content and approves (legislates) a federal bill that, when signed by the president, becomes federal statutory law. Federal executive departments (that is, the president's cabinet) such as the Department of Health and Human Services (HHS) and subordinate administrative agencies (for example, FDA, the National Institutes of Health [NIH]) declare and publish (promulgate) regulations/administrative law to codify and put into practical motion the governing federal statutory laws. The verb "promulgate" is interchangeable with the verb "codify," the verb associated with the Code (that is, collection) of Federal Regulations. The United States Justice Department indicts and prosecutes based on alleged violation of one or more specific regulations/administrative laws promulgated by a federal agency (which has its basis in a federal statutory law). A federal court verdict or injunction references the prevailing federal statutory law, while prosecutors present to the court evidence that is shown to violate one or more specific administrative laws/regulations.

United States Federal Food, Drug, and Cosmetic Act

The Food, Drug, and Cosmetic Act (FDCA) of 1938 is the United States federal statute governing the marketing, manufacture, and distribution of finished pharmaceuticals, medical devices, foods, dietary supplements, and cosmetics. It has never been replaced or renamed, but rather supplemented or amended through subsequent acts. The FDCA replaced the 1906 Federal Safe Food and Drugs Act after revelations in Upton Sinclair's book *The Jungle* of intentional inclusion of filth by the United States meat packing industry. The 1938 FDCA improvements over the 1906 Act included the following:

- Extending control to cosmetics and therapeutic devices
- Requiring new drugs to be shown safe before marketing
- Providing that safe tolerances be set for unavoidable poisonous substances
- Authorizing standards of identity, quality, and fill-of-container for foods
- Authorizing factory inspections
- Adding the remedy of court injunctions (to the previous Act's seizures and prosecutions)

Following are some amendments to the FDCA along with significant dates in U.S. food and drug law history:

Food and Drug Administration amendments to the Food, Drug, and Cosmetic Act: Significant dates in U.S. food and drug law history. PL = Public Law

Infant Formula Act of 1980, PL 96–359 (Oct 26, 1980).

Orphan Drug Act, PL 97–414 (Jan 4, 1983).

Drug Price Competition and Patent Term Restoration Act of 1984, PL 98–417 (Sept 24, 1984).

Prescription Drug Marketing Act of 1987, PL 100–293 (Apr 22, 1988).

Generic Animal Drug and Patent Term Restoration Act of 1988, PL 100–670 (Nov 16, 1988).

Nutrition Labeling and Education Act of 1990, PL 101–535 (Nov 8, 1990).

Safe Medical Devices Act of 1990, PL 101–629 (Nov 28, 1990).

Medical Device Amendments of 1992, PL 102–300 (June 16, 1992).

Prescription Drug Amendments of 1992; Prescription Drug User Fee Act of 1992, PL 102–571 (Oct 29, 1992).

Animal Medicinal Drug Use Clarification Act (AMDUCA) of 1994, PL 103–396 (Oct 22, 1994).

Dietary Supplement Health and Education Act of 1994, PL 103–417 (Oct 25, 1994).

FDA Export Reform and Enhancement Act of 1996, PL 104–134 (Apr 26, 1996).

Food Quality Protection Act of 1996, PL 104–170 (Aug 3, 1996).

Animal Drug Availability Act of 1996, PL 104–250 (Oct 9, 1996).

Food and Drug Administration Modernization Act (FDAMA) of 1997, PL 105–115 (Nov 21, 1997).

Best Pharmaceuticals for Children Act, PL 107–109 (Jan 4, 2002).

Medical Device User Fee and Modernization Act (MDUFMA) of 2002, PL 107–250 (Oct 26, 2002).

Animal Drug User Fee Act of 2003, PL 108–130 (Nov 18, 2003).

Pediatric Research Equity Act of 2003, PL 108–155 (Dec 3, 2003).

Minor Use and Minor Species Animal Health Act of 2004, PL 108–282 (Aug 2, 2004).

Dietary Supplement and Nonprescription Drug Consumer Protection Act, PL 109–462 (Dec 22, 2006).

Food and Drug Administration Amendments Act (FDAAA) of 2007, PL 110–85 (Sept 27, 2007).

Family Smoking Prevention and Tobacco Control Act, PL 111–31 (June 22, 2009).

FDA Food Safety Modernization Act, PL 111–353 (Jan 4, 2011).

FDA Safety and Innovation Act (FDASIA), PL 112–144 (July 9, 2012).

United States Public Health Service Act

The United States Public Health Service (PHS) originated in name in 1798; however, the PHS Act of 1944 consolidated many previous federal laws relating to public health services and newly mandated the licensing of human biologic products, or their suspension if they endangered public health. The PHS Act is codified and enforced by several federal agencies under the HHS.

The Virus-Serum-Toxin Act

The Virus-Serum-Toxin Act (VSTA) (as amended) requires licensing of animal vaccine products and establishments, and requires permits for the importation of animal biologic products. The Veterinary Biologics Program of the United States Department of Agriculture's (USDA) Animal and Plant Health Inspection Service (APHIS) enforces the VSTA by codifying regulations in 9 CFR 101–118.

It should be noted that the term "pharmaceuticals" is interchangeable with "drugs" in the United States and with the term "medicinal products" in the rest of the world. It is an umbrella term that includes biologics (usually large-molecule therapeutic entities greater than 5 kilodaltons [kDa]) and nonbiologic drugs (usually small-molecule entities less than 5 kDa). Biologics, like other drugs, are used for the treatment, prevention, or cure of disease in humans; however, they are generally derived from living material and are complex in structure and often less definitively characterized than nonbiologic drugs. Biologics are not characterized down to every element, ionic state, water of hydration, three-dimensional structure (including disulfide bond), hydrogen bond, or protein quaternary subunit.

EUROPEAN UNION'S LEGAL SYSTEM RELEVANT TO PHARMACEUTICALS

Within the member countries of the European Union (EU), regulations, directives, and decisions are the principal forms of legislation governing pharmaceutical manufacturing and pharmaceutical marketing. Specifically, a "regulation" is a binding legislative act that must be applied in its entirety across the EU. A "directive" is a legislative act that sets out a goal that all EU countries must achieve; however, it is up to the individual countries to decide how to achieve that goal. A "decision" only deals with a particular issue and specifically mentioned persons or organizations.

Member states of the EU enact national laws to enforce the provisions of EU directives. Directive 2001/83/EC (similar to the United States FDCA) continues to be amended over time but maintains its original structure and general content (comprising Articles and Annexes). This directive requires each holder of a medicinal product manufacturing authorization to have "permanently and continuously at his disposal the services of at least one qualified person [QP]." The specific duties of the QP, as well as the guidelines for pharmaceutical *good manufacturing practices* (GMP), are detailed in the EudraLex Volume 4, Part I.

JAPAN'S LEGAL SYSTEM RELEVANT TO PHARMACEUTICALS

The National Diet is the legislature of Japan. It functions similarly to a parliament in the Westminster model, as it is bicameral, and the lower house majority forms the government. However, it also resembles the American congressional model in that the upper house is elected, and both houses must generally consent to pass legislation. Bills are brought before the Diet for debate and passing. Once a bill is passed by both houses, the prime minister and relevant minister(s) of state countersign the bill, and then the emperor seals the bill and it becomes law. Bureaucratic regulations come in several types: binding regulations, executive orders, mandate orders, non-binding regulations, directives, circulars, standards of review, and administrative guidance.

The Ministry of Health, Labour, and Welfare (MHLW) was established by a merger of the Ministry of Health and Welfare and the Ministry of Labour on January 6, 2001, as part of the government program for reorganizing government ministries. The MHLW, originally established in 1938, was in charge of the improvement and promotion of social welfare, social security, and public health, and the new organization has the same tasks. The department consists of the ministry proper, affiliated institutions, councils, local branches, and an external organization. The MHLW is in charge of pharmaceutical regulatory affairs in Japan (veterinary drugs are under the jurisdiction of the Ministry of Agriculture, Forestry, and Fisheries). The Pharmaceutical and Food Safety Bureau (PFSB) undertakes the main duties and functions of the Ministry: it handles clinical studies, approval reviews, and post-marketing safety measures, that is, approvals and licensing for ensuring food and drug safety.

CHINA'S LEGAL SYSTEM RELEVANT TO PHARMACEUTICALS

The highest and ultimate source of legal norms in the People's Republic of China (PRC) is nominally the Constitution. It establishes the framework and principles of government, and lists the fundamental rights and duties of Chinese citizens. PRC governmental directives exist in a hierarchy, which is defined by the Legislation Law of the PRC. The hierarchy of regulations is:

1. The Constitution
2. National laws, which are issued by the National People's Congress (NPC)
3. Administrative regulations, which are issued by the State Council

The State Council is authorized to promulgate administrative regulations on social and economic sectors and affairs consistent with the laws adopted by the NPC.

The China Food and Drug Administration (CFDA) administers the Drug Administration Law of the People's Republic of China (No. 45). This law was enacted to strengthen drug administration, to ensure drug quality and safety for human beings, and to protect the health of people and their legitimate rights and interests in the use of drugs.

INDIA'S LEGAL SYSTEM RELEVANT TO PHARMACEUTICALS

Legislative power in India is exercised by the Parliament, a bicameral legislature consisting of the president of India, the Rajya Sabha, and the Lok Sabha. Of the two houses of Parliament, the former is considered to be the upper house, or the Council of States, and consists of members appointed by the president and elected by the state and territorial legislatures. The latter is considered the lower house or the house of the people.

The Director General of Health Services, Ministry of Health and Family Welfare, Government of India, is in charge of the Central Drugs Standard Control Organization (CDSCO). CDSCO is the central drug authority for discharging functions assigned to the central government under the Drugs and Cosmetics Act. CDSCO has six zonal offices, four sub-zonal offices, 13 port offices, and seven laboratories under its control. Major functions of CDSCO are regulatory control over the import of drugs, approval of new drugs and clinical trials, meetings of the Drugs Consultative Committee (DCC) and Drugs Technical Advisory Board (DTAB), and approval of certain licenses as central license approving authority is exercised by CDSCO.

Chapter 2

Regulations and Guidances

> Interpret frequently used regulations and guidelines/guidances/drafts including those published or administered by the Pharmaceutical Inspection Convention and Pharmaceutical Inspection Cooperation Scheme (PIC/S), the International Conference on Harmonization (ICH), the World Health Organization (WHO), the European Medicines Agency (EMA), the Food & Drug Administration (FDA), Health Canada, USDA 9CFR, and other national regulatory agencies and international standards (e.g., ISOs). (Understand)
>
> **Body of Knowledge I.B**

It is necessary to possess the ability to interpret regulations and guidances as published or administered by the Pharmaceutical Inspection Convention and Pharmaceutical Inspection Co-operation Scheme (PIC/S), Health Canada, the World Health Organization (WHO), the International Conference on Harmonization (ICH), the European Medicines Agency (EMA), the United States Food and Drug Administration (FDA), the Therapeutic Goods Administration (TGA), United States Department of Agriculture (USDA) 9 CFR (Code of Federal Regulations), USDA Veterinary Service, and the International Pharmaceutical Excipients Council (IPEC).

PHARMACEUTICAL INSPECTION CONVENTION AND PHARMACEUTICAL INSPECTION CO-OPERATION SCHEME (PIC/S)

PIC/S are two authorities that provide guidance in the field of *good manufacturing practices* (GMP). The goal of PIC/S is to lead the international development, implementation, and maintenance of harmonized GMP standards and quality systems in the field of medicinal products. This goal is to be attained by developing and promoting harmonized GMP standards and guidance documents. Additionally, PIC/S promotes the training of authorities and inspectors to ensure competent GMP assessments. Currently, 58 regulatory bodies participate in PIC/S, including representatives from Europe, Africa, America, Asia, and Australasia.

All PIC/S documents are publicly available and can be found at https://picscheme.org/en/publications.

HEALTH CANADA

A federal department known as Health Canada is responsible for helping Canadians maintain and improve their health while respecting individual choices and circumstances. The goal of the department is to maintain a balance between the potential health benefits and risks posed by all drugs and health products. Health Canada comprises the following branches, offices, bureaus, and agencies:

- Audit and Accountability Bureau
- Chief Financial Officer Branch
- Communications and Public Affairs Branch
- Corporate Services Branch
- Departmental Secretariat
- First Nations and Inuit Health Branch
- Health Products and Food Branch
- Healthy Environments and Consumer Safety Branch
- Legal Services
- Pest Management Regulatory Agency
- Regions and Programs Bureau
- Strategic Policy Branch
- Canadian Institutes of Health Research
- Patented Medicines Prices Review Board
- Public Health Agency of Canada

Branch-wide compliance and enforcement activities are the responsibility of the Inspectorate, enabling consistency of approach across the spectrum of regulated products. Core functions include compliance monitoring, compliance verification, and investigation—supported by establishment licensing of drugs, medical devices, and laboratory analysis.

Quality assurance (QA) that ensures that drugs are consistently produced and controlled in such a way as to meet the quality standards appropriate to their intended use, as required by the marketing authorization, includes GMP. Inspections are conducted to verify the compliance with GMP (Part C, Division 2 of the Food and Drugs Regulations), which is a requirement for the issuance of an establishment license. The Inspectorate has developed the GMP guidelines and a series of guides and other related documents to ensure a uniform application of the established requirements and to help the industry comply.

WORLD HEALTH ORGANIZATION (WHO)

WHO is responsible for providing leadership on global health matters, including post-marketing surveillance and eliminating substandard and falsified medicines. One of its core functions is to direct and coordinate international health work by promoting collaboration, mobilizing partnerships, and activating efforts of different health sectors to respond to national and global health challenges.

The role of WHO in the area of medicines regulatory support is twofold. One aspect relates to the development of internationally recognized norms, standards, and guidelines. The second aspect relates to providing guidance, technical assistance, and training in order to enable countries to implement global guidelines to meet their specific medicines regulatory environment and needs.

National governments are responsible for establishing strong national *medicines regulatory authorities* (MRAs). WHO partners with countries, the United Nations system, international organizations, civil society, foundations, academia, and research institutions. The World Health Assembly is the supreme decision-making body for WHO. It is attended by delegations from all 194 member states. Its main function is to determine the policies of the Organization.

INTERNATIONAL CONFERENCE ON HARMONIZATION

In 1990 the pharmaceutical industries of Europe, Japan, and the United States began to work with drug regulatory authorities with the goal of harmonizing the interpretation and application of technical guidelines and requirements for product registration. Regulatory and pharmaceutical industries benefit from this harmonization, as illustrated by the reduction of duplicate clinical trials and the minimization of the use of animal testing without compromising safety and effectiveness in drug development. ICH Tripartite Guidelines were developed through scientific consensus with regulatory and industry experts to achieve harmonization. The guidelines focus on four areas: quality, safety, efficacy, and multidisciplinary. The guidelines can be downloaded from https://www.ich.org/page/ich-guidelines.

Quality guidelines harmonization achievements in the quality area include pivotal milestones such as the conduct of stability studies, defining relevant thresholds for impurities testing, and a more flexible approach to pharmaceutical quality based on GMP risk management. Quality guidances include:

- Q1A–Q1F Stability
- Q2 Analytical Validation
- Q3A–Q3D Impurities
- Q4–Q4B Pharmacopoeias
- Q5A–Q5E Quality of Biotechnological Products
- Q6A–Q6B Specifications

- Q7 Good Manufacturing Practice
- Q8 Pharmaceutical Development
- Q9 Quality Risk Management
- Q10 Pharmaceutical Quality System
- Q11 Development and Manufacture of Drug Substances
- Q12 Lifecycle Management

Safety guidelines include a comprehensive set of guidelines for uncovering potential risks like carcinogenicity, genotoxicity, and reprotoxicity. A recent breakthrough has been a nonclinical testing strategy for assessing the QT interval prolongation liability—the single most important cause of drug withdrawals in recent years. Safety guidances include:

- S1A–S1C Carcinogenicity Studies
- S2 Genotoxicity Studies
- S3A–S3B Toxicokinetics and Pharmacokinetics
- S4 Toxicity Testing
- S5 Reproductive Toxicology
- S6 Biotechnological Products
- S7A–S7B Pharmacology Studies
- S8 Immunotoxicology Studies
- S9 Nonclinical Evaluation for Anticancer Pharmaceuticals
- S10 Photosafety Evaluation
- S11 Nonclinical Safety Testing

Efficacy guidelines are concerned with the design, conduct, safety, and reporting of clinical trials. They also cover novel types of medicines derived from biotechnological processes and the use of pharmacogenetics/genomics techniques to produce better targeted medicines. Efficacy guidances include:

- E1 Clinical Safety for Drugs Used in Long-Term Treatment
- E2A–E2F Pharmacovigilance
- E3 Clinical Study Reports
- E4 Dose-Response Studies
- E5 Ethnic Factors
- E6 Good Clinical Practice
- E7 Clinical Trials in Geriatric Population
- E8 General Considerations for Clinical Trials

- E9 Statistical Principles for Clinical Trials
- E10 Choice of Control Group in Clinical Trials
- E11 Clinical Trials in Pediatric Population
- E12 Clinical Evaluation by Therapeutic Category
- E14 Clinical Evaluation
- E15 Definitions in Pharmacogenetics/Pharmacogenomics
- E16 Qualification of Genomic Biomarkers
- E17 Multi-Regional Clinical Trials
- E18 Genomic Sampling Methodologies

Multidisciplinary guidelines are the cross-cutting topics that do not fit uniquely into one of the quality, safety, and efficacy categories. They include the ICH medical terminology (MedDRA), the Common Technical Document (CTD), and the development of Electronic Standards for the Transfer of Regulatory Information (ESTRI). Multidisciplinary guidances include:

- M1 MedDRA Terminology
- M2 Electronic Standards
- M3 Nonclinical Safety Studies
- M4 Common Technical Document
- M5 Data Elements and Standards for Drug Dictionaries
- M6 Gene Therapy
- M7 Genotoxic Impurities
- M8 Electronic Common Technical Document (eCTD)

EUROPEAN MEDICINES AGENCY (EMA)

The European Medicines Agency is a decentralized agency of the European Union; it is located in London and began operating in 1995. The Agency is responsible for the scientific evaluation of medicines developed by pharmaceutical companies for use in the European Union. This responsibility includes the scientific evaluation of applications for EU marketing authorizations for human and veterinary medicines. Pharmaceutical companies must submit a single marketing authorization application to the Agency and can only start to market a medicine after receiving marketing authorization.

Most of the Agency's scientific evaluation work is carried out by its scientific committees, which are made up of members from European Economic Area (EEA) countries, as well as representatives of patient, consumer, and healthcare professional organizations. Committees are responsible for evaluating the development, assessment, and supervision of medicines in the EU. The Agency constantly monitors the safety of medicines and can take action if information indicates that the benefit–risk balance of a medicine has changed since authorization.

The EMA is headed by the executive director, supported by the deputy executive director, advisory functions, corporate governance, and eight divisions including:

- Human Medicines Research and Development Support
- Human Medicines Evaluation
- Procedure Management and Committees Support
- Inspections and Human Medicines Pharmacovigilance
- Veterinary Medicines
- Stakeholders and Communication
- Information Management
- Administration

THE UNITED STATES FOOD AND DRUG ADMINISTRATION

FDA is an agency within the Department of Health and Human Services and consists of nine Center-level organizations and 13 Headquarter (HQ) Offices.

The high-level structure of the agency consists of the Office of the Commissioner, Center for Biologics Evaluation and Research (CBER), Center for Devices and Radiological Health (CDRH), Center for Drug Evaluation and Research (CDER), Center for Food Safety and Applied Nutrition (CFSAN), Center for Tobacco Products (CTP), Center for Veterinary Medicine (CVN), Office of Regulatory Affairs (ORA), and Oncology Center of Excellence (OCE).

The Office of the Commissioner is responsible for ensuring the safety, effectiveness, and security of human and veterinary drugs, vaccines, and other biological products for human use, as well as medical devices; the safety and security of our nation's food supply, cosmetics, dietary supplements, and products that give off electronic radiation; and the regulation of tobacco products.

Center for Biologics Evaluation and Research (CBER) is the Center within FDA that regulates biological products for human use under applicable federal laws, including the Public Health Service Act and the Federal Food, Drug and Cosmetic Act. CBER protects and advances the public health by ensuring that biological products are safe and effective and available to those who need them. CBER also provides the public with information to promote the safe and appropriate use of biological products.

Center for Devices and Radiological Health (CDRH) is responsible for protecting and promoting the public health. CDRH ensures that patients and providers have timely and continued access to safe, effective, and high-quality medical devices and safe radiation-emitting products. CDRH provides consumers, patients, their caregivers, and providers with understandable and accessible science-based information about the products we oversee. CDRH facilitates medical device innovation by advancing regulatory science, providing industry with predictable, consistent, transparent, and efficient regulatory pathways, and assuring consumer confidence in devices marketed in the United States.

Center for Drug Evaluation and Research (CDER) ensures that safe and effective drugs are available to improve the health of people in the United States. As part of the FDA, CDER regulates over-the-counter and prescription drugs, including biological therapeutics and generic drugs. This work covers more than just medicines. For example, fluoride toothpaste, antiperspirants, dandruff shampoos, and sunscreens are all considered drugs.

Center for Food Safety and Applied Nutrition (CFSAN) is one of six product-oriented centers, in addition to a nationwide field force, that carry out the mission of the FDA. CFSAN provides services to consumers, domestic and foreign industry, and other outside groups regarding field programs; agency administrative tasks; scientific analysis and support; and policy, planning, and handling of critical issues related to food and cosmetics.

Center for Tobacco Products (CTP) oversees the implementation of the Family Smoking Prevention and Tobacco Control Act. CTP responsibilities include setting performance standards, reviewing premarket applications for new and modified risk tobacco products, requiring new warning labels, and establishing and enforcing advertising and promotion restrictions.

Center for Veterinary Medicine (CVN) ensures animal drugs are safe and effective, properly made, and adequately labeled and packaged. CVN carries out research to support policies and regulatory decisions about animal food, drugs, and medical devices.

Office of Regulatory Affairs (ORA) is the lead office for all agency field activities. ORA inspects regulated products and manufacturers, conducts sample analyses of regulated products, and reviews imported products offered for entry into the United States. ORA also works with its state, local, tribal, territorial, and foreign counterparts.

Oncology Center of Excellence (OCE) conducts expedited review of medical products for oncologic and hematologic malignancies. The OCE also leads a variety of research and educational outreach projects and programs to advance the development and regulation of medical products for patients with cancer.

THERAPEUTIC GOODS ADMINISTRATION (TGA)

As part of the Australian Government Department of Health, the TGA safeguards and enhances the health of the Australian community through effective and timely regulation of therapeutic goods using a risk-based approach. TGA is responsible for regulating the supply, import, export, manufacturing, and advertising of therapeutic goods. TGA regulates therapeutic goods through premarket assessment, post-market monitoring and enforcement of standards, licensing of Australian manufacturers, and verifying overseas manufacturers' compliance with the same standards as their Australian counterparts. TGA's approach to therapeutic product vigilance is to continually monitor and evaluate the safety and efficacy (performance) profile of therapeutic products and to manage any risks associated with individual products.

The Medicines Regulation Division evaluates applications to approve new medicines for supply in Australia. The Division is also responsible for monitoring medicines approved for supply in Australia after they are on the market. The Division includes the following branches: Prescription Medicines Authorization is responsible for evaluating new prescription medicines, leading to an approval or rejection decision. Complementary and Over-the-Counter Medicines is responsible for regulating over-the-counter medicines as well as complementary medicines, which include traditional and herbal medicines and vitamin and mineral supplements. Scientific Evaluation is responsible for evaluating generic prescription medicines and biologicals. This branch includes experts in toxicology, biological sciences, and pharmaceutical chemistry, and it administers the regulation of internationally controlled drugs. Pharmacovigilance and Special Access Branch provides oversight of medicines and vaccines to ensure they maintain an appropriate level of quality, safety, and efficacy following entry into the Australian marketplace. The branch also evaluates and authorizes certain clinical trials and special access arrangements for all types of therapeutic products.

The Medical Devices and Product Quality Division monitors medical devices approved for supply in Australia and works to ensure that Australian and international therapeutic goods manufacturers meet specified standards. Medical Devices Branch is responsible for evaluating medical devices, including in vitro diagnostic tests, and monitoring them throughout their life cycle to ensure they continue to meet an appropriate level of quality, safety, and performance. Laboratories Branch is responsible for conducting laboratory testing, quality assessment, and test procedure development in scientific disciplines such as microbiology, immunobiology, molecular biology, biochemistry, chemistry, and biomaterials engineering. This branch also contributes to the evaluation of a range of therapeutic products for market authorization. Manufacturing Quality Branch is responsible for ensuring that manufacturers of medicines and medical devices—as well as blood, tissue, and cellular therapies—meet appropriate quality standards. This involves the physical inspection of manufacturing facilities in Australia and abroad as well as provision of clearances for facilities where suitable inspections have been carried out by comparable overseas regulators. The branch also coordinates product recalls when necessary and provides technical advice to support Medicines Regulation Division's decisions, particularly on matters relating to manufacturing practice and quality management.

The Regulatory Practice and Support Division provides operational regulatory policy advice and specific support services that ensure efficient, best-practice regulatory operations to the Regulatory Services Group. Regulatory Services and Improvement Branch is responsible for activity-based pricing, regulatory billing and industry assistance, project and change management, and regulatory decision reviews. The branch also provides regulatory information assistance and develops regulatory guidance for industry, and leads the Business Improvement Program to streamline and automate regulatory processes. Regulatory Practice, Education, and Compliance Branch is responsible for regulatory compliance activities, including education, advertising, enforcement, and investigations into illegal and counterfeit therapeutic goods. This branch also undertakes regulatory impact analysis and supports good regulatory practice through research and impact analysis of proposed regulatory reforms. Regulatory Engagement and Planning Branch is responsible for managing engagement with international regulators, planning and performance reporting, parliamentary processes, media responses, coordination of technical input for Freedom of Information requests, stakeholder engagement, and committee support.

UNITED STATES DEPARTMENT OF AGRICULTURE (USDA) 9 CFR—ANIMALS AND ANIMAL PRODUCTS

USDA works to support the American agricultural economy in order to strengthen rural communities, to protect and conserve our natural resources, and to provide a safe, sufficient, and nutritious food supply for the American people. The Department's wide range of programs and responsibilities touches the lives of every American every day.

The Animal and Plant Health Inspection Service (APHIS) makes a significant contribution to the value of the nation's food supply by protecting US agricultural resources from pests and diseases, managing wildlife damage, regulating genetically engineered organisms, and administering the Animal Welfare Act. APHIS programs integrate plant and animal disease surveillance, epidemiology, emergency response, and information delivery to ensure the marketability of US agricultural products. APHIS also works to resolve and manage trade issues related to animal or plant health. CFR Title 9 concerns animals and animal products.

USDA VETERINARY SERVICE MEMORANDUM NO. 800.65

This memorandum provides guidance on preparing veterinary biological products that use embryonated chicken eggs or chicken tissue as an ingredient. It is meant to assist licensees, permittees, and applicants in meeting the purity and quality requirements specified in 9 CFR 113.50. Specific ingredient requirements for each licensed product are found in the applicable Outline of Production or Special Outline. This memorandum defines the infectious agents that should be excluded from eggs and chickens used in the production of veterinary biological products without specifying the methods by which a source flock should be established or maintained. It also recommends procedures for the management and disposition of eggs and chickens following a disqualifying disease outbreak in a source flock.

INTERNATIONAL PHARMACEUTICAL EXCIPIENTS COUNCIL

The IPEC Federation is a global organization that promotes quality in pharmaceutical excipients. The IPEC Federation represents the four existing regional International Pharmaceutical Excipient Councils (IPECs)—IPEC Americas, IPEC Europe, IPEC Japan, and IPEC China—and provides a unified voice to promote the best use of excipients in medicines as a means of improving patient treatment and safety by helping to preserve the efficacy, safety, and stability of active pharmaceutical ingredients (APIs), and helping to ensure that they deliver their promised benefits to patients. Optimal use of excipients provides pharmaceutical manufacturers with drug development cost savings and enhanced functionality capability, and can also assist in drug formulation innovation.

IPEC India is in the formation stage. It provides a unified voice that promotes the proper use of excipients in medicines as a means of improving patient treatment and public safety. Although each association focuses its attention on the applicable law, regulations, science, and business practices within its region to accomplish its member's goals, they also cooperate on common excipient-related safety and public health issues, in connection with international trade matters, and pharmacopeial specifications.

Over 250 national and multinational excipient manufacturers, distributors, and finished drug product makers are members of at least one IPEC regional unit, and many are members of two or more. Creation of affiliated IPEC organizations in other regions is also in process.

JAPANESE MINISTRY OF HEALTH, LABOUR, AND WELFARE (MHLW)

The MHLW (the *Koseirodosho*) was established by a merger of the Ministry of Health and Welfare and the Ministry of Labour on January 6, 2001, as part of the government program for reorganizing government ministries. The MHLW, which was originally established in 1938, was in charge of the improvement and promotion of social welfare, social security, and public health, and the new organization has the same tasks. The department consists of the ministry proper, affiliated institutions, councils, local branches, and an external organization. The MHLW is in charge of pharmaceutical regulatory affairs in Japan (veterinary drugs are under the jurisdiction of the Ministry of Agriculture, Forestry and Fisheries). The Pharmaceutical and Food Safety Bureau (PFSB) undertakes the main duties and functions of the Ministry: it handles clinical studies, approval reviews, and post-marketing safety measures, that is, approvals and licensing for ensuring food and drug safety.

The Health Policy Bureau plans and proposes policies for the realization of a high-quality, effective system for offering medical services. The Health Policy Bureau includes the General Affairs Division, Regional Medical Care Planning Division, Medical Professions Division, Dental Health Division, Nursing Division, Economic Affairs Division, Research and Development Division, and Medical Institution Management Support Division.

The Health Services Bureau uses regional health centers to improve health, and takes measures against infectious diseases such as Ebola hemorrhagic fever, AIDS, and tuberculosis, as well as diseases caused by living practices such as diabetes and cancer, while promoting appropriate organ transplantation. Thus, the Bureau endeavors to improve the health of each Japanese national. The Health Service Bureau includes the General Affairs Division, Cancer Measures and Health Promotion Division, Specific Disease Control Division, Tuberculosis and Infectious Diseases Control Division, Environmental Health Division, and Water Supply Division.

The Pharmaceutical and Food Safety Bureau implements measures for securing the efficacy and safety of drugs/quasi-drugs, cosmetics, and medical devices, safety measures for medical institutions, and measures against narcotics and stimulants, in addition to handling blood banking services. Thus, the Bureau addresses various issues directly related with people's life and health. Pharmaceutical and Food Safety Bureau includes the General Affairs Division, Evaluation and Licensing Division, Safety Division, Compliance and Narcotics Division, and Blood and Blood Products Division.

Chapter 3
Mutual Recognition Agreements

> Interpret requirements that govern product registration, import or export of raw material or finished product, and the sharing of inspection findings. (Understand)
>
> **Body of Knowledge I.C**

MUTUAL RECOGNITION AGREEMENTS

Mutual recognition agreements (MRAs) are international agreements that specify the conditions under which the participating countries will accept or recognize one another's conformity assessments, and that identify how the parties will cooperate on other activities, as specified. The primary objective of MRAs is to provide effective and efficient market access throughout the territories of the MRA parties.

MRAs Established by the European Union and Partner Countries

The European Union (EU), Australia, Canada, New Zealand, and Switzerland are referred to as MRA *partner countries,* whereas other countries outside the European Economic Area (EEA) are referred to as *third countries.* Each agreement has slight differences in coverage; however, the broad general purpose of each MRA is fourfold:

- Facilitate market access while safeguarding consumer health
- Grant mutual acceptance of reports, certificates, and authorizations issued by the regulated authorities of the signatories, and provide certification of conformity to the requirements
- Exchange information concerning procedures used to ensure that the conformity bodies comply with the general principles of designation
- Encourage greater international harmonization

The premise of the MRA is that the MRA partner country and EU member states mutually recognize relevant manufacturing authorizations granted by the competent authorities of the other party, conclusions of inspections of manufacturers carried out by the relevant inspection services of the other party, and manufacturers' certification of the conformity of each batch to its specifications by the respective party without re-control at import.

The scope of the agreements covers medicinal products that have undergone one or a series of manufacturing operations in the MRA partner country and/or the EU. For the purposes of these agreements, manufacturing operations include manufacturing, labeling, testing, and wholesaling activities.

Key Elements of Mutual Recognition Agreements

Issuance of Certificate of GMP Compliance of Manufacturers. The purpose of the certificate is to confirm that the manufacturer or laboratory is authorized to carry out activities and is regularly inspected and complies with national *good manufacturing practices* (GMP) requirements that are deemed equivalent to EU GMP. Authorities responsible for supervision of the entity issue the certificate. A manufacturer can request an inspection by the local competent authority to obtain a certificate.

Batch Certification. Each batch being transferred between MRA countries must be accompanied by a batch certificate issued by the manufacturer in the exporting country. The importer of the batch needs to receive and to maintain this certificate, which attests that the batch has been manufactured in accordance with the requirements of the marketing authorization of the importing country and that it meets specifications. The certificate includes a statement that batch processing and packaging records have been reviewed and comply with GMP. The content of the batch certificate has been harmonized and agreed on by MRA partners.

Responsibilities of the Qualified Person. The *qualified person* (QP) may rely on the written confirmation of the manufacturer's batch certificate without further testing. Responsibilities of the various parties should, however, be clearly defined in a written agreement.

Pharmaceutical Inspection Co-operation Scheme

The Pharmaceutical Inspection Convention and the Pharmaceutical Inspection Co-operation Scheme (jointly referred to as PIC/S) provide a means for international cooperation in the field of GMP. PIC/S is a cooperation scheme. Membership, however, does not mean that an MRA exists. PIC/S's mission is to lead the international development, implementation, and maintenance of harmonized GMP standards and quality systems of inspectorates in the field of medicinal products. The purpose of PIC/S is to:

- Pursue and strengthen the cooperation established between the participating authorities in the field of inspection and related areas with a view to maintaining mutual confidence and promoting quality assurance of inspections
- Provide the framework for all necessary exchange of information and experience

- Coordinate mutual training for inspectors and for other technical experts in related fields
- Continue common efforts toward the improvement and harmonization of technical standards and procedures regarding the inspection of the manufacture of medicinal products and the testing of medicinal products by official control laboratories
- Continue common efforts for the development, harmonization, and maintenance of GMP
- Extend cooperation to other competent authorities having the national arrangements necessary to apply equivalent standards and procedures with a view to contributing to global harmonization

Currently, there are 54 member countries, including countries in Europe, Africa, America, Asia and Australasia. The United States Food and Drug Administration (FDA) is a participating authority.

Before a regulatory authority can become a member of PIC/S, a detailed assessment is undertaken to determine whether the authority has the arrangements and competence necessary to apply an inspection system comparable to that of current PIC/S members. This assessment involves an examination of the authority's inspection and licensing system, quality system, legislative requirements, inspector training, and so on, and is followed by a visit from a PIC/S delegation to observe inspectors carrying out actual GMP inspections.

Chapter 4
Regulatory Inspections

> Define and describe various types of inspections including preapproval, system-based, for-cause, and license renewal, and describe the frequency for each. Describe how to prepare for and host in-person and remote evaluations and how to accommodate record requests and digital reporting. (Understand)
>
> **Body of Knowledge I.D**

INSPECTION AUTHORITY

The Food, Drug, and Cosmetic Act (FDCA) grants the United States Food and Drug Administration (FDA) the authority to conduct establishment inspections and collect samples to enforce the Act. FDA, in the Center for Biologics Evaluation and Research (CBER) Compliance Program Guidance Manual, Inspection of Biological Products, 7345.848, outlines three reasons for conducting inspections:

- To safeguard the public health by reducing the risk of adulterated or misbranded drugs reaching the marketplace
- To increase communication between the industry and FDA
- To provide timely input to firms during inspections to improve the firm's compliance with *current good manufacturing practices* (cGMP) requirements

Countries that import drugs manufactured in the United States can inspect drug manufacturing establishments in the United States. For example, Health Canada performs inspections for on-site monitoring and assessment of applicable requirements in the Canadian Food and Drugs Act and associated regulations. Health Canada inspections routinely are conducted on a predetermined cycle or as required to assess compliance.

"The FDA has the authority to inspect 'records, files, papers, processes, controls, and facilities' bearing on whether prescription drugs are adulterated or misbranded or are otherwise violative." [21 USC. 374 (a) (1)] This means that FDA can enter and inspect

any facility that engages in FDA-regulated activities. In the course of an inspection, the agency has the authority to examine documentation, equipment, facilities, products, and materials that pertain to regulated activities in order for them to determine whether the company is compliant with the applicable regulations. If a company outside of the country wants to distribute its FDA-regulated product in the United States, it is subject to the same regulations as US domestic manufacturers and is subject to inspection.

Elements or areas outside the scope of the FDA's inspection authority include:

- Financial data and information, sales data, *except* for shipping data and metrics
- Pricing information, personnel records *except* for training records
- Research data except for product being inspected

As stated in the Medical Devices; Current Good Manufacturing Practice (cGMP) Final Rule quality system regulation:

> *Generally, the FDA will neither request nor review the actual results of audits and management reviews; however, the firm must be able to demonstrate that audits were conducted. A firm may provide an internal audit log or schedule or certification that an audit was performed on a particular date, covering a specific area.*

TYPES OF INSPECTIONS

Regulatory inspections for good manufacturing practice (GMP) compliance in the pharmaceutical industry typically are categorized as preapproval inspection, routine surveillance (post-approval), and compliance (one of which is "for cause"). Pre-Approval Inspections (PAIs) are used by the FDA to review information for drug approval. PAIs entail a systematic review of the of the application and the manufacturing process. PAIs will usually not have a pre-announcement. The inspection will focus on three issues: Chemistry, Manufacturing, and Controls (CMC) (Pharmaceutical Quality/CMC); conformance to application; and data integrity. The FDA may conduct a PAI for manufacturing sites listed in the application (foreign and domestic), including API manufacturers, testing laboratories, packaging facilities, finished dosage form manufacturers, container/closure system manufacturers, and specialty laboratories. If major issues are observed during a PAI, a follow-up inspection may be required before drug approval to review corrective actions implemented by the firm.

Routine surveillance inspections, conducted after a drug product is approved by the FDA, are intended to be conducted every two years. The FDA uses a risk management approach to assess which products pose a high risk and, as a result, require more frequent oversight. A firm is considered to be a higher risk if the company has a history of cGMP compliance issues or is closer to the end of the processing of a drug product, or if issues have occurred with similar categories of products (for example, heparin API contamination).

A *compliance inspection* is conducted to examine the effectiveness of corrective actions implemented by a firm due to a regulatory action. A compliance inspection may be referred to as a "for cause" inspection when an issue has occurred, such as contamination, adverse events (for example, patient deaths or severe side effects), or lack of drug effectiveness (perhaps due to a counterfeit drug).

SHARING OF INSPECTION FINDINGS

FDA is now a PIC/S member. PIC/S is the abbreviation and logo used to describe both the Pharmaceutical Inspection Convention (PIC) and the Pharmaceutical Inspection Co-operation Scheme (PIC Scheme) operating together in parallel. Members of the PIC/S organization have a tremendous impact by focusing on a set of key issues, importantly including international cooperation and collaboration in critical areas such as GMP standards, technical training, and quality systems of inspectorates—all essential to protecting the health of individuals and the public health of nations throughout the world.

FDA also works together with the European Medicines Agency (EMA) and its member state inspectorates on drug product inspections, and with EMA and the Australian Therapeutic Goods Administration (TGA) on active pharmaceutical ingredient manufacturer inspections in various parts of the world. They also collaborate with many other counterparts.

The majority of FDA's global engagement is accomplished through a variety of robust bilateral and multilateral relationships, including their three posts in Latin America—Mexico, Costa Rica, and Chile. With some countries, they actively collaborate on inspection report exchanges, joint inspection programs, and parallel assessment of applications.

FDA is currently exploring the idea of a single audit pilot program for medical devices with ANVISA, Health Canada, and TGA. The goal of the pilot would be to determine whether an inspection done by one regulator can count for multiple regulatory agencies. It will reduce redundant requirements for firms and save regulatory resources by focusing on higher-risk inspections in each respective region.

FDA INSPECTIONS OF OUT-OF-US MANUFACTURERS

Part of FDA's mission is to assure that drug, medical device, biologic, and food products manufactured in foreign countries and intended for US distribution are in compliance with the law and regulations, that noncompliance is identified and corrected, and that any unsafe or unlawful products are removed from the marketplace. FDA performs inspections of products at the US border using a risk-based approach. Materials that do not meet requirements are destroyed or exported back to the source. FDA typically conducts inspections of foreign drug manufacturing firms when:

- The firm is identified in an application or submission as the main supplier or alternate supplier of materials, products, or services.
- The firm has had a violative inspectional history and/or problems.
- The firm was prioritized using a tiered approach based on factors such as risk, volume of products, or complexity of processes.
- The firm had problems related to adverse reactions or was involved in recalls.

The foreign inspection program is under the management and direction of the International and Technical Operations Branch (ITOB) headquarters staff of the Division of Emergency and Investigational Operations/ORO. This branch schedules all foreign inspection trips and provides all the resources necessary for the program activities. The field offices evaluate and endorse establishment inspection reports (EIRs) and forward them to ITOB.

The foreign inspection program is crucial to the successful management of the program, and that can be achieved only through maintenance of consistency and uniformity of inspection and enforcement activities. There has been a significant increase in the number of foreign inspections of pharmaceutical manufacturing plants in the past few years.

Inspections at foreign drug facilities are expected to be approached in the same manner as domestic inspections. There are two factors that are unique to foreign inspections:

1. One main difference that poses a significant challenge to the inspection team abroad is the logistics of the program itself. The majority of foreign inspections are pre-scheduled and on a relatively tight time frame. Unless the inspection is prepared in advance and sharply focused, it is difficult to meet the objectives of the program satisfactorily.

2. Another factor to keep in mind is that the authority to inspect foreign drug facilities does not come from Section 704 of the Food, Drug, and Cosmetic Act (the Act,) but from the agency's ability to exercise Section 801 of the Act and commitments made by the sponsors of applications, if applicable. For that reason, the agency is not required to provide stringent documentary evidence to establish violations of the Act. However, the inspection team is expected to collect sufficient records to substantiate its findings and to aid in the further review process by the agency.

As a general rule, sample collection is not required during inspections at foreign facilities. However, some compliance programs, such as the drug preapproval and post-approval programs, may require sample collection as part of the inspection process. When significantly violative conditions are observed during an inspection, the team leader should alert ITOB with its findings either by phone or fax for timely evaluation of the situation. At the end of each inspection, the inspection team leader is expected to communicate the result of the inspection to ITOB, if possible, by fax transmission of the form FDA 483, a brief summary of findings, and a proposed recommendation.

Reviewing Documents

Foreign drug inspections will generally involve *bulk pharmaceutical chemicals* (BPCs) or dosage forms. The end product may be sterile or nonsterile. Contract processors or contract laboratories may also be assigned. The investigators/analysts will inspect these facilities in the same manner as domestic facilities.

While at the firm, the inspection team should evaluate all process failures, product failures, failures in laboratory tests, and process changes. Process failures and process changes are important because they have a significant impact on the adequacy of process validation. Attention should be given to all reports of failure investigations of any batches requiring rework or reprocessing.

Sometimes it becomes necessary to review documents that are not part of data filed with the agency prior to the scheduled inspection. Either ITOB or the inspection team may request these data from the domestic sponsor or the foreign facility. They are documents that would be routinely inspected but may be too bulky or lengthy to cover during the limited on-site inspection. Examples of such documents are manufacturing process validations, cleaning validations, and stability data.

For sterile products, the following should be reviewed: microbial test results for all batches, all initial positive sterility test results and reports of investigation, all organisms isolated and their source, environmental monitoring results and investigations, and monitoring of water for injection (WFI) systems for microbial and endotoxin qualities.

For aseptic processes, media fills need to be reviewed. Also, keep in mind that dosage forms other than injectables may have microbiological requirements. Complaint files should be reviewed for all products being covered.

Identify and report all products that the firm manufactures, which products of those are shipped to the United States, and the amounts being shipped to the United States since the last inspection, or over the last two years.

For an application product, the inspection team should, at least, obtain and review the *chemistry, manufacturing, and controls* (CMC) section of the application prior to the inspection. This section contains information regarding bio-batches, process and controls, analytical methods, and supporting analytical data. Also worth reviewing are reviewers' comments and correspondence between the sponsor and the agency (deficiency letters and written responses). They usually contain application-specific issues, but occasionally provide indications of GMP concerns to focus on. For an approved application, annual reports should be reviewed to identify any potential problems.

For bulk drug substances, most of which are covered by *drug master files* (DMFs), the inspection team should review the section describing the final purification steps of the manufacturing process.

Validation

All drug manufacturing processes are expected to have been validated prior to shipping. This may mean that a manufacturer is not required to have documentation of a validated manufacturing process at the time of the inspection. However, this should not deter the inspection team from making an assessment of the firm's manufacturing process(es). Especially for a bulk drug substance that has been manufactured for years, employing an established process for many years, and is not expected to change, the firm should have had sufficient experience with the respective process to have their critical process parameters under control to attain reliable and consistent product quality. During the inspection, the inspection team should attempt to evaluate whether there are any significant variations in critical process parameters by examining a series of batch production records.

If the process needs to be constantly fine-tuned from batch to batch, that is a sign of an invalidated process.

For the majority of bulk drug processes, the single most important quality attribute is the substance purity. Therefore, particular attention must be paid to the evaluation of final purification processes. For many applications submitted by foreign sponsors, the products have been marketed outside the United States for years and their processes usually are established. The inspection team should first determine whether the product and the process being inspected are new at the time of the inspection. If the application process is substantially similar to the existing process, the inspection team should attempt to find out if the firm has validation data and whether other available production data support a validated process.

Cultural Aspects

The inspection team is there to make an assessment of each foreign facility on FDA's terms. Foreign firms are under no obligation to comply with US regulations except for their commitments in applications filed with the agency and/or for their desire to market their products in the United States. In dealing with a foreign firm, therefore, it is important not to impose our views on it during the inspection. Observe its manufacturing operations and verify them with the filed processes. Review the production records and document any deficiencies and discrepancies. The purpose of the inspection team, which is similar to domestic inspections, is to gather as much information as possible to allow the agency to make an informed decision. Be assertive and clear about what information and/or documentation the firm should provide to the inspection team.

Views of the inspectional findings should be discussed with the firm in a constructive manner during the exit interview. Regulatory processes of other countries are usually different from US regulations, and the inspection team may need to explain to the firm the implication of the findings and ways to remedy the problems, including possible corrective actions.

Documentation of Violations of cGMPs

During the inspection of a foreign drug manufacturer, obtain the documentation to establish evidence of cGMP violations or data integrity problems. The agency has the authority under the FDCA to administratively restrict the importation of a product without demonstrating the adulteration of the product. The burden of proof is placed on the importing party. However, the inspection report should contain sufficient information and documentation to support a conclusion by the reviewing office that significant violations of the law exist to warrant restricting importation of the commodity and/or non-approval of affected application(s). Where data integrity problems are suspected, an attempt should be made to establish a pattern of practice. If the inspection team determines that extensive evidence documentation is required, and fears that the evidence might be destroyed, ITOB should be contacted at the earliest possible time so as to develop a prompt logistical support plan.

INSPECTION TECHNIQUES

During a regulatory inspection, FDA may evaluate the systems in place to assess the firm's cGMP compliance. If the system is determined to be in a state of control for the product or products examined during the inspection, the system should be adequate for the same class of products made by that firm.

FDA can conduct a comprehensive inspection, often referred to as a *full inspection,* for a broad and deep evaluation of the firm's cGMP compliance. The full inspection determines compliance status of every system at the facility subject to FDA jurisdiction. A full inspection is required for a compliance inspection and for the initial FDA inspection of a facility.

Alternatively, FDA can conduct a *directed inspection,* also known as an *abbreviated inspection,* to provide an efficient update of the firm's cGMP compliance. The abbreviated inspection covers specific areas to the depth necessary to address the purpose of the inspection.

FDA evaluates an organization by inspecting its cGMP systems. The FDA review of a system includes the systems being used as well as specific examples that illustrate the effectiveness of system implementation. The systems and evidence of implementation are covered in sufficient detail so that FDA can determine whether the system is in a state of control. FDA divides systems into six categories: quality, facilities and equipment, materials, production, packaging and labeling, and laboratory control. Full inspections require assessment of a minimum of four systems, one of which must be the quality system. Abbreviated inspections require a minimum of two systems, one of which must be the quality system. During subsequent abbreviated inspections, the systems evaluated are rotated so that all six systems are periodically addressed. Depth of coverage of each system depends on the findings during an inspection.

FDA INSPECTION PROCESS FOR A US FIRM

Upon arrival at the facility being inspected, and regardless of the type of inspection conducted, the FDA investigator displays credentials and gives documentation of the inspection on form FDA 482, Notice of Inspection, to the top management official at the facility. The FDA 482 usually is presented during an opening meeting with the inspection team and company management.

Inspections can be conducted by one investigator or a team of experts from within the firm's FDA district, from other districts, and/or from FDA headquarters, with one member designated as team leader. The inspection may consist of any combination of observing processes and the facility, reviewing procedures and records, and asking employees of the firm questions about their responsibilities and the tasks they perform.

Frequently, the inspection begins with a walk through the facility while it is in operation. FDA investigators take notes during all parts of the inspection to document what they have seen, the personnel encountered, and any concerns they have. Daily debriefings may occur so that FDA can indicate their concerns as they are encountered.

At the close of the inspection, the investigator provides the top management official with any inspection findings on form FDA 483, Inspectional Observations, usually during a closing meeting with company management. Observations listed on the FDA 483 are cGMP deficiencies related to a requirement in the cGMP regulations. If no observations of significance were identified, no FDA 483 is issued. Form FDA 484, Receipt for Samples, is presented to top management if samples were taken during the inspection. Routine surveillance samples are often collected during regulatory inspections. Samples also may be collected as a result of enforcement action.

After the close of the on-site inspection, the FDA investigator(s) are obligated to complete an *establishment inspection report* (EIR). Once the inspection is considered closed and the EIR is available, it is typically forwarded to the company for inclusion in their inspection files.

Preparing for and Hosting a Regulatory Inspection

Preparing for and hosting a regulatory inspection is an ongoing process. Organizations should conduct themselves to always be inspection-ready, as there may be times when an unannounced regulatory inspection may happen.

As part of the preparation process for a regulatory inspection, organizations should establish and maintain a regulatory inspection procedure that defines roles and responsibilities and determine the required physical resources, including where in the facility the inspection will be hosted and what information technology is required, such as copiers and printers.

The key to hosting a regulatory inspection is practice! Organizations should periodically conduct "mock" regulatory inspections to ensure the individuals involved know their responsibilities and expectations.

During the inspection, the Management Representative responds to all questions and requests from the inspector(s) or ensures a Subject Matter Expert (SME) is present to respond. A Scribe should be available to participate in the inspection and take accurate notes of the inspection dialog as well as records document and records requests requested by the inspector.

During an inspection the Management Representative and Scribe will host the inspector(s) in the Front Room. The Front Room is the area where the organization's representatives will directly interface with an investigator during a regulatory inspection. The Back Room is the area where the organization's representatives gather and review information and documentation for a regulatory inspection. The Mid Room serves as the staging area for SME preparation.

Depending on the regulatory jurisdiction and the circumstances for the inspection, some records are allowed to be kept from inspector review. These records include personnel records not directly related to training and competence, financial records, management review, internal audits, supplier audits, and external audits. However, regulatory agencies may request management review and audit agendas, dates, and attendees.

Upon arrival of a regulatory inspector, whomever greets the inspector(s), should ask for identification, register their attendance in the guest registry, ask the inspector(s) to have a seat, and immediately notify the Management Representative. All employees should be notified that a regulatory inspection is in process and the organizations' management team should proceed to the Front Room to attend the opening meeting.

The Management Representative should confirm the scope and the dates of the inspection, determine what guidance to follow based on the type and scope of inspection, and provide the inspector(s) assistance with the inspection logistics. It is also advisable to inform the inspectors what records are available on-site and what records are stored off-site and the time frame for requesting records from off-site storge facilities.

The Management Representative should offer to present the inspector(s) a summary of company, products, and management structure (PowerPoint presentation), and offer to provide a tour of the facility upon conclusion of the opening meeting. The Back Room completes document records requests by reviewing and making three copies (Quality Manager, Inspector, File Copy), stamping/marking each document or record using red ink/font "COPY" and "CONFIDENTIAL." If there is an issue with any of the requests, the Management Representative must be informed prior to taking the documents or records to the Front Room. (All copies are handed directly to the Management Representative).

The Management Representative assembles appropriate management for the closing meeting. All participants (excluding the regulatory inspectors) should sign the organizations meeting attendance sheet. Participants in the meeting should ensure that all observations are discussed and clearly understood.

Every effort should be made to verify the completeness of training records, ensure calibration and maintenance activities are up to date, and ensure the facility is neat, clean, and orderly.

Chapter 5

Enforcement Actions

> Define and describe various global enforcement actions and consequences (e.g., warning letters, consent decree, license withdrawals, product seizure, and import alerts). (Understand)
>
> **Body of Knowledge I.E**

Pharmaceutical companies and the regulatory agencies that provide oversight of their activities share responsibility for ensuring the manufacture and distribution of quality drug products and for ensuring compliance with *good manufacturing practices* (GMP) regulations. When these agencies identify areas of noncompliance within a company (for example, during a regulatory inspection or as a result of mandatory problem reporting), they have an obligation to the public to take appropriate enforcement action to remedy the situation. The agencies have a number of enforcement tools at their disposal and will take action consistent with the seriousness of the violation (or violations) and the risk of harm to the public.

The most common enforcement tool used to compel compliance with regulatory requirements is a notice of observations/violations (for example, form FDA 483) that is provided to a company after an on-site inspection by the applicable regulatory agency. If a company fails to adequately address all of the issues in the notice, or if the compliance issues identified during the inspection are more serious or widespread, the regulatory agency may issue a warning letter to the company. In more serious cases, where there is an immediate risk to patient health and safety, the regulatory agency may choose to seize impacted product (product seizure) or prevent its continued distribution (through an injunction). In the United States, a company may be compelled to enter into a consent decree that outlines, in specific detail, the steps required to bring the organization back into compliance, and the reporting and oversight obligations necessary to fulfill the requirements of the consent decree. Finally, in those instances where previous enforcement action has been unsuccessful in compelling compliance, or where there are indications that compliance will not be possible, the regulatory agency may choose to withdraw a marketing authorization (specific to a product) or withdraw an establishment license (applicable to the company as a whole). This stepwise approach allows the regulatory agency to use a risk-based methodology to determine where to focus its efforts toward ensuring compliance with applicable regulations while continuing to meet its obligation to protect the public health.

TYPES OF ENFORCEMENT ACTIONS

Notice of Observations/Violations

If, during an inspection by FDA, significant compliance issues are observed, the investigator may choose to document those observations on form FDA 483, Inspectional Observations. Form 483 is used to notify the inspected establishment's top management, in writing, of significant objectionable conditions that were observed during the course of the inspection. If an observation related to a previous inspection has not been corrected (FDA is very good about verifying commitments made to correct previous Form 483 items) or is a recurring observation, it may also be noted on the Form 483 for the current inspection. It is important to note, however, that additional observations of questionable significance may be discussed verbally with a firm's top management at the inspection closeout meeting. FDA's intent in making top management aware of these issues is to let them know that, if left uncorrected or unaddressed, these issues could become a violation. The *establishment inspection report* (EIR) that is provided after the inspection is closed provides documentation of both the written observations (on Form 483) and verbal observations documented during the inspection.

A similar process is used by Health Canada. The Health Products and Food Branch Inspectorate (HPFBI) will document GMP-related observations on the inspection exit notice that is shared with the firm's management at the formal exit meeting. Differently than in the United States, a specific rating is assigned to the firm at the end of the inspection. Either a "C" rating (firm is recommended for the continuation or issuance of the establishment license) or an "NC" rating (firm is not recommended for the continuation or issuance of the establishment license) is assigned based on the results of the inspection.

In the United Kingdom (UK), after an inspection by the Medicines and Healthcare Products Regulatory Agency (MHRA), a post-inspection letter outlining the deficiencies noted during the inspection is sent to the inspected firm to provide written confirmation of the deficiencies reported verbally during the closing meeting with management. In the case of the post-inspection letter, all deficiencies are classified as critical, major, or other.

In all cases where an inspection report is issued, or areas of noncompliance are discussed verbally with the firm, the expectation of the regulatory agency is that the firm will take the appropriate corrective and preventive action to prevent future noncompliance. A written response from the firm, outlining the planned or completed actions, is necessary to preclude further enforcement action and to close the inspection.

These enforcement actions might include warning letter or notice, caution, criminal prosecution, and asset confiscation and civil action (injunction). Application of a sanction depends on several factors, including the nature of the noncompliance (critical or major), duration of the noncompliance, and actual or potential impact on public health.

In Europe, serious deficiencies that potentially could result in risk to public health are typically discussed in the inspection closeout (final) meeting, and immediate action is taken, as necessary, before issuing an inspection report. After the inspection, a draft inspection report or post-inspection letter is sent to the company for its comments to allow the report to be finalized. Any response received from the manufacturing firm is considered in the final report. Issuance of the final report is needed for the issuance of a GMP certificate, where applicable.

FDA Untitled Letters

An Untitled Letter is an initial correspondence with regulated industry that cites violations that do not meet the threshold of regulatory significance for a Warning Letter. Therefore, the format and content of an Untitled Letter should clearly distinguish it from a Warning Letter.

Any appropriate agency compliance official may issue an Untitled Letter. All agency components responsible for issuing Warning Letters and Untitled Letters must follow the agency's "Procedures for Clearing FDA Warning Letters and Untitled Letters, Identification of Untitled Letters."

All Untitled Letters must be entered into the Compliance Management System (CMS), whether they are generated by a district or center, and whether they are approved and issued or not. Every Untitled Letter that is issued bears the CMS-assigned number or a sequential code number assigned by the issuing district or center. If a district- or center-assigned number is used, this number should be recorded in the CMS to facilitate tracking.

When a profitable firm (that is, domestic or foreign drug, biologics, or medical device facility) undergoes a current good manufacturing practices (cGMP) or *quality system* (QS) inspection, the inspected profile classes should be updated by the action office (that is, the district for domestic inspections or the district and the center for foreign inspections) at each stage in the review process. When an Untitled Letter is issued as a result of the inspection, the date and type of letter issued should be entered in the *remarks* field for the relevant profile classes.

Warning Letters

FDA may choose to issue a *Warning Letter* if a company fails to adequately address all of the issues in the form FDA 483, or if the compliance issues identified during the inspection are more serious or widespread. Warning Letters typically are issued for violations of regulatory significance, that is, those that may lead to further enforcement action if not corrected promptly and adequately.

FDA will label the correspondence as a "Warning Letter." Warning Letters generally are addressed both to the highest known official at the corporation and to the highest known official at the site that was inspected, and include the dates of the inspection, a brief description of the violations, and the specific sections of the law that are in violation. The correspondence also will acknowledge any corrections promised during the inspection or in the firm's written response, and the Agency's position related to those corrections. Finally, it will request a prompt response to the Warning Letter, typically within 15 working days.

Unlike form FDA 483, Warning Letters are posted publicly on the FDA website and can result in significant harm to the firm's reputation. It is in a firm's best interest to correct identified areas of noncompliance quickly and completely to preclude the issuance of a Warning Letter. A Warning Letter may include a recommendation to withhold approval for any pending applications, which could result in significant financial harm to the company. It is important to note, however, that FDA is under no legal obligation to issue such a letter before taking additional and more serious enforcement action.

FDA may issue a Warning Letter closeout letter once it has completed an assessment of the corrective actions taken in response to the Warning Letter. Closeout letters apply only to those Warning Letters issued after September 1, 2009. The closeout letter can only be issued after FDA has verified (typically through a follow-up inspection) that the corrective actions have been taken. If the letter contains GMP violations that cannot be corrected, no closeout letter will be issued.

Product Seizures

A *product seizure* is a judicial (court-approved) action for removing violative products from the marketplace. Seizures generally fall into one of three categories:

- *Mass seizure.* The seizure of all FDA-regulated products at an establishment or facility. This type of seizure is indicated when all of the suspect products have been held in the same environment (for example, dirty warehouse) or were produced under the same conditions (for example, nonconformance with GMP requirements).
- *Open-ended seizure.* The seizure of all units of a specific product or products, regardless of lot or batch number, where the violation is anticipated to be continuous.
- *Batch-specific seizure.* The seizure of a batch or batches of a regulated product where the violation is limited to a period of time or unique to a specific set of batches.

Product seizures typically are initiated when a company has not voluntarily recalled product that the regulatory agency believes is in violation of the law and where there is a perceived or known health risk to the public.

Injunctions

An *injunction* is another type of judicial action, initiated to stop or prevent a violation of the law. It is typically invoked to halt the flow of violative products and to correct the conditions that caused the violations to occur. An injunction is considered for any significant out-of-compliance circumstance, but particularly when a health hazard has been identified. When an injunction is granted, FDA has a duty to monitor the injunction and to advise the courts if the company fails to follow through on their obligations. The four most common types of injunctions are the temporary restraining order, the preliminary injunction, the permanent injunction, and the consent decree of permanent injunction.

Temporary Restraining Order. A *temporary restraining order* (TRO) is a court-ordered cease-and-desist order that is used in emergency situations. FDA, for example, may recommend a TRO when the Agency believes that a serious violation has occurred and the situation must be controlled or the flow of product stopped immediately.

Preliminary Injunction. A motion for preliminary injunction is subject to a full hearing before a court. Once the motion is granted, the preliminary injunction is in effect. A preliminary injunction may stand indefinitely until the case is settled or a permanent injunction has been entered, after trial.

Permanent Injunction. A decree of permanent injunction may be entered at any time after the complaint is filed, either after a hearing or as a result of a negotiated settlement between the government and the defendant. A decree of permanent injunction remains in effect until it is dissolved by an order of the court.

Consent Decree of Permanent Injunction. In the United States, a consent decree is a court-ordered agreement between FDA and a company that outlines the steps that a company needs to take to resume normal operations. Although each situation is unique, the consent decree will likely include the need for changes within the company (for example, to bring the organization into compliance with applicable regulations) and a requirement to involve outside consultants to oversee the changes and ensure ongoing compliance. FDA will typically seek a consent decree only when serious issues have been identified with a drug product manufacturer and other enforcement mechanisms have been unsuccessful in encouraging the company to address the issues in a timely fashion.

Loss of Marketing License

Regulatory agencies such as FDA, MHRA, and the European National Competent Authorities are charged with ensuring that the drugs that are marketed are safe for their intended use and that their benefits outweigh the risks involved with any drug product. As a result, regulatory agencies constantly monitor and evaluate the safety data that are transmitted to them, either through periodic safety reports (by the manufacturer) or spontaneous reports (by healthcare practitioners or consumers). They weigh the benefits of existing drugs against any new emerging drug products or treatment therapies to determine whether older drugs still provide a medical benefit to patients.

When the regulatory agencies identify serious safety issues with a drug product, or deem the safety risks to outweigh the benefits provided by the drug product, they may choose to suspend or revoke the existing marketing license.

According to the World Health Organization (WHO), a marketing authorization may be suspended or revoked in any of the following circumstances:

- The product has proved to be ineffective for the approved indication(s).
- It is strongly suspected that the product is unsafe under the normal conditions of use.
- The quantitative or qualitative composition is not as agreed on in the marketing authorization.

- The product is not in compliance with the conditions of the marketing authorization.
- The product is being promoted in an inappropriate or unethical manner.

The marketing authorization holder is typically notified of the suspension or revocation along with the reasons for the decision and any appeal mechanisms available to them.

Loss of Establishment License

In the most serious cases of noncompliance, a regulatory agency may choose to suspend or revoke a pharmaceutical company's establishment license. A firm's repeated failure and unwillingness to comply with cGMP requirements, or the submission of false data in support of an application, are instances where a regulatory agency may be compelled to take this type of enforcement action.

Debarment

FDA has the authority to prohibit specific individuals and companies from participating in certain aspects of the drug industry. If an individual is convicted of certain felonies or misdemeanors, such as those related to the drug approval process or FDA's regulation of drugs in general, FDA may debar that individual. *Debarment* means that the individual may no longer work for a firm holding a pending or approved drug product application at FDA. Moreover, FDA will not accept or review abbreviated drug applications submitted by debarred individuals. Companies are subject to debarment for felonies and some misdemeanors if the offenses are related to the development or approval of an abbreviated drug product application, which are related to generic drugs. A debarred company cannot submit, or assist others in submitting, future abbreviated drug product applications.

Voluntary Actions

Recalls. Under the Food, Drug, and Cosmetic Act, FDA does not have the statutory authority to mandate that a firm recall or remove drug products from distribution. In those cases where the Agency believes that the product may pose a threat to consumer health and safety, FDA may recommend or request a field action, but companies are under no regulatory obligation, in the United States, to comply. In reality, however, most companies recall or remove product voluntarily from the field, and it is only in rare situations that FDA is required to take more immediate and serious actions (for example, court-ordered recall or seizure).

Like Warning Letters, information related to the recall of drug products is made public. FDA publishes a weekly Enforcement Report that identifies product recalls across the industries that they oversee. For more serious recalls, a formal press release, either by the company or FDA, may be initiated to alert the public of the potential health hazard.

Pharmaceutical companies need to have a procedure in place to handle recalls, and should have systems in place that maintain traceability of product distributed to their customers. Procedures should identify the person or persons responsible for the recall decision and the individual or group that is responsible for carrying out the recall action. In the absence of an actual recall, it is recommended that a pharmaceutical company conduct mock recalls to exercise their procedures and ensure that their systems are adequate to reconcile 100% of their distributed product.

There are three classifications of recalls:

- Class I is a situation in which there is a reasonable probability that the use of, or exposure to, a violative product will cause serious adverse health consequences or death.
- Class II is a situation in which use of, or exposure to, a violative product may cause temporary or medically reversible adverse health consequences or where the probability of serious adverse health consequences is remote.
- Class III is a situation in which use of, or exposure to, a violative product is not likely to cause adverse health consequences.

The recall strategy will specify the method(s) to be used and the level of effectiveness checks that will be conducted, as follows:

- Level A—100% of the total number of consignees to be contacted
- Level B—Some percentage of the total number of consignees to be contacted; which percentage is to be determined on a case-by-case basis, but is greater than 10% and less than 100% of the total number of consignees
- Level C—10% of the total number of consignees to be contacted
- Level D—2% of the total number of consignees to be contacted
- Level E—No effectiveness checks

Field Corrections and Market Withdrawals. Field corrections and market withdrawals are two other voluntary mechanisms to repair or remove questionable product that has left the control of the pharmaceutical company. In the case of a field correction, the repair, modification, adjustment, relabeling, destruction, or inspection of a product is conducted without physically removing it to some other location. In the case of a market withdrawal, the company removes or corrects a distributed product that involves a minor violation that would not be subject to legal action by FDA.

Chapter 6
Regulatory Agency Reporting

Once a pharmaceutical product is introduced into commercial distribution, a number of changes or events can trigger a requirement to submit information to the applicable regulatory authorities. These events can be planned (for example, a predetermined change to the product or manufacturing process) or unplanned (for example, failure of a drug product on stability), but, in either case, the burden of responsibility for reporting these events falls to the pharmaceutical manufacturer. Professionals working in a *good manufacturing practices* (GMP) environment must have an understanding of the types of events (both planned and unplanned) that result in a regulatory reporting obligation, and the specific requirements for submitting the required reports. Failure to submit the required reports can result in enforcement action against the firm.

POST-MARKETING CHANGES

> Describe how post-marketing changes to specifications, processes, and methods are assessed for impact to determine the appropriate reporting method [e.g., scale up and post-approval changes (SUPAC)]. (Understand)
>
> **Body of Knowledge I.F.1**

It is not uncommon for changes (or variations) to marketing applications/authorizations to be necessary throughout the life cycle of the drug product. Changes resulting from increased technical or process knowledge, or additional safety and quality information resulting from consumer experience or post-marketing studies, need to be effectively assessed and communicated to the appropriate regulatory agencies for inclusion in the approved application.

In the United States, the Food and Drug Administration (FDA) seeks to provide clarity around the types of changes and the reporting requirements through their guidance document *Changes to an Approved NDA or ANDA*. In this guidance document, FDA outlines three categories of changes based on their potential to have an adverse effect on the identity, strength, quality, purity, or potency of a drug product (Table 6.1).

In Europe, the marketing authorization holder is required to take into account technical and scientific progress and to submit amendments that incorporate changes (variations) based on that progress. Additionally, marketing authorization holders may wish to alter or improve the drug product, and those changes or variations also require submission of supplements or amendments to their approved application. The regulations in Europe have classified these changes or variations into two categories:

- *Minor variations.* Type IA (administrative changes or simple changes with no potential impact on the safety of the drug product) and Type IB (other minor changes as specified in Annex I to Commission Regulation 1085/2003)
- *Major variations.* All changes (other than Type IA or IB) that require approval before the marketing authorization holder can move forward with the change

EVALUATING POST-MARKETING CHANGES

The holder of the marketing application or authorization is responsible for assessing the effects of any change before distributing a drug product made with the change in place. Drug manufacturers should have a formal change control process in place to identify what needs to be done to complete the assessment and to document the outcome of the assessment process. While the assessment may vary depending on the type of change, some of the typical steps that may be taken to evaluate the impact of a change on the drug product are standard.

Table 6.1 Three categories of changes based on potential for adverse effect on identity, strength, quality, purity, or potency of a drug product.

Category	Criteria
Major	Change has a substantial potential to have an adverse effect on the identity, strength, quality, safety, purity, or potency of a drug product. A submission to the Agency, and approval of the submission by the Agency, are required before distribution of the drug product made using the change.
Moderate	Change has a moderate potential to have an adverse effect on the identity, strength, quality, safety, purity, or potency of a drug product. A submission to the Agency before the distribution of the drug product made using the change is required. Two types of submissions have been identified for moderate changes.
Minor	Change has minimal potential to have an adverse effect on the identity, strength, quality, safety, purity, or potency of a drug product. A submission to the Agency before the distribution of the drug product made using the change is not required.

Agency = United States Food and Drug Administration

Technical Assessment

It is important to evaluate the impact of the change on the manufacturing process, and any impact on the finished dosage form. This evaluation should pay close attention to the impact of the change on the *critical quality attributes* (CQA) of the product and the *critical processing parameters* (CPP) for the drug manufacturing process. The technical assessment should determine the impact on the validated state of the process and determine whether additional validation activities will be necessary to demonstrate that the process remains in a state of control.

Analytical Assessment

First, it is important to determine whether the change has an impact on the approved regulatory specifications. Changes to incoming raw materials or changes in the manufacturing process may have an impact on the ability of the finished product to meet the regulatory specifications identified in the application.

Depending on the type of change proposed, a change to the regulatory specifications might be necessary. A change that requires a change in the approved specifications will almost always require approval by the regulatory agency involved before product made with the change can be distributed.

Second, it may be important to perform additional analytical testing to demonstrate that the change has not had, or will not have, an adverse effect on the bioavailability, stability, or any of the chemical, physical, microbiological, and biological properties of the drug product. Providing a scientific argument or data that demonstrate that the product manufactured with the change will maintain an appropriate level of quality over time is often required before making a change.

Equivalence Assessment

It may be prudent, depending on the type of change, to conduct an assessment of product manufactured before the change versus product manufactured after the change to demonstrate that they are equivalent. Simply stated, is the finished drug product made after the change equivalent to the drug product made before the change? Demonstrating in vitro equivalence may be necessary to preclude the need for in vivo studies.

Adverse Effect

The change needs to be evaluated to determine whether the particular change in question has an adverse effect on the identity, strength, quality, purity, or potency of the drug product. While it seems counterintuitive that a company would move forward with a change that has an adverse impact, an application holder may be forced to do so as a result of changes beyond their control. For example, a change made by an *active pharmaceutical ingredient* (API) supplier may result in a change in the impurity profile for the finished drug product. If the change is necessary for the API supplier, the pharmaceutical company may be forced to move forward with the change to be able to continue to market the product. While these types of situations are rare, they do occur, and the pharmaceutical company must have systems in place to effectively evaluate these types of changes.

Reporting Strategy

The required reporting strategy may differ depending on the type of change. It is critical that the group responsible for determining the regulatory strategy (typically Regulatory Affairs) is made aware of the change(s) early on so that the requirements for submitting the change(s) to the regulatory agency can be evaluated and the appropriate timing for implementing the change(s) can be determined. It is often through this assessment that the need for additional testing or manufacture of batches made with the change in place is identified.

REGULATORY REPORTING REQUIREMENTS

> Describe global reporting requirements including supplements, field alerts, biological product deviation reports, adverse events, product recalls, annual reports, and variations to dossiers and applications. (Understand)
>
> **Body of Knowledge I.F.2**

Planned Changes

Requirements in the United States. For drug products approved in the United States, FDA identifies four reporting categories depending on the classification of the change:

- *Prior Approval Supplement* (PAS). A PAS is recommended for anything classified as a major change. In the case of a PAS, the company must wait to distribute product made with the change until after they receive approval from FDA on the supplement.
- *Supplement—Changes Being Effected in 30 Days* (CBE-30). A CBE-30 is recommended for certain types of moderate changes. In the case of a CBE-30, the supplement needs to be submitted to FDA, and the application holder must wait at least 30 days before the distribution of the drug product. If FDA informs the applicant within 30 days of receipt of the supplement that information is missing, or that a PAS is required, distribution of product must be delayed until the missing information is provided or the PAS is approved. If, after review of the CBE-30, FDA disagrees with the change, they may request that the manufacturer cease distribution of the drug products made using the change.
- *Supplement—Changes Being Effected* (CBE-0). A CBE-0 is recommended for certain types of moderate changes. In the case of a CBE-0,

distribution of product manufactured with the change can occur once FDA receives the supplement. Similarly to what can transpire with the CBE-30, if FDA disagrees with the change, they may request that the manufacturer cease distribution of the drug products made using the change.

- *Annual reports.* The submission of annual reports for marketed drug products covered by an approved application should include a description of minor changes made to the drug product over the time period covered by the annual report. As a reminder, these minor changes are those changes that have a minimal potential to have an adverse impact on the identity, strength, quality, safety, purity, or potency of the drug product. It is not appropriate to include changes with either a substantial potential or moderate potential for adverse impact in a drug product annual report.

It is important to note that a single change can require more than one change to the approved application. For example, a change in manufacturing site may involve a change in the equipment used to manufacture the product, or a change in the API may involve a change in the processing parameters for the manufacturing process or necessitate a change in the approved regulatory specifications. For multiple related changes where the recommended reporting category for the individual changes differs, FDA recommends that the submission strategy use the most restrictive of the categories recommended for the individual changes. When the multiple related changes all have the same recommended reporting category, FDA recommends that the submission use the reporting category for the individual changes.

Requirements in Europe. For drug products approved in Europe, a reporting system similar to that of FDA is used depending on the type of variation (change) to the marketing authorization being proposed. The submission requirements differ depending on the type of variation, as does the ability of the manufacturer to move forward with implementing the change in commercial product. Three types of variations are recognized:

- *Type IA variations.* The application holder should verify that all the conditions for a type IA notification are met and that all the documentation requirements are complete. Once the type IA notification is received by the applicable member state (reference member state), the clock is started. By day 14, the reference member state is required to notify the applicant regarding the status of the notification, that is, the notification is acknowledged (accepted) or it is not accepted along with the reason for nonacceptance. If it is acknowledged, or if the application holder has not received communication from the reference member state that the notification has not been approved, the application holder may move forward with the change and distribution of product manufactured with the change.

- *Type IB variations.* Similar to a type IA notification, the application holder must ensure that all the conditions for the type IB notification are met and that all documentation requirements are complete and included with the notification. In the case of a type IB notification, the clock does not start immediately upon receipt. The reference member state has 10 working days to verify that the requirements for a type IB notification have been met and that all the supporting data are included. After that 10-day period, the reference member state will inform the applicant of the start date. By day 30 after the clock start date, the reference member state will communicate the status of the notification (approval or non-approval). Again similarly to type IA notifications, if it is acknowledged, or if the application holder has not received communication from the reference member state that the notification has not been approved, the application holder may move forward with the change and distribution of product manufactured with the change.

- *Type II variations.* These types of variations are typically processed within 60 days of the clock start date; however, the regulations do allow for a reduced or extended assessment period. As with the type IB variations, the reference member state has 10 working days to validate the submission, and they must notify the application holder of the clock start date. Where there is agreement between member states regarding the outcome of the assessment, the reference member state will send a formal approval or refusal letter to the applicant. The application holder must wait until they receive a formal approval letter before implementing the change.

PRODUCT SURVEILLANCE

> Describe monitoring requirements for risk evaluation and mitigation strategy (REMS) and pharmacovigilance. (Understand)
>
> **Body of Knowledge I.F.3**

Most often, unplanned events related to product quality issues or failure of the drug product to meet regulatory specifications throughout its intended shelf life trigger regulatory agency reporting. In those situations where the event is unplanned, strict and very tight deadlines exist for providing the reports to the appropriate regulatory agencies. It is critical to understand the reporting requirements specific to the country or countries in which the products are marketed and the deadlines for submission of the initial reports. Once the issue or event is made known to the pharmaceutical manufacturer, the clock starts, and it is incumbent on the individuals responsible for the reporting to act quickly and to gather as much relevant information as possible in a short period of time.

FDA Field Alert Reports. The FDA Field Alert reporting requirements, as outlined in 21 CFR 314.81, first became effective in May 1985. This regulation requires holders of New Drug Applications (NDA) and Abbreviated New Drug Applications (ANDA) to submit information about distributed drug products to their jurisdictional FDA district office within three business days of receipt of the information by the application holder. These reports, unlike the Adverse Drug Experience (ADE) Reports required under 21 CFR 314.80, are required when a manufacturer is made aware of specific quality issues affecting batches of product that remain in commercial distribution.

For example, generation of an out-of-specification (OOS) result for a finished drug product would be considered the type of "information concerning any failure" described in this regulation. Unless the OOS result is found, through investigation, to be invalid within three days, an initial Field Alert Report should be submitted. Follow-up reports are submitted as new information is identified during the investigation; a final Field Alert Report is submitted once the investigation is finalized and closed.

Biological Product Deviation Reports. Under 21 CFR 600.14, manufacturers of licensed biologic products are required to promptly notify FDA of deviations in the manufacturing of products that may affect their safety, purity, or potency. Manufacturers must report any event, as well as any information relevant to the event, associated with the manufacturing of the biologic product if that event:

- Represents a deviation from GMP, applicable regulations, applicable standards, or established specifications that may affect the safety, purity, or potency of that product, or represents an unexpected or unforeseeable event that may affect the safety, purity, or potency of that product
- Occurs in the sponsor's facility or a facility under contract with the sponsor
- Involves a distributed biologic product

It is important to note that the expectation regarding the decision to report should be based on whether the event has the potential to affect the safety, purity, or potency of a product.

These reports should be submitted as soon as possible to FDA but, in all cases, no more than 45 calendar days from the date the deviation was first discovered. Like the Field Alert Reports, follow-up reports are required to update the Agency on any actions taken since the initial report. A final report is also required to close these reports.

Rapid Alert Reports. The alert system agreed on in the framework of the *mutual recognition agreement* (MRA) ensures that batch recalls and other safety measures resulting from quality defects of medicinal products discovered by one party are transmitted to the MRA partner without delay. It is based on the current *rapid alert system* (RAS) operational in the European Economic Area (EEA) and in the respective MRA partner country (Australia, Canada, Switzerland, and New Zealand). The alert system provides for communication about counterfeiting and fraud. Suspension and withdrawals of manufacturing authorization are communicated through the alert system as well.

The alert is classified from 1 to 3 depending on the expected risk presented to the public or animal health by the defective product. This classification is internationally agreed on for medicinal product recalls (Table 6.2). In the case of class 1 recalls, a rapid alert notification must be sent to all European Economic Area

Table 6.2 The rapid alert system alerts are classified from 1 to 3 depending on the expected risk presented to the public or animal health by the defective product. This classification is internationally agreed on for medicinal product recalls.

Class	Criteria
1	The defect presents a life-threatening or serious risk to health.
2	The defect may cause mistreatment or harm to the patient or animal, but it is not life threatening or serious.
3	The defect is unlikely to cause harm to the patient, and the recall is carried out for other reasons, such as noncompliance with the marketing application or specification.

member states, and the Pharmaceutical Inspection Convention and Pharmaceutical Inspection Co-operation Scheme (PIC/S), European Directorate for the Quality of Medicines & HealthCare (EDQM), World Health Organization (WHO), FDA, and MRA partners, irrespective of whether the batch was exported to a particular country. In most cases, class 1 recalls are to patient level; however, this action may not be appropriate if alternative medicines are not available, so that an assessment must be conducted by the reporter of the overall risk to patients/animals. Consideration must be given to the difficulties of communicating to patients since marketing authorization holders may need to arrange press releases and advertising campaigns.

Class 2 rapid alert notifications should be sent only to those EEA member states and PIC/S, EDQM, WHO, and MRA partners to which it is known, or believed, that the batch has been distributed. In identifying those countries, due consideration should be given to parallel distribution and import arrangements and the free trade between wholesale distributors within the EEA. In the case of parallel imports where there is difficulty in establishing the traceability of batches, consideration should be given to notifying all member states through the RAS.

Class 3 recalls are not notified through the RAS. For class 2 and class 3 recalls, recall to patient level is rarely required as lack of the product may present a greater risk to the patient than continuing treatment. Occasionally, class 2 or class 3 recalls can be accomplished solely at the wholesale level in circumstances such as when stocks are unlikely to be found farther down the supply chain and the level of risk is sufficiently low.

Chapter 7

Site Master File (SMF), Validation Master Plan (VMP), Drug Master File (DMF), and Site Reference File (SRF)

> Describe the purpose and content of Site Master Files (SMFs), Validation Master Plans (VMPs), Drug Master Files (DMFs), and Site Reference Files (SRFs). (Understand)
>
> **Body of Knowledge I.G**

SITE MASTER FILE

A *site master file* (SMF) is a succinct document that provides information about the control of pharmaceutical manufacturing operations for a production site. The document is prepared by the manufacturer and contains information about the specific *good manufacturing practices* (GMP) operations carried out at the named site and any closely integrated operations. If only part of a pharmaceutical operation is done at the site (for example, testing, packaging), the SMF describes only those operations performed at the site.

When submitted to a regulatory authority, the SMF provides information on the manufacturer's operations and procedures that can be useful in the efficient planning and undertaking of a GMP inspection. The GMP inspection report issued for inspections conducted by the European Medicines Agency (EMA) includes a notation on whether an SMF was available before the inspection, and the date of the SMF is included in the report.

SMFs are required by the European Union (EU), Medicines and Healthcare Products Regulatory Agency (MHRA), Pharmaceutical Inspection Convention and the Pharmaceutical Inspection Co-operation Scheme (PIC/S), and the World Health Organization (WHO). United States Food and Drug Administration (FDA) regulations currently do not require submission or maintenance of an SMF. The *mutual recognition agreements* (MRAs) between the European Community (EC) and Canada, New Zealand, and Australia include the SMF as part of a full inspection report that may be shared upon request.

The structure, formatting, and length of the SMF are prescribed by several guidance documents, including those published by WHO (guidelines for drafting SMF), PIC/S (explanatory notes), and the MHRA (guidance notes) (Figure 7.1). The

- Each page is a loose and individually numbered A4 sheet.
- Each page has an edition (revision) number and an effective date.
- Each chapter (1 to 9) starts on a new page so that updates can be provided and the relevant sheets replaced.
- The format and headings follow those given in the guidance notes.
- Wherever possible, simple plans, outline drawings, or schematic layouts are used instead of a narrative. These plans, drawings, or layouts must fit on A4 sheets of paper.
- If narrative or page limitations are not included in a section, information may be included in a table, tabular list, or bulleted format.

Figure 7.1 Formatting requirements for a site master file.

SMF should be succinct and, where possible, not more than 25 to 30 A4 pages. For small companies, particularly those with a small product range, the SMF may be fewer than 25 pages. It is expected that the SMF will be maintained and improved to facilitate inspections. The SMF contains nine chapters (Table 7.1).

Content of Site Master Files

SMF Chapter 1. General Information

- *Name, address, and company description* (C.1.1): Includes a brief description of the company, relation to other sites, and any information relevant to understanding the manufacturing operations.
- *Licensed pharmaceutical manufacturing activities* (C.1.2): Provides the relevant document issued by a national authority, and, if applicable, the period of validity for the manufacturing license. This section includes any conditions/restrictions of the manufacturing license and, where applicable, the *drug master file* (DMF) for each product.
- *Description of other pharmaceutical and nonpharmaceutical activities* (C.1.3).
- *Name and address of the site* (C.1.4): Includes the name of the site, the complete street address and postal address, the telephone number, fax number, and e-mail address of a contact person, and a 24-hour contact telephone number.
- *Products manufactured at the site* (C.1.5): Lists the types of products manufactured at the site and provides information about specific toxic or hazardous substances handled, including their manufacturing. This section provides information on whether the products are manufactured in a dedicated facility or on a campaign basis and states whether both human and veterinary products are manufactured on the site. For contract manufacturing or analytical sites, this section defines whether the firm is the contract giver or acceptor.

Table 7.1 Site master files contain nine chapters.

Chapter number	Title
1	General Information
2	Personnel
3	Premises and Equipment
4	Documentation
5	Production
6	Quality Control
7	Contract Manufacturing and Analysis
8	Distribution, Complaints, and Product Recall
9	Self-Inspection

- *Description of the site* (C.1.6): Provides the location and immediate environment, and the size of the site, types of buildings, and their ages.
- *Number of employees engaged in production, quality control, storage, and distribution* (C.1.7): Provides the total number of employees and the number of employees engaged in production, *quality control* (QC), *quality assurance* (QA), storage and distribution, and technical and engineering support services.
- *Listing and description of outside scientific, analytical, or other technical assistance in relation to manufacture and analysis* (C.1.8): Provides the company name, address, telephone number, and fax number of each outside contractor, and includes a brief outline (not more than 100 words) of the activity being undertaken. Also includes the establishment license number, where applicable.
- *Description of the quality management system* (QMS) (C.1.9): Provides a short description (750 words or three A4 pages) of the QMS of the firm responsible for manufacture. Describes the quality policy and the responsibilities of the QA function, including audit programs, approval and starting, primary packaging material suppliers, batch release, and document control. The responsibilities specific to the head of quality, QA manager, and QC manager are included. Describes the elements of the QA system (for example, organizational structure, responsibilities, procedures, and processes, specifications, test methods, and other quality-related data collection). It is important to show that quality management is independent from production management. Briefly describes the self-inspections or audits undertaken by external organizations. A cross-reference to SMF Chapter 9 can be added for additional details regarding the self-inspection program. Describes how quality system results are

reviewed to demonstrate the adequacy of the quality system in relation to the objective. Identifies the standards used to assess or audit the company's quality system, or used by the company to assess suppliers, such as ISO 9001, 21 CFR 211, and EU GMP. When suppliers of critical materials and packing materials are assessed, this section details how this is done, and describes the supplier selection process.

SMF Chapter 2. Personnel

- *Organization charts* (C.2.1): Organization charts define the overall organizational structure for QA, production, and QC, with detailed charts for each area depicting senior managers and supervisors only. As this information also is included in Section 2.1, a cross-reference to that section can be used.
- *Qualifications, experience, and responsibilities of key personnel* (C.2.2): Includes brief details of academic qualifications, work-related qualifications, and years of relevant experience since qualifying.
- *Training program* (C.2.3): Describes how training needs are identified and by whom. Gives details of training relative to GMP requirements, including new employee training and continuous GMP training. Describes job-related skills training and in-house and external training, and how practical experience is gained. Explains how retraining needs are identified and details how training is recorded and records are maintained.
- *Health requirements for production personnel* (C.2.4): Describes who is responsible for checking the health of employees. Describes requirements for pre-employment medical examinations and how employees are checked depending on the nature of their work. Describes the system for reporting sickness or contact with sick people before working in a critical area. Provides information on any system of reporting back to work after illness, and the requirements for personnel who work in sterile areas and any additional monitoring to which they may be subjected.
- *Personnel hygiene and clothing requirements* (C.2.5): States the washing, changing, and rest areas and the clothing/gowning requirements for the production areas. Gives instructions on how clothing should be used and when it should be changed. Describes gowning room/change room procedures. States the rules on eating, drinking, smoking, and use of chewing gum or tobacco. It is helpful to inspectors to include references to the applicable procedures for hygiene and clothing requirements.

SMF Chapter 3. Premises and Equipment

- *Site plan and manufacturing areas* (C.3.1): Provides a site plan highlighting the production areas, and a simple plan of each production area with indication of scale. For sterile product areas, indicates room and area classification(s) and pressure differentials between adjoining areas of different classifications.

- *Nature of construction and finishes* (C.3.2): A narrative format is preferred for this section. To reduce the length of the narrative for a large, complex plant, the details should be limited to critical areas, including processing and packaging areas. Describes wall construction, nature of finishes, floors, ceilings, doors and windows, lighting, piping, and drainage system(s).

- *Brief description of ventilation system* (C.3.3): To reduce the length of the narrative, schematic drawings should be used where possible. Details should be given for critical areas with potential risks of airborne contamination, including sterile product areas and areas for processing powders, granulation, and tableting. For manufacture of sterile or aseptic products, classification of the rooms used for production should be mentioned. A summary of the results of the most recent qualification/requalification should also be included. The MHRA guidance specifies that room classification should be given in accordance with the grading system outlined in the *EC/PIC Guide to GMP.* Outline the *heating, ventilation, and air conditioning* (HVAC) system, areas with different classes of air, pressure differential principles to prevent cross-contamination, dedicated *air handling units* (AHUs), and forced ventilation systems. The limits for changing the filters should be given. If a *drop-off point* (DOP) is introduced, the point must be shown.

- *Special areas for the handling of highly toxic, hazardous, and sensitizing materials* (C.3.4): This section is similar to section C.3.1, and includes information such as whether materials are handled only in the QC laboratory, whether solids or waste are neutralized and collected in separate containers, and whether extracted gases from the fume hood are neutralized and scrubbed before they are allowed to exit into the external environment.

- *Description of water and steam systems* (C.3.5): This section provides a schematic drawing back to the main city supply system or source of raw water, and the capacity of the system (maximum quantity produced per hour). Construction materials of the vessels and distribution system (pipework) are provided, along with specification of any filters in the system. If water is stored and circulated, the temperature at the point of return is provided, and the specifications of the water produced (that is, chemical, conductivity, and microbiological descriptions) are provided. The sampling points and frequency of sampling should be included. If a steam system is needed, the same sort of information must be provided for the system.

- *Description of planned preventive maintenance program and recording system* (C.3.6): "Maintenance" is done by the manufacturer and "servicing" by an outside contractor. This section describes the planned preventive maintenance program, including written procedures and suitable reporting forms for maintenance and servicing.

- *Brief description of major production and control laboratories equipment* (C.3.7): This section provides a list of equipment. Although the make and model numbers of the equipment are not required, several points should be addressed, including appropriateness and validity of construction material and ease of cleaning. For specific pieces of machinery (for example, a rotary tablet press), only a general description is required. If the equipment has additional devices, however, these devices should be recorded (for example, automatic weighing machines with printer, a labeler incorporating a bar code reader for the label). Further information is required for QC laboratory and microbiology laboratory and for computers and microprocessors in the manufacturing facility.
- *Maintenance (description of planned preventive maintenance program and recording system)* (C.3.8): Identifies who is responsible for maintenance and servicing and whether there are written procedures and contractual details for servicing. Maintenance routines that could affect product quality must be identified, and annual planned preventive maintenance programs for equipment must be in accordance with machine/instrument manufacturers' requirements. Documentation must be provided that records are kept on the type and frequency of service or check, details of service repairs and modifications, and reports provided to the users.
- *Qualification, validation, and calibration, including recording system and arrangements for computerized systems validation* (C.3.9): This section briefly describes the company's general policy and protocols for qualification, prospective validation, and retrospective validation, and describes and outlines the essential steps of the manufacturing process validation and cleaning validation. Any revalidation policy must be described. An outline of process validation may be given here or cross-referenced to production paragraph C.5.4. A description of the system for the release for sale or supply of development and validation batches should be provided, as well as the arrangements for computer validation, including software validation. The equipment calibration policy and description of how calibration records are kept should be included in this section.
- *Availability of written specifications and procedures for cleaning manufacturing areas and equipment* (C.3.10): This section summarizes the cleaning strategy, schedules, and documentation for manufacturing areas and equipment. If cleaning agents are changed periodically, it should be noted, along with any validation of cleaning processes, including the cleaning methods (and their frequency) for the water supply system, air handling system, and dust extraction system.

SMF Chapter 4. Documentation

- *Arrangements for the preparation, revision, and distribution of necessary documentation for manufacture* (C.4.1): This section includes a description of the documentation system and who is responsible for the preparation, revision, and distribution of documents. The storage place for the master documents is described, as well as type of documents stored (that is, product/process specifications, raw material specifications, packaging component specifications, standard process instructions including packaging, batch records including packaging, analytical methods, and QA release procedures). Information on how documentation is prepared and controlled and retention policy is provided, as well as the arrangements for any electronic or microfilmed records.

- *Other documentation related to product quality* (C.4.2): This section includes any other relevant documentation related to product quality that is not described elsewhere (for example, microbiological controls on air and water). This section should list and briefly explain the use of any additional standard documentation used routinely, including documentation for equipment specifications, specifications for disposables, *standard operating procedures* (SOPs), QC procedures, training procedures, computer program specifications, documentation control of process deviations, other validation documents, and reconciliation of batches of raw materials, bulk product, and major packing components (for example, product-contact and printed materials).

SMF Chapter 5. Production

- *Brief description of production operations using, wherever possible, flow sheets and charts specifying important parameters* (C.5.1): This section should describe the operations capable of being done at the site with the existing facilities, and specify the categories of medicinal products produced. In facilities where only packaging is undertaken, it is necessary only to provide a brief description of labeling and filling and the nature of the containers used (for example, tamper-evident glass containers). More information is required in this section if cytotoxic or radioactive substances are handled. Flowcharts are useful in describing the production operations; technical details are not required.

- *Arrangements for the handling of starting materials, packaging materials, bulk and finished products, including sampling, quarantine, release, and storage* (C.5.2): Purchasing procedures (approval of suppliers of active pharmaceutical ingredients, excipients, and primary packing materials) and material requisition procedures from stores to manufacturing plant and vice versa, including sampling, quarantine, release, and storage, are all described in this section. Identification of the supplier's lot number with the company's lot number is provided, along with sampling plans. Other details to be included are status labeling (for example, by using labels or by computer), issue of materials to manufacture and package, control of weighing, checking methods, and how materials being used for manufacture are identified and confirmed. Information for the control of bulk manufacturing (that is, records of key parameters and in-process checks) and packing (that is, release of bulk, semifinished product, and packing materials, along with in-process checks) is provided.

- *Arrangements for the handling of rejected materials and products* (C.5.3): Information must be given to show that rejected materials are clearly labeled and that they are stored in a dedicated area. This section also describes the arrangements for sentencing the materials and their disposal.
- *Brief description of general policy for process validation* (C.5.4): An outline of process validation protocol only is required. An outline may be given here or cross-referenced to SMF Chapter 3 (C.3.9).

SMF Chapter 6. Quality Control

Chapter 6 provides a description of the QC system and activities, particularly the procedures for the release of finished products. The information includes a description of the analytical testing, packaging, component testing, and microbiological testing, and outlines the arrangements for the preparation, revision, and distribution of documents for specification test methods and release criteria, if not mentioned elsewhere (see also paragraph C.1.9 and SMF Chapter 4, Documentation).

SMF Chapter 7. Contract Manufacture and Analysis

This chapter describes how GMP compliance of the contract acceptor is assessed, as well as the technical contract between the contract giver and acceptor, and the way in which the GMP compliance is assessed to ensure product compliance with the marketing authorization.

SMF Chapter 8. Distribution, Complaints, and Product Recalls

- *Arrangements and recording system for distribution* (C.8.1): A description of the storage and distribution practices, ways of securing the warehouse, environmental controls of the warehouse, and how materials are stored is provided in this section. It is important to describe how the status of products is controlled (for example, by computer or by label), and how rejected materials are securely isolated. The authorities should be comfortable that the distribution records permit full batch traceability from the factory to the customer, in terms of the date of sale, customer details, and quantity dispatched, and that these records are readily available to permit an effective recall if required.
- *Arrangements for the handling of complaints and product recalls* (C.8.2): This section refers to the procedures for handling complaints and product recalls. Quality defects should be investigated and appropriate measures taken to prevent recurrence.
- *Complaints* (C.8.3): The person or persons responsible for logging, classifying, and investigating complaints is provided, along with

information about written reports, their review and retention, and how causes of quality defects are investigated.

- *Product recalls* (C.8.4): Who is responsible for product recalls? Is there a written procedure that describes the sequence of actions to be followed (for example, retrieval of distribution data, notification to customers, receipt/segregation/ inspection of returned product, investigation/reporting of cause, and reporting corrective action)? Who notifies the authorities of complaints and recalls? Is the authority involved in complaints and the decision to recall? Can recalls be effective below wholesale level? List products recalled over the last two years. If none has been recalled, record "none."

SMF Chapter 9. Self-Inspection

- Include references to documented procedures that describe the self-inspection program and follow-up activities to self-inspections.
- Describe how the self-inspection system verifies that activities that have a bearing on quality comply with planned arrangements (that is, are records reviewed and activities observed to verify compliance with documents, procedures, and regulations?).
- State whether the program assesses the effectiveness of the quality system.
- Describe how the results of the self-inspection system are documented and brought to the attention of the personnel having responsibility for the area and activities inspected.
- Describe how the system ensures that those responsible for the area or activity take timely corrective action on the deficiencies found.

VALIDATION MASTER PLAN (VMP)/MASTER VALIDATION PLAN (MVP)

The VMP, or MVP, document defines the overall structure, philosophy, and approach of the phases of the validation life cycle. The VMP should define the structure and required documentation for design, qualification, and validation activities, to include prerequisites, and stage-gate requirements to move on to the next phase in the validation life cycle. The VMP should detail key criteria within each stage and phase, and other qualification and validation requirements and considerations, including lists of products, processes, and systems to be validated, documentation format, required standard operating procedures (SOPs), and planning and scheduling. Consideration should also include triggers (times, cycles, changes, trends, and so on) for revalidation and/or retrospective validation activities.

DRUG MASTER FILE

A *drug master file* (DMF) is a submission to FDA of information, usually concerning the *chemistry, manufacturing, and controls* (CMC) of a component of a drug product, to permit FDA to review this information in support of a third party's submission. A DMF may be used to provide confidential, detailed information about facilities, processes, or articles used in the manufacturing, processing, packaging, and storing of one or more human drugs. Drug product information or other non-CMC information may be filed in a DMF.

The submission of a DMF is not required by law or FDA regulation. A DMF is submitted solely at the discretion of the holder to maintain confidentiality of proprietary information (for example, manufacturing procedures) and to permit access to information by reviewers in the Center for Drug Evaluation and Research (CDER) to support applications submitted by one or more applicants. The information contained in the DMF may be used to support an Investigational New Drug Application (IND), a New Drug Application (NDA), an Abbreviated New Drug Application (ANDA), another DMF, an export application, or amendments and supplements to any of these.

A DMF is not a substitute for an IND, NDA, ANDA, or export application. It is not approved or disapproved. Technical contents of a DMF are reviewed only in connection with the review of an IND, NDA, ANDA, or export application. The DMF is reviewed only when it is referenced in an application or another DMF. The DMF is reviewed using the same regulatory and scientific criteria as review of an application.

Inspections of drug substance manufacturers usually are triggered when there is an application under review that references a DMF for the manufacture of that drug substance.

Confidentiality of information in DMFs is assured through 21 CFR 314.420(e). As DMFs usually cover manufacturing information, they usually are not considered releasable under a Freedom of Information Act (FOIA) request. Applications and DMFs are different (Table 7.2).

Types of Drug Master Files

A drug master file (DMF) is a submission to the Food and Drug Administration (FDA) that may be used to provide confidential detailed information about facilities, processes, or articles used in the manufacturing, processing, packaging, and storing of one or more human drugs (Table 7.3). The submission of a DMF is not

Table 7.2 Differences between applications and drug master files.

Application	Drug master file
Submitted to a particular review division.	Submitted to Center for Drug Evaluation and Research
Each submission (including supplement) is entered into the application database and assigned to a reviewer, and an acknowledgment letter sent.	Each submission is entered into a database (different from application database) and no acknowledgment letter sent.
Each submission has a due date.	Reviewed only when referenced. No assignment to a reviewer, no due date.

required by law or FDA regulation. A DMF is submitted solely at the discretion of the holder. The information contained in the DMF may be used to support an Investigational New Drug Application (IND), a New Drug Application (NDA), an Abbreviated New Drug Application (ANDA), another DMF, an export application, or amendments and supplements to any of these.

A DMF is *not* a substitute for an IND, NDA, ANDA, or export application. It is not approved or disapproved. Technical contents of a DMF are reviewed only in connection with the review of an IND, NDA, ANDA, or an export application.

The recommendations in the guidance Drug Master Files: Guidelines are, in general, still applicable. However, additional information or clarification of the recommendations in the guidance provided falls into four categories:

Category 1: Recommendations that are no longer applicable due to changes in regulations or guidances.

Category 2: Additional clarification of recommendations in the guidance.

Category 3: New information for aspects of DMF filing that was not in effect when the guidance was written.

Category 4: Information related to the Generic Drug User Fee Act (GDUFA)

See www.fda.gov/drugs/forms-submission-requirements/drug-master-files-dmfs.

Changes to Drug Master Files

The DMF holder is expected to submit all changes to the DMF as amendments, notify FDA of change in holder name or address, notify FDA of change in agent/representative, submit an annual report or update, submit a *letter of authorization* (LOA) for each item referenced for each customer, and notify authorized parties of changes. While annual updates to a DMF are not required by regulation, Section VII of the DMF Guidelines recommends reporting a list of authorized parties, what they are authorized to reference, the date of the LOA, and a list of changes reported during the past year (that is, not a list of changes made, but a list of changes already reported). Stability updates should be reported as amendments.

If the anniversary date is missed, FDA will not send a reminder (unlike applications). If no changes have been made, it is important to send an update with a statement to that effect. It should be remembered that an *amendment* is a report of a change or addition of technical or administrative information. A *supplement* is not an amendment; it only applies to approved applications. Pages that replace an already numbered page from a previous submission should also contain the page number in the current submission (for example, a page replacing page 10 in the original submission may be page 14 in the new submission). Finally, no pages are ever physically replaced in a DMF.

PREPARING A SITE MASTER FILE (SMF)

Pharmaceutical Inspection Convention/Pharmaceutical Inspection Co-operation Scheme PE 008-4 contains the Explanatory Notes for Pharmaceutical Manufacturers on the Preparation of a Site Master File. The *site master file* contains specific

Table 7.3 Types of drug master files.

Type I	Manufacturing site, facilities, operating procedures, and personnel (no longer applicable)
Type II	Drug substance, drug substance intermediate, and material used in their preparation, or drug product
Type III	Packaging material
Type IV	Excipient, colorant, flavor, essence, or material used in their preparation
Type V	FDA accepted reference information

information about the quality management policies and activities of the site, the production and/or quality control of pharmaceutical manufacturing operations carried out at the named site, and any closely integrated operations at adjacent and nearby buildings. If only part of a pharmaceutical operation is carried out on the site, a site master file need only describe those operations, for example, analysis or packaging.

The site master file should provide information on the manufacturer's GMP-related activities that can be useful in general supervision and in the efficient planning and undertaking of inspections by regulatory authorities. A site master file should contain adequate information but, as far as possible, not exceed 25–30 pages plus appendixes. Simple plans, outline drawings, or schematic layouts are preferred instead of narratives.

The site master file should be a part of documentation belonging to the quality management system of the manufacturer and kept updated accordingly. The site master file should have an edition number, the date it becomes effective, and the date by which it has to be reviewed. It should be subject to regular review to ensure that it is up to date and representative of current activities. Each appendix can have an individual effective date.

CONTENT OF SITE MASTER FILE

1. General information on the manufacturer
   - 1.1 Contact information on the manufacturer
   - 1.2 Authorized pharmaceutical manufacturing activities of the site
   - 1.3 Any other manufacturing activities carried out on the site
2. Quality management system of the manufacturer
   - 2.1 The quality management system of the manufacturer
   - 2.2 Release procedure of finished products
   - 2.3 Management of suppliers and contractors
   - 2.4 Quality risk management (QRM)
   - 2.5 Product quality reviews

3. Personnel
4. Premises and equipment
    4.1 Premises
    4.2 Equipment
5. Documentation
6. Production
    6.1 Type of products
    6.2 Process validation
    6.3 Material management and warehousing
7. Quality control (QC)
8. Distribution, complaints, product defects, and recalls
    8.1 Distribution
    8.2 Complaints, product defects, and recalls
9. Self-inspections

Appendixes

Appendix 1 Copy of valid manufacturing authorization

Appendix 2 List of dosage forms manufactured, including the INN-names or common name (as available) of active pharmaceutical ingredients (API) used

Appendix 3 Copy of valid GMP certificate

Appendix 4 List of contract manufacturers and laboratories including contact information and flowcharts of the supply chain for these outsourced activities

Appendix 5 Organizational charts

Appendix 6 Layouts of production areas, including material and personnel flows, and general flowcharts of manufacturing processes of each product type (dosage form)

Appendix 7 Schematic drawings of water systems

Appendix 8 List of major production and laboratory equipment

Part II

Quality Systems

Chapter 8
Quality Management System (QMS)

Describe key elements of the structure of a Quality Management System (QMS). Outline the requirements for the development, operation, and management review for suitability and effectiveness as defined in ICH Q7, ICH Q10, EU GMP, and other guidances. (Evaluate)

Body of Knowledge II.B

A *quality management system* (QMS) can be defined as a set of policies, processes, and procedures required for planning and execution (production/development/service) in the core business area of an organization. A QMS integrates the various internal processes within the organization and provides a process approach for project execution. The QMS enables organizations to identify, measure, control, and improve the various core business processes, which ultimately leads to improved business performance.

The design and implementation of a QMS will vary depending on the type, size, and products of the organization. Implementing a QMS within an organization needs to be a decision made by top management. Continued support by top management is critical to the ongoing success of the QMS.

The objectives of the QMS need to be clearly defined so that the system can be effective. QMS objectives must be measurable and reflect the overall company objectives. The QMS must be managed properly, and adequate resources must be allocated. The system must be practical and accessible to all employees within the organization. The system also must be reviewed regularly and measured for effectiveness, and adjustments must be made to reflect major changes to the organization and business practices.

Each company will have its own objectives for implementing a QMS; however, enhancing company effectiveness and profitability are often cited as primary drivers for implementation. A QMS can assist a company by managing costs and risks, increasing effectiveness and productivity, identifying improvement opportunities, and increasing customer satisfaction. A well-managed QMS will have an impact on customer loyalty and repeat business, market share, operational efficiencies, and flexibility, and the ability to respond to market opportunities. It also helps in the effective and efficient use of resources and reducing costs, provides competitive advantages, and can increase participation and motivation of human resources, industry reputation, and control over all processes.

ESTABLISHING A QUALITY MANAGEMENT SYSTEM

A QMS must ensure that products and services conform to customer needs and expectations and, at the same time, meet the objectives of the organization. Factors to be considered when setting up a QMS include design and build, deployment and implementation, control, measurement, review, and improvement.

Design and build includes the structure of the QMS, the process, and its implementation. The design should be led by senior managers and ideally is accomplished using a framework to lead the thinking. Design of the QMS should result from determining the organization's core processes and well-defined goals and strategies, and be linked to the needs of the stakeholders. The process for designing and building the QMS must be clear. The quality function plays a key role in the design and build of the QMS, but involvement and buy-in to the system must come from all other functions as well.

Deployment and implementation are best achieved using process packages, where each core process is broken down into subprocesses and described by a combination of documentation, education, training, tools, systems, and metrics.

Control of the QMS will depend on the size and complexity of the organization. Local control, where possible, is effective, and good practice is found where key stakeholders are documented within the process and where the process owner is allowed to control the entire process. Ideally, process owners/operators are involved in writing the procedures that document the QMS.

Measurement determines the effectiveness and efficiency of each process in attaining its objectives. It should include the contribution of the QMS to the organization's goals, and could be achieved by measuring policy definition completeness, business coverage, reflection of policies, deployment, usage, opinion regarding helpfulness of the QMS by staff, speed of change of the QMS, and relevance of QMS architecture to the job at hand. A form of scorecard deployed throughout the organization down to the individual objective level can be used, and the setting of targets at all levels is vital.

Review of the effectiveness, efficiency, and capability of the QMS is crucial, and the outcome of these reviews should be communicated to all employees, including senior management. Reviewing and monitoring should be conducted whether or not improvement activities have achieved their expected outcomes.

Improvement should follow as a result of the review process, with the aim of seeking internal best practice. It is part of the overall improvement activities and an integral part of managing change within the organization.

QUALITY MANAGEMENT SYSTEM PRINCIPLES

ISO 9001:2015 *Quality management systems—Requirements* is an internationally recognized standard that specifies the requirements for a QMS that may be used by organizations for internal application, certification, or contractual purposes. ISO 9001 emphasizes the need for management commitment, customer focus, a quality policy, quality planning, communication of responsibility and authority, and the need for a review of the organization's QMS.

Monitoring of customer satisfaction is necessary to evaluate and validate whether customer requirements have been met. Reference is made in the standard to a methodology known as *plan–do–check–act* (PDCA) that can be applied to all processes.

The standard suggests that an effective QMS must be a strategic tool designed to deliver business objectives and must have, at its core, a *process approach*, with each process transforming one or more inputs to create an output of value to the customer. The key business processes may be supported by procedures and work instructions in those cases where it is judged necessary to rigidly define what rules are to be followed when undertaking a task. Most organizations will have core business processes that define those activities that directly add value to the product or service for the external customer, and supporting processes that are required to maintain the effectiveness of the core processes.

The understanding of the many interrelationships between these processes demands that a *systems* approach to management is adopted. The processes must be thoroughly understood and managed so that the most efficient use is made of available resources and to ensure that the needs of all the stakeholders—customers, employees, shareholders, and the community—are met.

Customer satisfaction is a constantly moving target depending on changes in technology and the marketplace, so an effective QMS must be in a state of continual improvement. For this goal to be achieved, attention needs to be given to both the voice of the customer—through complaint analysis, opinion surveys, and regular contacts—and the voice of the processes—through measurement, monitoring, and analysis of both process and product data.

ISO 9001:2015 (*Quality management systems—Requirements*) was released in 2015. The updated version of ISO 9001 followed the Annex L structure taking the standard from eight clauses to 10. Figure 8.1 depicts the structure of ISO 9001:2015. The updated standard places an emphasis on risk-based management and increases emphasis on achieving value for the organization and its customers. *Organizational context* emphasizes responsiveness to the ever-changing business environment by

ISO 9001:2015

0. Introduction
1. Scope
2. Normative references
3. Terms and definitions
4. Context of the organization
5. Leadership
6. Planning
7. Support
8. Operations
9. Performance evaluations
10. Improvement

Figure 8.1 ISO 9001:2015 clause structure.

determining external and internal issues that are relevant to the organization's purpose and its strategic direction and that affect its ability to achieve the intended results of its QMS.

QUALITY MANAGEMENT SYSTEM PRACTICES

Audits, Reviews, and Assessments

A good QMS will not function or improve without adequate audits and management reviews. Audits are done to ensure that actual practices adhere to documented procedures. System reviews should be done periodically and methodically to ensure that the system achieves the required effect.

A schedule for audits, with different activities requiring different frequencies, should be established. An audit should not be conducted solely for the purpose of revealing defects or irregularities; it should be conducted to establish the facts rather than to find fault. Audits may indicate necessary improvement and corrective actions, but they also are intended to determine whether processes are effective and responsibilities have been correctly assigned. The emphasis on process improvement, enhancing customer satisfaction, and risk management. ISO 9001:2015 requires a thoughtful approach to auditing. Although periodic is not defined, the general expiration by standards auditors and regulatory inspectors is internal audits should be conducted at a minimum annually.

The procedures for conducting audits and the results from them, should be documented and be available for review. Internal system audits should be positive and conducted as part of the preventive strategy, and not as a matter of expediency resulting from problems.

The assessment of a quality system against a standard or set of requirements by internal audit and review is known as a *first-party* assessment or approval scheme. If an external customer makes the assessment of a supplier against its own requirements, or a national or international standard, a *second-party* scheme is in operation. The assessment by an independent organization, not connected with any contract between the customer and supplier, but acceptable to them both, is an independent, *third-party* assessment scheme.

ICH Q10: Pharmaceutical Quality System

ICH Q10 describes one comprehensive model for an effective pharmaceutical quality management system that is based on International Organization for Standardization (ISO) quality concepts, includes applicable *good manufacturing practices* (GMP) regulations, and complements ICH Q8, *Pharmaceutical Development*, and ICH Q9, *Quality Risk Management.*

ICH Q10 is a model for a pharmaceutical QMS that can be implemented throughout the different stages of a product life cycle. Much of the content of ICH Q10 applicable to manufacturing sites is currently specified by regional GMP requirements and ICH Q7, *Good Manufacturing Practice Guidance for Active Ingredients.*

ICH Q10 demonstrates industry and regulatory authorities' support of an effective pharmaceutical QMS to enhance the quality and availability of medicines globally in the interest of public health. Implementation of ICH Q10 throughout

Table 8.1 Technical activities included in the product life cycle for new and existing products.

Pharmaceutical development	Drug substance development
	Formulation development, including container/closure system
	Manufacture of investigational products
	Delivery system development (where relevant)
	Manufacturing process development and scale-up
	Analytic method development
Technology transfer	New product transfers during the development to manufacturing stages
	Transfers within or between manufacturing and testing sites for marketed products
Commercial manufacturing	Acquisition and control of materials
	Provision of facilities, utilities, and equipment
	Production, including packaging and labeling
	Quality control and assurance
	Release
	Storage
	Distribution (excluding wholesaler activities)
Product discontinuation	Retention of documentation
	Sample retention
	Continued product assessment and reporting

the product life cycle should facilitate innovation and continual improvement and strengthen the link between pharmaceutical development and commercial manufacturing activities. The elements of ICH Q10 should be applied in a manner that is appropriate and proportionate to each of the product life cycle stages, recognizing the differences between, and the different goals of, each stage (Table 8.1).

ICH Q10 provides a harmonized model for a pharmaceutical QMS throughout the life cycle of a product, and is intended to be used together with regional GMP requirements. The QMS elements and management responsibilities described in this guideline encourage the use of science and risk-based approaches at each life cycle stage, thereby promoting continual improvement across the entire product life cycle.

ICH Q10 Objectives

Implementation of the Q10 model should result in achievement of three main objectives that complement or enhance regional GMP requirements:

- *Achieve product realization.* To establish, implement, and maintain a system that allows the delivery of products with the quality attributes appropriate to meet the needs of patients, healthcare professionals, regulatory authorities (including compliance with approved regulatory filings), and other internal and external customers.

- *Establish and maintain a state of control.* To develop and use effective monitoring and control systems for process performance and product quality, thereby providing assurance of continued suitability and capability of processes. Quality risk management can be useful in identifying monitoring and control systems.
- *Facilitate continual improvement.* To identify and implement appropriate product quality improvements, process improvements, variability reduction, innovations, and pharmaceutical quality system enhancements, thereby increasing the ability to fulfill quality needs consistently. Quality risk management can be useful for identifying and prioritizing areas for continual improvement.

Enablers: Knowledge Management and Quality Risk Management

Use of knowledge management and quality risk management enables a company to implement ICH Q10 effectively and successfully. These enablers facilitate achievement of the three regional GMP requirements objectives by providing the means for science- and risk-based decisions related to product quality.

Knowledge Management. *Knowledge management* is a systematic approach to acquiring, analyzing, storing, and disseminating information related to products, manufacturing processes, and components. Sources of knowledge include, but are not limited to, public domain or internally documented knowledge, pharmaceutical development studies, technology transfer activities, process validation studies over the product life cycle, manufacturing experience, innovation, continual improvement, and change management activities. Product and process knowledge should be managed from development through the commercial life of the product up to and including product discontinuation. Development activities using scientific approaches provide knowledge for product and process understanding.

Quality Risk Management. Quality risk management is integral to an effective pharmaceutical QMS. It can provide a proactive approach to identifying, scientifically evaluating, and controlling potential risks to quality. It facilitates continual improvement of process performance and product quality throughout the product life cycle. ICH Q9 provides principles and examples of tools for quality risk management that can be applied to different aspects of pharmaceutical quality.

Design and Content Considerations, Quality Manual

The design, organization, and documentation of the pharmaceutical QMS should be well structured and clear to facilitate common understanding and consistent application. The elements of ICH Q10 should be applied in a manner that is appropriate and proportionate to each of the product life cycle stages, recognizing the different goals and knowledge available for each stage. The size and complexity of the company's activities should be taken into consideration when developing a new pharmaceutical QMS or modifying an existing one.

- Quality policy
- Quality objectives
- Scope of the pharmaceutical QMS
- Identification of the pharmaceutical QMS processes, including sequences, linkages, and interdependencies, as well as process maps and flow
- Management responsibilities within the pharmaceutical QMS

- Quality policy
- Quality objectives
- Scope of the pharmaceutical QMS
- Identification of the pharmaceutical QMS processes, including sequences, linkages, and interdependencies, as well as process maps and flow
- Management responsibilities within the pharmaceutical QMS

Figure 8.2 Components of a quality manual.

The design of the pharmaceutical QMS should incorporate appropriate risk management principles. While some aspects of the pharmaceutical QMS can be companywide and others site specific, the effectiveness of the pharmaceutical QMS normally is demonstrated at the site level. The pharmaceutical QMS should include appropriate processes, resources, and responsibilities to provide assurance of the quality of outsourced activities and purchased materials. The pharmaceutical QMS should include process performance and product quality monitoring, corrective and preventive action, and change management and management review. Performance indicators should be identified and used to monitor the effectiveness of processes within the pharmaceutical QMS. Finally, management responsibilities should be identified within the pharmaceutical QMS. A quality manual or equivalent documentation approach should be established and should contain a description of the pharmaceutical QMS (Figure 8.2).

MANAGEMENT RESPONSIBILITY

Leadership is essential in establishing and maintaining a companywide commitment to quality and for the performance of the pharmaceutical QMS.

Management Commitment

Senior management holds the ultimate responsibility to ensure that an effective pharmaceutical QMS is in place to achieve the quality objectives. They are responsible for ensuring that roles, responsibilities, and authorities are defined, communicated, and implemented throughout the company. Management should:

- Participate in the design, implementation, monitoring, and maintenance of an effective pharmaceutical QMS.

- Demonstrate strong and visible support for the pharmaceutical QMS and ensure its implementation throughout their organization.
- Ensure that a timely and effective communication and escalation process exists to raise quality issues to the appropriate levels of management.
- Define individual and collective roles, responsibilities, authorities, and interrelationships of all organizational units related to the pharmaceutical QMS.
- Ensure that these interactions are communicated and understood at all levels of the organization. An independent quality unit or structure with authority to fulfill certain pharmaceutical quality system responsibilities is required by regional regulations.
- Conduct management reviews of process performance and product quality, and of the pharmaceutical QMS.
- Advocate continual improvement.
- Commit appropriate resources.

Quality Policy

Senior management should establish a *quality policy* that describes the overall intentions and direction of the company related to quality. The quality policy should include an expectation to comply with applicable regulatory requirements and should facilitate continual improvement of the pharmaceutical QMS. The quality policy should be communicated to, and understood by, personnel at all levels in the company. The quality policy should be reviewed periodically to ensure its continued effectiveness.

Quality Objectives and Planning

Senior management should ensure that the quality objectives needed to implement the quality policy are defined and communicated. Quality objectives should be supported by all relevant levels of the company. Quality objectives should align with the company's strategies and be consistent with the quality policy. Management should provide the appropriate resources and training to achieve the quality objectives. Performance indicators that measure progress against quality objectives should be established, monitored, communicated regularly, and acted on as appropriate.

Resource Management

Management should determine and provide adequate and appropriate resources (human, financial, materials, facilities, and equipment) to implement and maintain the pharmaceutical quality system and continually improve its effectiveness. Management should ensure that resources are appropriately applied to a specific product, process, or site.

Internal Communication

Management should ensure that appropriate communication processes are established and implemented within the organization. Communication processes should ensure the flow of appropriate information between all levels of the company and should ensure the appropriate and timely escalation of critical product quality and pharmaceutical quality system related issues.

Management Review

Senior management should be responsible for pharmaceutical QMS governance through management review to ensure its continued suitability and effectiveness. Management should assess the conclusions of periodic reviews of process performance and product quality, and of the pharmaceutical QMS. Although periodic is not defined, the general expiration by standards auditors and regulatory inspectors is management reviews should be conducted at a minimum annually.

Typically, a management review should cover:

- Results of audits
- Customer feedback
- Process and product conformity
- Status of preventive and corrective actions
- Follow-up actions from previous management reviews
- Changes that could affect the QMS
- Recommendations for improvements

Management review outputs should include improvements to the QMS and processes, product improvements related to customer requirements, and identification of resource needs. In addition, the procedures for conducting management reviews, and the results from them, should be documented and be available for review. Management reviews should be conducted as part of the preventive strategy, and not as a matter of expediency resulting from problems.

Management of Outsourced Activities and Purchased Materials

The pharmaceutical QMS, including the management responsibilities, extends to the control and review of any outsourced activities and quality of purchased materials. The pharmaceutical company is responsible to ensure that processes are in place to assure the control of outsourced activities and quality of purchased materials (Figure 8.3).

Management of Change in Product Ownership

When product ownership changes (for example, through acquisitions or divestitures), management should consider the complexity of these events and ensure

- Assessing, before outsourcing operations or selecting material suppliers, the suitability and competence of the other party to carry out the activity or provide the material using a defined process (for example, audits, material evaluations, qualification).
- Defining the responsibilities and communication processes for quality-related activities of the involved parties. For outsourced activities, this information should be included in a written agreement between the contract giver and contract acceptor.
- Monitoring and review of the performance of the contract acceptor or the quality of the material from the provider, and the identification and implementation of any needed improvements.
- Monitoring incoming ingredients and materials to ensure they are from approved sources using the agreed-on supply chain.

Figure 8.3 Management of outsourced activities and purchased materials.

that ongoing responsibilities are defined for each company involved, and the necessary information is shared or transferred.

CONTINUAL IMPROVEMENT OF PROCESS PERFORMANCE AND PRODUCT QUALITY

Life cycle stage goals and four specific pharmaceutical quality system elements—pharmaceutical development, technology transfer, commercial manufacturing, and product discontinuation—augment regional requirements to achieve the ICH Q10 objectives.

Life Cycle Stage Goals

Pharmaceutical Development. The goal of pharmaceutical development activities is to design a product and its manufacturing process to consistently deliver intended performance and meet the needs of patients, healthcare professionals, and regulatory authorities, and internal customers' requirements. Approaches to pharmaceutical development are described in ICH Q8. The results of exploratory and clinical development studies, while outside the scope of this guidance, are inputs to pharmaceutical development. Figure 8.4 depicts the pharmaceutical life cycle.

Technology Transfer. The goal of technology transfer activities is to transfer product and process knowledge between development and manufacturing and within or between manufacturing sites to achieve product realization. This knowledge forms the basis for the manufacturing process, control strategy, process validation approach, and ongoing continual improvement.

Commercial Manufacturing. The goals of manufacturing activities include achieving product realization, establishing and maintaining a state of control, and facilitating continual improvement. The pharmaceutical quality system should assure that the desired product quality is routinely met, suitable process performance is achieved, the set of controls is appropriate, improvement

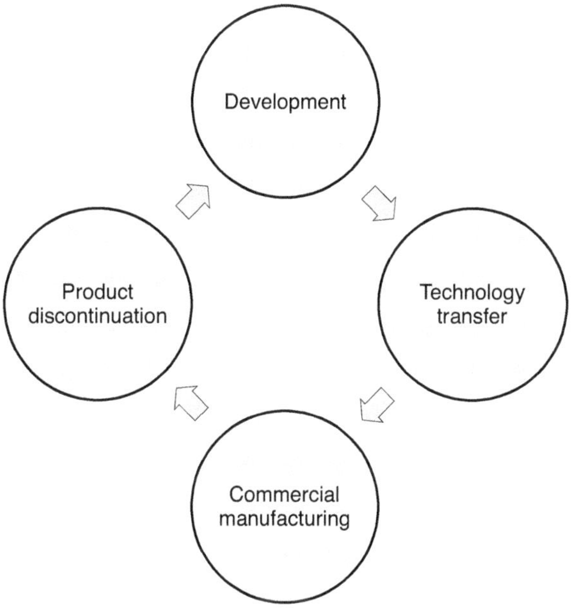

Figure 8.4 Pharmaceutical life cycle.

opportunities are identified and evaluated, and the body of knowledge is continually expanded.

Product Discontinuation. The goal of product discontinuation activities is to manage the terminal stage of the product life cycle effectively. For product discontinuation, a predefined approach should be used to manage activities such as retention of documentation and samples and continued product assessment (for example, complaint handling, adverse event reporting, and stability) and reporting in accordance with regulatory requirements.

Pharmaceutical Quality Management System Elements

Four elements are required, in part, under regional GMP regulations; however, the Q10 model's intent is to enhance these elements to promote the life cycle approach to product quality. These four elements are process performance and product quality monitoring system, corrective action and preventive action (CAPA) system, change management system, and management review of process performance and product quality. These elements should be applied in a manner that is appropriate and proportionate to each of the product life cycle stages, recognizing the differences between, and the different goals of, each stage. Throughout the product life cycle, companies are encouraged to evaluate opportunities for innovative approaches to improve product quality.

Process Performance and Product Quality Monitoring System. Pharmaceutical companies should plan and execute a system for the monitoring of process performance and product quality to ensure that a state of control is maintained

Table 8.2 Application of process performance and product quality monitoring system throughout the product life cycle.

Pharmaceutical development	Technology transfer	Commercial manufacturing	Product discontinuation
Process and product knowledge generated and process and product monitoring conducted throughout development can be used to establish a control strategy for manufacturing.	Monitoring during scale-up activities can provide a preliminary indication of process performance and the successful integration into manufacturing. Knowledge obtained during transfer and scale-up activities can be useful in further developing the control strategy.	A well-defined system for process performance and product quality monitoring should be applied to assure performance within a state of control and to identify improvement areas.	Once manufacturing ceases, monitoring such as stability testing should continue to completion of the studies. Appropriate action on marketed product should continue to be executed according to regional regulations.

Source: From *Guidance for Industry: Q10 Pharmaceutical Quality System*, April 2009.

(Table 8.2). An effective monitoring system provides assurance of the continued capability of processes and controls to produce a product of desired quality and to identify areas for continual improvement. The process performance and product quality monitoring system should:

- Use quality risk management to establish the control strategy, which can include parameters and attributes related to drug substance and drug product materials and components, facility and equipment operating conditions, in-process controls, finished product specifications, and the associated methods and frequency of monitoring and control. The control strategy should facilitate timely feedback/feed-forward and appropriate corrective action and preventive action.
- Provide the tools for measurement and analysis of parameters and attributes identified in the control strategy (for example, data management and statistical tools).
- Analyze parameters and attributes identified in the control strategy to verify continued operation within a state of control.
- Identify sources of variation affecting process performance and product quality for potential continual improvement activities to reduce or control variation.
- Include feedback on product quality from both internal and external sources, for example, complaints, product rejections, nonconformances, recalls, deviations, audits, and regulatory inspections and findings.

- Provide knowledge to enhance process understanding, enrich the design space (where established), and enable innovative approaches to process validation.

Corrective Action and Preventive Action (CAPA) System. The pharmaceutical company should have a system for implementing corrective actions and preventive actions resulting from the investigation of complaints, product rejections, nonconformances, recalls, deviations, audits, regulatory inspections and findings, and trends from process performance and product quality monitoring (Table 8.3). A structured approach to the investigation process should be used with the objective of determining the root cause. The level of effort, formality, and documentation of the investigation should be commensurate with the level of risk, in line with ICH Q9. CAPA methodology should result in product and process improvements and enhanced product and process understanding. Additionally, verification of effectiveness should be performed on all CAPAs.

Change Management System. Innovation, continual improvement, the outputs of process performance and product quality monitoring, and CAPA drive change (Table 8.4). To evaluate, approve, and implement these changes properly, a company should have an effective change management system. Generally, a difference in formality of change management processes exists before the initial regulatory submission and after submission, where changes to the regulatory submission might be required under regional requirements. The change management system ensures that continual improvement is undertaken in a timely and effective manner. It should provide a high degree of assurance that no unintended consequences will occur because of the change. As part of the change management system, as appropriate for the stage of the life cycle:

- Quality risk management should be used to evaluate proposed changes. The level of effort and formality of the evaluation should be commensurate with the level of risk.

Table 8.3 Application of corrective action and preventive action system throughout the product life cycle.

Pharmaceutical development	Technology transfer	Commercial manufacturing	Product discontinuation
Product or process variability is explored. CAPA methodology is useful where corrective actions and preventive actions are incorporated into the iterative design and development process.	CAPA can be used as an effective system for feedback, feed-forward, and continual improvement.	CAPA should be used, and the effectiveness of the actions should be evaluated.	CAPA should continue after the product is discontinued. The impact on product remaining on the market should be considered, as well as other products that might be affected.

Source: From *Guidance for Industry: Q10 Pharmaceutical Quality System*, April 2009.

Table 8.4 Application of change management system throughout the product life cycle.

Pharmaceutical development	Technology transfer	Commercial manufacturing	Product discontinuation
Change is an inherent part of the development process and should be documented; the formality of the change management process should be consistent with the stage of pharmaceutical development.	The change management system should provide management and documentation of adjustments made to the process during technology transfer activities.	A formal change management system should be in place for Commercial manufacturing. Oversight by the quality unit should provide assurance of appropriate science and risk-based assessments.	Any changes after product discontinuation should go through an appropriate change management system.

Source: From *Guidance for Industry: Q10 Pharmaceutical Quality System*, April 2009.

- Proposed changes should be evaluated relative to the marketing authorization, including design space, where established, and/or current product and process understanding. An assessment determines whether a change to the regulatory filing is required under regional requirements. As stated in ICH Q8, working within the design space is not considered a change (from a regulatory submission perspective); however, from a pharmaceutical QMS standpoint, all changes should be evaluated by the company's change management system.
- Proposed changes should be evaluated by subject matter expert teams contributing the appropriate expertise and knowledge from relevant areas (for example, Pharmaceutical Development, Manufacturing, Quality, Regulatory Affairs, and Medical) to ensure that the change is technically justified. Prospective evaluation criteria for a proposed change should be set.
- After implementation, an evaluation of the change should be undertaken to confirm that the change objectives were achieved and that there was no deleterious impact on product quality.

Management Review of Process Performance and Product Quality. Management review should provide assurance that process performance and product quality are managed over the product life cycle (Table 8.5). Depending on the size and complexity of the company, management review can be a series of reviews at various levels of management, and should include a timely and effective communication and escalation process to raise appropriate quality issues to senior levels of management for review. The management review system should include:

- The results of regulatory inspections and findings, audits, and other assessments, and commitments made to regulatory authorities.

Table 8.5 Application of management review of process performance and product quality throughout the product life cycle.

Pharmaceutical development	Technology transfer	Commercial manufacturing	Product discontinuation
Aspects of management review can be performed to ensure adequacy of the product and process design.	Aspects of management review should be performed to ensure the developed product and process can be manufactured at commercial scale.	Management review should be a structured system, and should support continual improvement.	Management review should include such items as product stability and product quality complaints.

Source: From *Guidance for Industry: Q10 Pharmaceutical Quality System*, April 2009.

- Periodic quality reviews that can include measures of customer satisfaction such as product quality complaints and recalls, conclusions of process performance and product quality monitoring, and the effectiveness of process and product changes, including those arising from corrective action and preventive actions.
- Any follow-up actions from previous management reviews. The management review system should identify appropriate actions, such as improvements to manufacturing processes and products, provision, training, and/or realignment of resources, and capture and dissemination of knowledge.

CONTINUAL IMPROVEMENT OF THE PHARMACEUTICAL QUALITY SYSTEM

Management Review

Management should have a formal process for reviewing the pharmaceutical QMS on a periodic basis. The review should include measurement of achievement of pharmaceutical quality system objectives, and assessment of performance indicators that can be used to monitor the effectiveness of processes within the pharmaceutical quality system (that is, complaints, deviations, CAPAs, and change management processes, feedback on outsourced activities, self-assessment processes, including risk assessments, trending, and audits, and external assessments such as regulatory inspections and findings, and customer audits).

Monitoring of Internal and External Factors

Management can monitor several factors that impact the QMS, including emerging regulations, guidance, quality issues, innovations, changes in business environment and objectives, and changes in product ownership.

Outcomes of Management Review and Monitoring

The outcome of management review of the pharmaceutical QMS and monitoring of internal and external factors can include improvements to the pharmaceutical quality system and related processes, allocation or reallocation of resources and/or personnel training, revisions to quality policy and quality objectives, and documentation and timely and effective communication of the results of the management review and actions, including escalation of appropriate issues to senior management.

CONCLUSION

With so many similar yet different quality requirements and regulations, it is very difficult and somewhat complicated to ensure compliance. Appendix A At-a-Glance Comparisons Tables A.1 and A.2, provided by José Rodríguez-Pérez, map the US cGMP, the EU/PIC-S GMP, Health Canada, and WHO pharmaceutical requirements. These tables provide a framework of similarities and of the regulations and what Rodríguez-Pérez refers to as the "missing subparts" (management review, CAPA, and process validation) related to cGMPs.

Chapter 9

Quality Unit (Site) Management

> Describe quality management elements for individual sites or units including responsibilities for the quality unit management such as Qualified Person, Management Representative, batch release (disposition) requirements for investigational and commercial products, and the need for quality units to be independent from operations. (Understand)
>
> **Body of Knowledge II.B**

Regulations define the *quality unit* (QU) as the concept that is consistent with modern quality systems in ensuring that the various operations associated with all systems are appropriately planned, approved, conducted, and monitored. Assigned responsibilities for the QU include ensuring that controls are implemented and completed satisfactorily during manufacturing operations and that developed procedures and specifications are appropriate and followed; approving and rejecting incoming materials, in-process materials, and drug products; and reviewing production records and investigating any unexplained discrepancies.

Responsibilities are divided between *quality control* (QC) and *quality assurance* (QA) functions. QC typically assesses suitability of components and products, evaluates the performance of the manufacturing process with respect to specifications and limits, and determines the acceptability of each batch for release. QA typically reviews and approves procedures, reviews records, and performs audits and trend analyses.

The United States Food and Drug Administration (FDA) has not moved away from the term *quality control unit* (QCU), which is defined in 21 CFR 210.3(b) (15) and described in 211.22. According to 21 CFR 210.3, the QCU is responsible for the duties relating to QC, whereas according to 211.22, the QCU is an independent group with unique responsibilities and authority to manage product excellence, consumer safety, compliance, and company liability. The QU encompasses nearly all manufacturing processes, including setting ingredient standards and ensuring they are met, qualifying third-party testing laboratories, packaging components and labeling, calculating practical cost considerations, and ensuring finished product quality.

Management Representative

Top management is required to appoint a member of management who, irrespective of other responsibilities has responsibility and authority for ensuring that processes needed for the QMS are documented reporting to top management on the effectiveness of the QMS, any need for improvement, the promotion of awareness of regulatory and QMS requirements with the organization.

QUALIFIED PERSON

The legal basis for the *qualified person* (QP) is defined in the European Directives. These Directives specify that European Union (EU) member states take all appropriate measures to ensure that the holder of the manufacturing authorization has permanently and continuously at its disposal the services of at least one QP. Additionally, member states must ensure that the QP fulfills the minimum conditions of qualification.

Qualifications for the role of QP include possession of a diploma, certificate, or other evidence of formal qualifications awarded upon completion of a university course of study or a course recognized as equivalent by the member state concerned, obtained over a period of at least four years of theoretical and practical study in one of six scientific disciplines (pharmacy, medicine, veterinary medicine, chemistry, pharmaceutical chemistry and technology, or biology). Alternatively, the minimum duration of the university course may be 3.5 years where the course is followed by a period of theoretical and practical training for a minimum of one year, which includes a training period of at least six months in a pharmacy open to the public, corroborated by an examination at university level. The course of studies includes both theoretical and practical usage in a number of subjects, including applied physics, general and inorganic chemistry, organic chemistry, analytical chemistry, pharmaceutical chemistry (including analysis of medicinal products), general and applied biochemistry (medical), physiology, microbiology, pharmacology, pharmaceutical technology, toxicology, and pharmacognosy (study of the composition and effects of natural active substances of plant and animal origin).

In addition to the formal training, a QP must have acquired practical experience over at least two years, in one or more facilities that are authorized to manufacture medicinal products, in the activities of qualitative analysis of medicinal products, quantitative analysis of active substances, and the testing and checking necessary to ensure the quality of medicinal products. The duration of practical experience may be reduced by one year when a university course lasts for at least five years and by 1.5 years when the course lasts for at least six years.

Responsibilities of the Qualified Person

The QP has several important responsibilities. In the case of medicinal products, the QP must ensure that each batch of medicinal products has been manufactured and checked in compliance with the laws in force in the member state and in accordance with the requirements of the marketing authorization; and in the case of medicinal products coming from third countries, the QP must ensure that each production batch has undergone in the importing member state a full qualitative analysis, a quantitative analysis of all the active constituents, and all the other tests or checks necessary to ensure the quality of medicinal products in accordance with the requirements of the marketing authorization. The batches of medicinal products that have undergone such controls in a member state are exempt from the controls if they are marketed in another member state and are accompanied by the control reports signed by the QP. In the case of medicinal products imported from a third country, where appropriate arrangements have been made by the European Community (EC) with the exporting country, the QP must ensure that the manufacturer of the medicinal product applied standards of *good manufacturing practices* (GMP) at least equivalent to those of the EC, and must ensure that the controls have been carried out in the exporting country. The QP may then be relieved of responsibility for carrying out those controls. In all cases, and particularly where the medicinal products are released for sale, the QP must certify in a register or equivalent document provided for that purpose that each production batch satisfies the provisions of the regulations, and ensure that the register or equivalent document is kept current as operations are completed and remains at the disposal of the agents of the competent authority for the period specified in the provisions of the member state concerned, and in any event, for at least five years.

Specific Requirements for the Qualified Person

Qualification of Suppliers of Active Pharmaceutical Ingredients. EU Directives require that *active pharmaceutical ingredients* (APIs) used for the manufacture of medicinal products have been manufactured according to the GMP for starting materials. The ICH Q7A guide *Good Manufacturing Practice Guidance for Active Pharmaceutical Ingredients* has been transferred to Part II of the *EU Guide to GMP*. It is expected that these standards are implemented in API manufacturing.

The QP responsible for the final production batch must be satisfied that the provisions of the Directive(s) are met and that a systematic approach to supplier qualification is in place to provide the assurance that only GMP-compliant APIs will be used for the manufacture of medicinal products.

Contract Manufacture and Analysis. Chapter 7, "Contract Manufacture and Analysis," of Part 1 of the *EU Guide to GMP* states that the contract manufacture and analysis must be correctly defined, agreed on, and controlled to avoid misunderstandings that could result in a product or work of unsatisfactory quality. To realize this objective, the QP needs to have the assurance that a written contract is in place that clearly defines the duties of each party, and that the contract contains the requirements. Any proposed changes in technical or other arrangements should be in accordance with the marketing authorization for the product concerned. The QP must ensure that a system to assess the competence of the contract acceptor to fulfill the principles of GMP is in place, and the contract acceptor must be aware of any problems associated with the product or the work that might pose a hazard to the premises, equipment, personnel, other materials, or other products. All processed products and materials delivered by the contract acceptor comply with their specification or have been released by the responsible quality unit.

Manufacture of Investigational Medicinal Products. Annex 13 of the GMP guide *Manufacture of Investigational Medicinal Products* defines in detail the requirements for the manufacture and testing of *investigational medical products* (IMPs), as well as ordering, shipping, and returning clinical supplies. In the case of IMPs, the QP is specifically responsible for ensuring that systems are in place that meet the requirements of Annex 13. The QP must, therefore, have a broad knowledge of pharmaceutical development and clinical trial processes. The QP is charged with making sure that release of IMPs does not occur until after the QP has certified that relevant requirements have been met.

Shipping. Investigational medicinal products should remain under the control of the sponsor until completion of a two-step release procedure: certification by the QP and release after fulfillment of the requirements of Article 9 ("Commencement of a Clinical Trial") of Directive 2001/20/EC. The sponsor should ensure that these are consistent with the details actually considered by the QP. Both releases should be recorded and retained in the relevant trial files held by or on behalf of the sponsor. Transfers of IMPs from one trial site to another should be an exception to the process. Such transfers should be covered by *standard operating procedures* (SOPs). The product history while outside of the control of the manufacturer, through trial monitoring reports and records of storage conditions at the original trial site, should be reviewed as part of the assessment of the product's suitability for transfer, and the advice of the QP should be sought. The product should be returned to the manufacturer or another authorized manufacturer for relabeling, if necessary, and certification by a QP. Records should be retained and full traceability ensured.

Complaints. The conclusions of any investigation carried out in relation to a complaint that arises from the quality of the product should be discussed between the manufacturer or importer and the sponsor (if different). This discussion should involve the QP and those personnel responsible for the relevant clinical trial to assess any potential impact on the trial, product development, or subjects.

Certification of Batch Release

The certification of batch release is defined in Annex 16 of the *EU Guide to GMP*. In addition, European Medicines Agency (EMA) published a paper on a proposed solution for dealing with minor deviations from the detail described in *Marketing Authorization for Human and Veterinary Medicinal Products* (including biological products).

MAINTENANCE OF A QUALITY MANAGEMENT SYSTEM (QMS)

Chapter 1 of the *EU Guide to GMP* requires that a comprehensively designed and correctly implemented system of QA incorporating GMP, and thus QC, is in place. It should be fully documented and its effectiveness monitored. The key requirement for the QP (specific to Chapter 1) is to ensure that medicinal products are not sold or supplied before a QP has certified that the batch has been produced and tested in accordance with the requirements of the marketing authorization and any other regulations relevant to the production, control, and release of medicinal products. To realize this objective, a QP should be involved in the implementation and maintenance of the *quality management system* (QMS).

The Qualified Person and the Product Quality Review

The requirement of a *product quality review* (PQR) was introduced with the addition of article 1.5 in Chapter 1 ("Quality Management") of the *EU Guide to GMP*. The aim of this requirement is to verify the consistency and appropriateness of the existing process, the adequacy of current specifications for starting material and finished product, and the presence of product and process improvements.

The PQR should be conducted and documented annually and should cover all aspects of the supply chain: starting materials, the process, process environment, and the final product. The manufacturer and marketing authorization holder, where different, should evaluate the results of this review, and an assessment should be made as to whether corrective and preventive action or any revalidation should be undertaken.

The QP responsible for final batch certification, together with the marketing authorization holder, should ensure that the quality review is performed in a timely manner and is accurate. It is necessary to understand the aim and basic principles of the PQR to assume responsibility. The generation of the PQR can be done by any of the functions, but QA, in association with the QP, should lead the way in interpreting data, and define and follow up any consequences.

The Qualified Person and Reference and Retention Samples

The requirements for the taking and holding of reference samples of starting materials, packaging materials, or finished products, and retention samples of finished products, are defined and included in the EU GMP as Annex 19 to the *EU Guide to GMP* Part 1. *Reference samples* are samples of starting materials or finished products intended to be reanalyzed in case of any complaint, dispute, or discrepancy; *retention samples* are samples of a fully packaged unit from a batch of finished product stored for identification purposes.

According to the different purposes, the amounts to be stored are different: reference samples are usually in a sample size sufficient to allow at least two full analytical controls on the batch (with some dispute still ongoing whether this would include sterility testing, if appropriate), and retention samples are usually only one package per packaging presentation.

Reference and retention samples from finished products must be retained for at least one year after the expiry date. If no longer period is required under national law, samples of starting materials (other than solvents, gases, or water used in the manufacturing process) must be retained for at least two years after the release of the product. Storage conditions must be in accordance with the conditions submitted in the marketing authorization. In case a product has been manufactured (partially) outside the EU, reference samples must be stored in the EU unless there is an operational *mutual recognition agreement* (MRA) in place.

External Qualified Persons

Some companies may use an external QP who provides an independent service. The duties and responsibilities do not differ from the ones that are applicable to QPs that are full-time employees of a company. To fulfill the tasks, it is required to have a detailed contract to ensure that all duties and responsibilities are met.

Delegation of Tasks/Absence of a Qualified Person

In general, the task of a QP, especially the process of batch certification and release, can only be delegated to another QP who is registered with the responsible authority for supervision of the marketing authorization and/or supervision.

Organizational Duties to Be Established by the Company of the Qualified Person

To perform their duties and responsibilities, a QP needs support from the company and its senior management. Three organizational tasks should be implemented:

- Senior management should be actively involved in the implementation, maintenance, and further development of the QMS.
- A clear job description should be in place to define the responsibilities of the QP and their relationship to other departments, responsible managers, and especially to the head of production and the head of QC, and, if appropriate, the head of QA.
- The QP should at all times have access to the marketing authorization dossier and should be informed about all changes related to the dossier.

ANNEX 16 TO THE EU GUIDE TO GMP CERTIFICATION AND BATCH RELEASE

Annex 16 covers those cases where a batch has had different stages of production or testing conducted at different locations or by different manufacturers, and where an intermediate or bulk production batch is divided into more than one finished product batch. It also covers the release of batches that have been imported both when there is and is not an MRA between the EC and the third country.

Principle

Each batch of finished product must be certified by a QP before being released for sale or supply or for export. The purpose of controlling batch release in this way is to ensure that the batch has been manufactured and checked in accordance with the requirements of its marketing authorization, the principles and guidelines of EC GMP or the GMP of a third country recognized as equivalent under an MRA, and any other relevant legal requirement before it is placed on the market.

In the event that a defect needs to be investigated or a batch recalled, this process ensures that the QP who certified the batch and the relevant records are readily identifiable.

Introduction

Manufacture, including QC testing, of a batch of medicinal products takes place in stages that may be conducted at different sites and by different manufacturers. Each stage should be conducted in accordance with the relevant marketing authorization, GMP, and the laws of the member state concerned, which should be taken into account by the QP who certifies the finished product batch before release to the market.

In an industrial situation, however, it is usually not possible for a single QP to be closely involved with every stage of manufacture. The QP who certifies a finished product batch may need therefore to rely in part on the advice and decisions of others. Before doing so, the QP should ensure that this reliance is well founded, either from personal knowledge or from confirmation by other QPs within a quality system.

When some stages of manufacture occur in a third country, it is still a requirement that production and testing are in accordance with the marketing authorization, that the manufacturer is authorized according to the laws of the country concerned, and that manufacture follows GMP of the EC.

General

One batch of finished product may have different stages of manufacture, importation, testing, and storage conducted at different sites before release. Each site should be approved under one or more manufacturing authorizations and should have at its disposal the services of at least one QP. The correct manufacture of a particular batch of product, regardless of how many sites are involved, should be the overall concern of the QP who certifies that finished product batch before release.

Different batches of a product may be manufactured or imported and released at different sites in the EC/European Economic Area (EEA). For example, an EC marketing authorization may name batch release sites in more than one member state, and a national authorization may also name more than one release site. In this situation, the holder of the marketing authorization and each site authorized to release batches of the product should be able to identify the site at which any particular batch has been released and the QP who was responsible for certifying that batch.

The QP who certifies a finished product batch before release may do so based on personal knowledge of all the facilities and procedures employed, the expertise of the persons concerned, and the quality system within which they operate. Alternatively, the QP may rely on the confirmation by one or more other QPs of the compliance of intermediate stages of manufacture within a quality system. This confirmation by other QPs should be documented and should identify clearly the matters that have been confirmed. The systematic arrangements to achieve this should be defined in a written agreement.

This agreement is required whenever a QP wishes to rely on a confirmation by another QP. The agreement should be in general accordance with SMF Chapter 7. The QP who certifies the finished product batch should ensure that the arrangements in the agreement are verified. The form of such an agreement should be appropriate to the relationship between the parties, for example, an SOP within a company or a formal contract between different companies, even if within the same group.

The agreement should include an obligation on the part of the provider of a bulk or intermediate product to notify the recipient(s) of any deviations, out-of-specification (OOS) results, noncompliance with GMP, investigations, complaints, or other matters that should be taken into account by the QP who is responsible for certifying the finished product batch.

When a computerized system is used for recording certification and batch release, particular note should be taken of the guidance in Annex 11. Certification of a finished product batch against a relevant marketing authorization by a QP in the EC/EEA need not be repeated on the same batch provided the batch has remained within the EC/EEA.

Whatever particular arrangements are made for certification and release of batches, it should always be possible to identify and recall without delay all products that could be rendered hazardous by a quality defect in a batch.

Batch Testing and Release of Products Manufactured in the EC/EEA

All Manufacture Occurs at a Single Authorized Site. When all production and control stages are carried out at a single site, the conduct of certain checks and controls may be delegated to others, but the QP at this site who certifies the finished product batch normally retains personal responsibility for these within a defined quality system. However, he or she may, alternatively, take account of the confirmation of the intermediate stages by other QPs on the site that are responsible for those stages.

Different Stages of Manufacture Are Conducted at Different Sites within the Same Company. When different stages of the manufacture of a batch are done at different sites within the same company (which may or may not be covered by the same manufacturing authorization), a QP should be responsible for each stage. Certification of the finished product batch should be performed by a QP of the manufacturing authorization holder responsible for releasing the batch to the market, who may take personal responsibility for all stages or may take account of the confirmation of the earlier stages by the relevant QP responsible for those stages.

Some Intermediate Stages of Manufacture Are Contracted to a Different Company. One or more intermediate production and control stages may be contracted to a holder of a manufacturing authorization in another company. A QP of the contract giver may take account of the confirmation of the relevant stage by a QP of the contract acceptor but is responsible for ensuring that this work is conducted within the terms of a written agreement. The finished product batch should be certified by the QP of the manufacturing authorization holder responsible for releasing the batch to the market.

Bulk Production Batch Is Assembled at Different Sites into Several Finished Batches Released under Single Marketing Authorization. This situation could occur, for example, under a national marketing authorization when the assembly sites are all within one member state, or under an EC marketing authorization when the sites are in more than one member state. One alternative is for a QP of the manufacturing authorization holder making the bulk production batch to certify all the finished product batches before release to the market. In doing so, he or she may either take personal responsibility for all manufacturing stages or take account of the confirmation of assembly by the QPs of the assembly sites. Another alternative is for the certification of each finished product batch before release to the market to be performed by a QP of the manufacturer who has conducted the final assembly operation. In doing so, he or she may either take personal responsibility for all manufacturing stages or take account of the confirmation of the bulk production batch by a QP of the manufacturer of the bulk batch.

In all cases of assembly at different sites under a single marketing authorization, one person—normally a QP of the manufacturer of the bulk production batch—should have overall responsibility for all released finished product batches that are derived from one bulk production batch. The duty of this person is to be aware of any quality problems reported on any of the finished product batches and to coordinate any necessary action arising from a problem with the bulk batch. While the batch numbers of the bulk and finished product batches are not necessarily the same, there should be a documented link between the two numbers so that an audit trail can be established.

Bulk Production Batch Is Assembled at Different Sites into Several Finished Batches Released under Different Marketing Authorizations. This situation could occur, for example, when a multinational organization holds national marketing authorizations for a product in several member states or when a generic manufacturer purchases bulk products and assembles and releases them for sale under its own marketing authorization. A QP of the manufacturer doing the assembly who certifies the finished product batch may either take personal responsibility for all manufacturing stages or may take account of the confirmation of the bulk production batch by a QP of the bulk product manufacturer. Any problem identified in any of the finished product batches that may have arisen in the bulk production batch should be communicated to the QP responsible for confirming the bulk production batch, who should then take any necessary action with respect to all finished product batches produced from the suspected bulk production batch. This arrangement should be defined in a written agreement.

A finished product batch may be purchased and released to the market by a manufacturing authorization holder, which could occur, for example, when a company supplying generic products holds a marketing authorization for products made by another company, purchases finished products that have not been certified against its marketing authorization, and releases them under its own manufacturing authorization in accordance with its own marketing authorization. In this situation, a QP of the purchaser should certify the finished product batch before release. In doing so, he or she may either take personal responsibility for all manufacturing stages or may take account of the confirmation of the batch by a QP of the vendor manufacturer.

Quality Control Laboratory and Production Site Are Authorized under Different Manufacturing Authorizations. A QP certifying a finished product batch may either take personal responsibility for the laboratory testing or may take account of the confirmation by another QP of the testing and results. The other laboratory and QP need not be in the same member state as the manufacturing authorization holder releasing the batch. In the absence of such confirmation, the QP should have personal knowledge of the laboratory and its procedures relevant to the finished product to be certified.

BATCH TESTING AND RELEASE OF PRODUCTS IMPORTED FROM A THIRD COUNTRY

General

Importation of finished products should be conducted by an importer. Each batch of imported finished product should be certified by a QP of the importer before release for sale in the EC/EEA. Unless an MRA is in effect between the EC and the third country, samples from each batch should be tested in the EC/EEA before certification of the finished product batch by a QP. Importation and testing need not necessarily be performed in the same member state. The guidance in this section should be applied as appropriate to the importation of partially manufactured products:

- A complete batch or the first part of a batch of a medicinal product is imported. The batch or part batch should be certified by a QP of the importer before release. This QP may take account of the confirmation of the checking, sampling, or testing of the imported batch by a QP of another manufacturing authorization holder (that is, within EC/EEA).
- Part of a finished product batch is imported after another part of the same batch has previously been imported to the same or a different site. A QP of the importer receiving a subsequent part of the batch may take account of the testing and certification by a QP of the first part of the batch. If this is done, the QP should ensure, based on evidence, that the two parts do come from the same batch, that the subsequent part has been transported under the same conditions as the first part, and that the samples that were tested are representative of the whole batch. The conditions are most likely to be met when the manufacturer in the third country and the importer(s) in the EC/EEA belong to the same organization operating under a corporate system of quality assurance. If the QP cannot ensure that the conditions are met, each part of the batch should be treated as a separate batch. When different parts of the batch are released under the same marketing authorization, one person, normally a QP of the importer of the first part of the batch, should take overall responsibility for ensuring that records are kept of the importation of all parts of the batch and that the distribution of all parts of the batch is traceable within the EC/EEA. The QP should

be made aware of any quality problems reported on any part of the batch and should coordinate any necessary action concerning these problems and their resolution. This should be ensured by a written agreement between all the importers concerned.

Location of Sampling for Testing in EC/EEA

Samples should be representative of the batch and be tested in the EC/EEA. To represent the batch, it may be preferable to take some samples during processing in the third country. For example, samples for sterility testing may best be taken throughout the filling operation; however, to represent the batch after storage and transportation, some samples should be taken after receipt of the batch in the EC/EEA.

When any samples are taken in a third country, they should either be shipped with and under the same conditions as the batch that they represent, or if sent separately, it should be demonstrated that the samples are still representative of the batch, for example, by defining and monitoring the conditions of storage and shipment. When the QP wishes to rely on testing of samples taken in a third country, this should be justified on technical grounds.

Batch Testing and Release of Products Imported from a Third Country with EC MRA

Unless otherwise specified in the agreement, an MRA does not remove the requirement for a QP within the EC/EEA to certify a batch before it is released for sale or supply within the EC/EEA. However, subject to details of the particular agreement, the QP of the importer may rely on the manufacturer's confirmation that the batch has been made and tested in accordance with its marketing authorization and the GMP of the third country, and need not repeat the full testing. The QP may certify the batch for release when he or she is satisfied with this confirmation and that the batch has been transported under the required conditions and has been received and stored in the EC/EEA by an importer under defined conditions.

Routine Duties of a Qualified Person

Before certifying a batch for release, the QP doing so should ensure that at least the following requirements have been met:

- The batch and its manufacture comply with the provisions of the marketing authorization (including the authorization required for importation, where relevant).
- Manufacture has been done in accordance with GMP or, in the case of a batch imported from a third country, in accordance with GMP standards at least equivalent to EC GMP.
- The principal manufacturing and testing processes have been validated, and account has been taken of the actual production conditions and manufacturing records.

- Any deviations or planned changes in production or QC have been authorized by the persons responsible in accordance with a defined system. Any changes requiring variation to the marketing or manufacturing authorization have been reported to and authorized by the relevant authority.
- All the necessary checks and tests have been performed, including any additional sampling, inspection, tests, or checks initiated because of deviations or planned changes.
- All necessary production and QC documentation has been completed and endorsed by the staff authorized to do so.
- All audits have been carried out as required by the QA system.
- The QP should in addition take into account any other factors of which he or she is aware that are relevant to the quality of the batch.

A QP may have additional duties in accordance with national legislation or administrative procedures. A QP who confirms the compliance of an intermediate stage of manufacture has the same obligations in relation to that stage as there would be with final release. A QP should maintain his or her knowledge and experience up to date in the light of technical and scientific progress and changes in quality management relevant to the products that he or she is required to certify.

If a QP is called on to certify a batch of a product type with which they are unfamiliar, for example, because the manufacturer for whom they work introduces a new product range or because they start to work for a different manufacturer, they should first ensure that they have gained the relevant knowledge and experience necessary to fulfill this duty. In accordance with national requirements, the QP may be required to notify the authorities of such a change and may be subject to renewed authorization.

Chapter 10

Risk Management and Assessment

Quality risk management is a systematic process for the assessment, control, communication, and review of risks to the quality of the product throughout its life cycle. ICH Q9, *Quality Risk Management*, provides the framework that may be applied to all aspects of pharmaceutical quality, including development, manufacturing, distribution, inspection, and submission/review processes across the life cycle of drug substances and drug products as well as biologic and biotechnological products. Quality risk management can be applied in the product life cycle to the use of raw materials, solvents, excipients, packaging, and labeling materials.

The two primary principles of quality risk management are that the evaluation of the quality risk should ultimately link back to the potential harm to the patient, and that the level of effort, formality, and documentation of the quality risk management process should be commensurate with the level of risk.

RISK MANAGEMENT

> Use various methods to apply risk management principles as described in ICH Q9, ICH Q12, and other guidance and regulatory documents. (Apply)
>
> **Body of Knowledge II.C.1**

Quality risk management should include systematic processes designed to coordinate, facilitate, and improve science-based decision making with respect to risk. Steps used to initiate and plan a quality risk management process include:

- Defining the process, which typically involves risk assessment, risk control, risk communication, and risk review elements at an appropriate level of detail
- Assembling background information and/or data on the potential hazard, harm, or human health impact relevant to the risk assessment

- Identifying a leader and critical resources
- Specifying a timeline, deliverables, and appropriate level of decision making for the risk management process
- Identifying the interrelationships of the elements of the quality risk management process

One development in monitoring a company's risk management is through the FDA Quality Management Maturity (QMM) assessment program. Initiated in 2022, this program allows companies to privately assess themselves based on the following rubric and identify areas that may need addressed based on the maturity and culture of the companies' quality management system (QMS):

- Leadership
- Business continuity
- Technical excellence
- Advanced pharmaceutical quality system (PQS)
- Employee empowerment and engagement

It is important to note that the QMM program is not intended to be used in lieu of or as a surrogate to establishment inspections, nor does it provide a measurement of product quality manufactured by a facility. Instead, it is an evaluation of an establishment's behavior, culture, and quality practices regarding the implementation and maturity of a QMS.

Reference: FDA Office of Pharmaceutical Quality White Paper, "Quality Management Maturity: Essential for Stable US Supply Chains of Quality Pharmaceuticals," 2022

RISK ASSESSMENT

> Use the Quality Management Maturity (QMM) assessment program to assess the effectiveness of the Quality Management System to ensure process and product quality. (Apply)
>
> **Body of Knowledge II.C.2**

Risk assessment consists of the identification of hazards and the analysis and evaluation of risks associated with exposure to those hazards. Quality risk assessments begin with a well-defined problem description or risk question. When the risk in question is well defined, an appropriate risk management tool and the types of information required to address the risk question will be readily identifiable. As an aid to clearly defining the risk(s) for risk assessment purposes, three fundamental questions are often used:

- What might go wrong?
- What is the likelihood (probability) it will go wrong?
- What are the consequences (severity)?

The output of a risk assessment is either a quantitative estimate of risk or a qualitative description of a range of risk. When risk is expressed quantitatively, a numerical probability is used. In quantitative risk assessments, a risk estimate provides the likelihood of a specific consequence given a set of risk-generating circumstances. Thus, quantitative risk estimation is useful for one particular consequence at a time. Alternatively, risk can be expressed using qualitative descriptors, such as "high," "medium," or "low," that should be defined in as much detail as possible. Sometimes, a *risk score* is used to further define descriptors in risk ranking. In addition, some risk management tools use a relative risk measure to combine multiple levels of severity and probability into an overall estimate of relative risk. The intermediate steps within a scoring process can sometimes employ quantitative risk estimation.

Risk identification provides the basis for further steps in the quality risk management process. It systematically uses information to identify hazards associated with the risk question or problem description. Information can include historical data, theoretical analysis, informed opinions, and the concerns of stakeholders. It addresses the "What might go wrong?" question, including identifying the possible consequences. *Risk analysis* is the estimation of the risk associated with the identified hazards. It is a qualitative or quantitative process of linking the likelihood of occurrence with the severity of harms. The extent to which the harm can be detected also factors into the estimation of risk. *Risk evaluation* compares the identified risk against given risk criteria and considers the strength of evidence for the three fundamental questions.

RISK CONTROL

The purpose of *risk control* is to reduce the risk to an acceptable level. The amount of effort used for risk control should be proportional to the significance of the risk. Decision makers might use different processes, including benefit–cost analysis, for understanding the optimal level of risk control.

Risk control might focus on the following questions:

- Is the risk estimated in the risk assessment above an acceptable level?
- What can be done to reduce or eliminate risk(s)?
- What is the appropriate balance between benefits, risks, and resources?
- Are new risks introduced due to the identified risks being controlled?

Risk reduction focuses on mitigation and avoidance of quality risk when it exceeds a specified (acceptable) level. It includes actions taken to mitigate the severity, and reduce the probability, of harm. It may also identify processes that improve the detectability of hazards.

The implementation of risk reduction measures can introduce new risks into the system or increase the significance of other existing risks. Hence, it might be appropriate to revisit the risk assessment to identify and evaluate any possible change in risk after implementing a risk reduction process.

Risk acceptance is a decision to accept risk. It can be a formal decision to accept the residual risk, or it can be a passive decision in which residual risks are not specified. For some types of harms, even the best quality risk management practices might not entirely eliminate risk. In these circumstances, it might be agreed that an appropriate quality risk management strategy has been applied, and that quality risk is reduced to a specified (acceptable) level. This specified and acceptable level will depend on many parameters and should be decided on a case-by-case basis.

RISK COMMUNICATION AND RISK REVIEW

Risk communication is the exchange or sharing of information about risk and its management between the decision makers and others. Communication should include the existence, nature, form, probability, severity, and detectability of the risk. This process can sometimes be formal. It may be internal (that is, occur within the organization) or external (that is, occur with suppliers, customers, or regulators). Formal processes should be documented. *Risk review* is a process that allows for the review or monitoring of events. It should incorporate new knowledge and experience and may be used for planned or unplanned events, and it should be performed periodically.

RISK MANAGEMENT METHODOLOGY

Quality risk management supports a scientific and practical approach to decision making. It provides documented, transparent, and reproducible methods to accomplish steps of the quality risk management process based on current knowledge about assessing the probability, severity, and, sometimes, the detectability of the risk. Figure 10.1 lists some risk management tools. Combined use of these tools provides flexibility that can facilitate the application of quality risk management principles.

INTEGRATION INTO INDUSTRY AND REGULATORY OPERATIONS

Quality risk management is the foundation for science-based decisions when integrated into a *quality management system* (QMS). It does not obviate industry's obligations to comply with regulatory requirements. It can facilitate better and more-informed decisions and may provide regulators with greater assurance of a company's ability to address potential risks. It may affect the extent and level of direct regulatory oversight.

The degree of rigor and formality used in quality risk management should be commensurate with the complexity and/or criticality of the issue. Quality risk management should be integrated into existing operations and documented appropriately. Finally, training of personnel provides for greater understanding of decision-making processes and builds confidence in quality risk management outcomes.

- Traditional informal ways (empirical and/or internal procedures) based on compilation of observations, trends, and other information
- Failure mode, effects, and criticality analysis (FMECA)
- Fault tree analysis (FTA)
- Hazard analysis and critical control points (HACCP)
- Hazard operability analysis (HAZOP)
- Preliminary hazard analysis (PHA)
- Risk ranking and filtering
- Supporting quality and statistical tools

Figure 10.1 Some risk management tools.

METHODS AND TOOLS

Basic Risk Management Facilitation Methods

Basic risk management facilitation methods include brainstorming, affinity diagrams, flowcharts, check sheets, process maps, and cause-and-effect diagrams (that is, the Ishikawa diagram, or fishbone diagram). These tools are used to generate and document ideas used to develop consensus.

Failure Mode and Effects Analysis

Failure mode and effects analysis (FMEA) includes the evaluation of potential failure modes and their effect on outcomes. It methodically breaks down the analysis of complex processes into manageable steps; risk reduction can then be used to eliminate, contain, reduce, or control failures. (See Figure 10.2.)

The potential areas of use for FMEA are to prioritize risks and monitor the effectiveness of risk control activities. It can be applied to equipment and facilities, and might be used to analyze a manufacturing operation and its effect on product or process. It identifies elements/operations within the system that render it vulnerable. The output or results of FMEA can be used as a basis for design or further analysis, or to guide resource deployment.

Failure Mode, Effects, and Criticality Analysis

Failure mode, effects, and criticality analysis (FMECA) is extended to investigate the degree of severity of the consequences, probabilities of occurrence, and detectability. It can identify places for additional preventive actions.

A potential area of use for FMECA in the pharmaceutical industry is with failures and risks associated with manufacturing processes; however, it is not limited to this application.

Process FMEA											
Filling Operation											
Process step	**Process/product parameters**	**SEV**	**OCC**	**DET**	**RPN**	**Evaluation**	**Actions Taken**	**SEV**	**OCC**	**DET**	**OCC**	**Evaluation**
Load Vials	Vial orientation	3	3	3	9	risk acceptability low						
Vial not sealed	Vial does not leak	4	5	3	120	risk acceptability is undesirable	Perform leak test	4	3	1	12	risk acceptability low

Figure 10.2 Partial FMEA example in which a leak test was added to reduce the risk as the initial RPN of 120 was deemed undesirable according to RPN action requirements for a five-point scale. With the improved gaging and vision system, the RPN was reduced to 12, which is low.

Severity, or *criticality*, is a measure of the seriousness of the possible consequences of a hazard. *Probability of occurrence* is the likelihood that a cause will occur and that it will induce the detrimental consequences of the associated hazard. *Detection* is the probability that control will catch the defect. Multiplying the severity by the probability of occurrence by the probability of detection yields the *risk priority number* (RPN). The RPN is the output of an FMECA, a relative risk score for each failure mode that is used to rank the modes on a relative risk basis. Table 10.1 is an example of an FMEA rating scheme.

Risk priority numbers are used to assess risk. To calculate the RPN, the team must rate the *severity* of each effect of failure, rate the likelihood of *occurrence* for each cause of failure, and rate the likelihood of *detection* for each cause of failure. Once this is completed, calculate the RPN by multiplying the three ratings:

$$\text{RPN} = \text{Severity} \times \text{Occurrence} \times \text{Detection}$$

Example: A process FMEA has been developed for a new drug product. A certain characteristic has a severity of 3, a probability of occurrence of 3, and a probability of detection of 3. Calculate the RPN.

$$\text{RPN} = 3 \times 3 \times 3 = 27$$

According to Table 10.2, an RPN of 27 is considered to be tolerable. However, the criticality and probability of occurrence should also be evaluated for this characteristic. Table 10.3 indicates a severity of 3 with a probability of occurrence of 3; the risk is considered to be as low as reasonably possible (ALARP).

Another consideration is evaluating the criticality and probability of occurrence. Table 10.3 provides an example of criticality and occurrence action requirements.

A drawback of RPN scores is there are multiple ways the same RPN can be achieved (see Tables 10.4–10.6). For example:

> scenario 1, an RPN of 50 could have a Severity (S) rating of 5, Occurrence (O) rating of 10, and Detection (D) rating of 1

or

> scenario 2, an RPN of 50 could have a Severity (S) rating of 1, Occurrence (O) rating of 5, and Detection (D) rating of 10

or

> scenario 3, an RPN of 50 could have a Severity (S) rating of 5, Occurrence (O) rating of 1, and Detection (D) rating of 10

Although the Risk Priority Number (RPN) is 50 for each of these scenarios, which of these scenarios pose the greatest risk? To answer this question, we must first look at severity. Scenarios 1 and 3 both have a Severity (S) rating of 5; however, scenario 1 has an Occurrence (O) rating of 5, making it higher risk. Scenario 2 has a Severity (S) rating of 1, making lower risk than scenarios 1 and 3.

Ranking the order RPN scenario risks in this example, the order 1, 3, 2 is the obvious answer.

Table 10.1 Typical FMEA rating scheme.

Severity	
5	*Catastrophic.* Failure affects safety and/or involves noncompliance with government regulations.
4	*Critical.* Customer dissatisfaction due to the nature of the failure, such as unusable product. May cause disruption to future processes. Process changes to be examined.
3	*Serious.* Customer is made uncomfortable or is bothered by failure. May require change in controls.
2	*Minor.* Customer may notice a slight deterioration of the product.
1	*Negligible.* Customer will probably not notice the failure.
Occurrence	
5	Certain to fail
4	High number of failures likely
3	Occasional failure likely
2	Few failures likely
1	Failure unlikely
Detection	
5	*Undetectable.* Controls very likely will not detect the existence of a failure. Very high risk that the product will be shipped with this defect.
4	*Poor.* Controls likely will not detect the existence of a failure. Likely that the product will be shipped with the defect.
3	*Moderate.* Controls may detect the existence of a failure. Moderate risk that the product will be shipped with this defect.
2	*Good.* Controls have a good chance of detecting the existence of a failure. Low risk the product will be shipped with this defect.
1	*Excellent.* Controls almost certainly will detect the existence of a failure. Remote risk the product will be shipped with this defect.

Table 10.2 Typical RPN action requirements.

RPN	Risk acceptability	Action
1 to 14	Low	Although low risk, continue mitigation process as far as possible
15 to 29	Tolerable	This should only be revisited in future design review if corrective action enhances the reliability or product appeal.
30 to 49	Undesirable	Risk is acceptable only if it cannot be further mitigated by organizational or technological solutions that do not reduce the clinical/functional utility of the product.
Above 49	Intolerable	Risk should be eliminated or reduced by protective measures. Justification is required for risk that is accepted.

Table 10.3 Typical criticality and occurrence action requirements.

		Criticality				
		Catastrophic	Critical	Serious	Minor	Negligible
Occurrence		5	4	3	2	1
Certain	5	Unacceptable	Unacceptable	Unacceptable	Unacceptable	ALARP
High	4	Unacceptable	Unacceptable	Unacceptable	ALARP	ALARP
Occasional	3	Unacceptable	Unacceptable	ALARP	ALARP	ALARP
Few	2	Unacceptable	ALARP	ALARP	ALARP	Acceptable
Unlikely	1	ALARP	ALARP	ALARP	Acceptable	Acceptable

ALARP = As low as reasonably possible

Table 10.4 Ten-point FMEA rating scheme.

Severity	
10	Failure could injure the customer or an employee.
9	Failure creates noncompliance with federal regulations.
8	Failure renders the unit inoperable or unfit for use.
7	Failure causes a high degree of customer dissatisfaction.
6	Failure results in a subsystem or partial malfunction of the product.
5	Failure creates enough of a performance loss to cause the customer to complain.
4	Failure can be overcome with modifications to the customer's process or product, but there is minor performance loss.
3	Failure creates a minor nuisance for the customer, but the customer can overcome it without performance loss.
2	Failure may not be readily apparent to the customer but has minor effects on the customer's process or product.
1	Failure is not noticeable to the customer and does not affect the customer's process or product.
Occurrence	
10	More than one occurrence per day or a probability of more than three occurrences in 10 events ($C_{pk} < 0.33$).
9	One occurrence every three to four days or a probability of three occurrences in 10 events ($C_{pk} \approx 0.33$).
8	One occurrence per week or a probability of five occurrences in 100 events ($C_{pk} \approx 0.67$).
7	One occurrence every month or one occurrence in 100 events ($C_{pk} \approx 0.83$).
6	One occurrence every three months or three occurrences in 1000 events ($C_{pk} \approx 1.00$).

Continued

Table 10.4 *Continued.*

5	One occurrence every six months to one year or five occurrences in 10,000 events ($C_{pk} \approx 1.17$).
4	One occurrence per year or six occurrences in 100,000 events ($C_{pk} \approx 1.33$).
3	One occurrence every one to three years or six occurrences in 10 million events ($C_{pk} \approx 1.67$).
2	One occurrence every three to five years or two occurrences in 1 billion events ($C_{pk} \approx 2.00$).
1	One occurrence in greater than five years or less than two occurrences in 1 billion events ($C_{pk} > 2.00$).
Detection	
10	The product is not inspected or the defect caused by failure is not detectable.
9	Product is sampled, inspected, and released based on acceptable quality level (AQL) sampling plans.
8	Product is accepted based on no defectives in a sample.
7	Product is 100% manually inspected in the process.
6	Product is 100% manually inspected using go/no-go or other mistake-proofing gages.
5	Some SPC is used in process and product is final inspected off-line.
4	SPC is used and there is immediate reaction to out-of-control conditions
3	An effective SPC program is in place with process capabilities (C_{pk}) greater than 1.33.
2	All product is 100% automatically inspected.
1	The defect is obvious or there is 100% automatic inspection with regular calibration and preventive maintenance of the inspection equipment.

Table 10.5 RPN action requirements for a 10-point scale.

RPN	Risk acceptability	Action
1 to 65	Low	Although risk is low, continue mitigation process as far as possible.
66 to 125	Tolerable	This should only be revisited in future design review if corrective action enhances the reliability or product appeal.
126 to 300	Undesirable	Risk is acceptable only if it cannot be further mitigated by organizational or technological solutions that do not reduce the clinical/functional utility of the product.
Above 300	Intolerable	Risk should be eliminated or reduced by protective measures. Justification required for risk that is accepted.

Table 10.6 Criticality and occurrence action requirements for a 10-point scale.

		Severity				
		Catastrophic	**Critical**	**Serious**	**Minor**	**Negligible**
Occurrence		**9–10**	**7–8**	**5–6**	**3–4**	**1–2**
Certain	10	Unacceptable	Unacceptable	Unacceptable	Unacceptable	ALARP
	9	Unacceptable	Unacceptable	Unacceptable	Unacceptable	ALARP
High	8	Unacceptable	Unacceptable	Unacceptable	ALARP	ALARP
	7	Unacceptable	Unacceptable	Unacceptable	ALARP	ALARP
Occasional	6	Unacceptable	Unacceptable	ALARP	ALARP	ALARP
	5	Unacceptable	Unacceptable	ALARP	ALARP	ALARP
Few	4	Unacceptable	ALARP	ALARP	ALARP	Acceptable
	3	Unacceptable	ALARP	ALARP	ALARP	Acceptable
Unlikely	2	ALARP	ALARP	ALARP	Acceptable	Acceptable
	1	ALARP	ALARP	ALARP	Acceptable	Acceptable

ALARP = as low as reasonably possible/practical.
AFAP = as far as possible/practical.
These terms both suggest that mitigations should be investigated and implemented to prevent or minimize risks to ensure safety and performance.

Fault Tree Analysis

Fault tree analysis (FTA) evaluates system (or subsystem) failures one at a time, represented pictorially in the form of a branching tree of fault modes. At each level in the tree, combinations of fault modes are described with logical operators (for example, AND, OR). FTA relies on experts' process understanding to find causes. A potential area of use for FTA can be to establish the pathway to the root cause of the failure. FTA can be used in investigating complaints or deviations to fully understand their root cause and to ensure that intended improvements will resolve the issue and not lead to other issues. FTA is an effective tool for evaluating how multiple factors affect a given issue. The output of an FTA includes a visual representation of failure modes. It is useful both for risk assessment and in developing monitoring programs. Figure 10.3 depicts a basic FTA.

Hazard Analysis and Critical Control Points

Hazard analysis and critical control points (HACCP) is used to manage risks with physical, chemical, and biological hazards. HACCP has seven steps:

1. Conduct a hazard analysis and identify preventive measures for each step of the process.
2. Determine the critical control points.

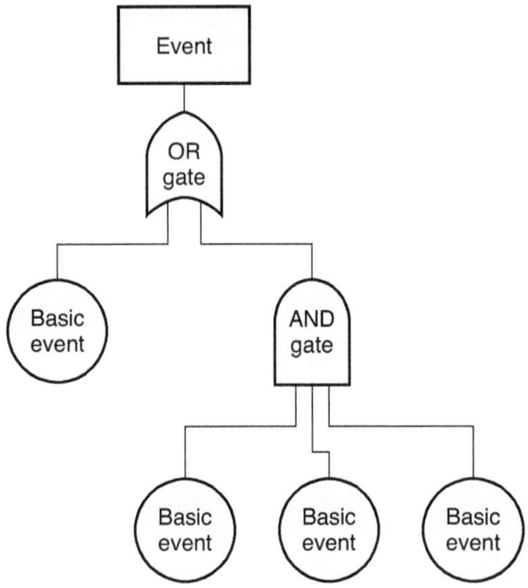

Figure 10.3 Fault tree analysis example.

3. Establish critical limits.
4. Establish a system to monitor the critical control points.
5. Establish the corrective actions to be taken when monitoring indicates that the critical control points are not in a state of control.
6. Establish a system to verify that the HACCP system is working effectively.
7. Establish a record-keeping system.

Potential areas of use for HACCP include identifying and managing risks associated with physical, chemical, and biologic hazards (including microbiological contamination). HACCP is useful when product and process understanding is sufficiently comprehensive to support identification of critical control points. The output of a HACCP analysis is risk management information that facilitates monitoring of critical points not only in the manufacturing process but also in other life cycle phases.

Hazard Operability Analysis

Hazard operability analysis (HAZOP) is used primarily to evaluate process safety hazards. It is a systematic brainstorming technique for identifying hazards using guide words (for example, *no, more, other than, part of*). HAZOP is for risk events caused by design or operating deviations. Potential areas of use for HAZOP are manufacturing processes, including outsourced production and formulation, and the upstream suppliers, equipment, and facilities for manufacture of drug substances and drug products. It has been used primarily in the pharmaceutical industry for evaluating process safety hazards. As is the case with HACCP, the output of a HAZOP analysis is a list of critical operations for risk management. This facilitates regular monitoring of critical points in the manufacturing process.

Preliminary Hazard Analysis

Preliminary hazard analysis (PHA) is the application of experience or knowledge of a hazard to identify future hazards. This tool consists of identification of the possibility that the risk event happens, qualitative evaluation of the extent of possible injury or damage to health that could result, relative ranking of the hazard using a combination of severity and likelihood of occurrence, and identification of possible remedial measures.

Potential areas of use for PHA might be analysis of existing systems or prioritizing hazards where circumstances prevent a more extensive technique from being used. It can be used for product, process, and facility design, as well as to evaluate the types of hazards for a general product type, then the product class, and finally the specific product. PHA is most commonly used early in the development of a project when little information on design details or operating procedures is available; thus, it will often be a precursor to further studies. Typically, hazards identified in the PHA are further assessed with other risk management tools.

Risk Ranking and Filtering

Risk ranking and filtering is a tool for comparing and ranking risks that includes filters (for example, cutoff scores). Potential areas of use include prioritizing manufacturing sites for inspection/audit by regulators or industry. Risk ranking methods are particularly helpful in situations in which the portfolio of risks and the underlying consequences to be managed are diverse and difficult to compare using a single tool. Risk ranking is useful to management in evaluating both quantitatively assessed and qualitatively assessed risks within the same organizational framework.

Supporting Quality and Statistical Tools

There are a number of tools used to support data assessment and facilitate decision making. Some of the more common tools are listed below.

Seven Basic Quality Tools

1. *Cause-and-effect diagram* (also called *Ishikawa diagram* or *fishbone diagram*). Identifies many possible causes for an effect or problem and sorts ideas into useful categories.
2. *Check sheet.* A structured, prepared form for collecting and analyzing data; a generic tool that can be adapted for a wide variety of purposes.
3. *Control charts.* Graphs used to study how a process changes over time.
4. *Histogram.* The most commonly used graph for showing frequency distributions, or how often each different value in a set of data occurs.

5. *Pareto chart.* Shows on a bar graph which factors are more significant.
6. *Scatter diagram.* Graphs pairs of numerical data, one variable on each axis, to look for a relationship.
7. *Stratification.* A technique that separates data gathered from a variety of sources so that patterns can be seen (some lists replace *stratification* with *flowchart* or *run chart*).

Seven New Quality Tools*

1. *Affinity diagram.* Organizes a large number of ideas into their natural relationships.
2. *Relations diagram.* Shows cause-and-effect relationships and helps with analysis of the natural links between different aspects of a complex situation.
3. *Tree diagram.* Breaks down broad categories into finer and finer levels of detail, helping to move thinking step by step from generalities to specifics.
4. *Matrix diagram.* Shows the relationship between two, three, or four groups of information and can give information about the relationship, such as its strength, the roles played by various individuals, or measurements.
5. *Matrix data analysis.* A complex mathematical technique for analyzing matrices, often replaced in this list by the similar *prioritization matrix.* One of the most rigorous, careful, and time-consuming of decision-making tools, a prioritization matrix is an L-shaped matrix that uses pairwise comparisons of a list of options to a set of criteria in order to choose the best option(s).
6. *Arrow diagram.* Shows the required order of tasks in a project or process, the best schedule for the entire project, and potential scheduling and resource problems and their solutions.
7. *Process decision program chart* (PDPC). Systematically identifies what might go wrong in a plan under development.

Suppliers, Inputs, Process, Outputs, Customers (SIPOC) Diagram. SIPOC diagrams define the scope of work for a team and identify at a high level the potential gaps (deficiencies) between what a process expects from its suppliers and what customers expect from the process, thus defining the scope for process improvement activities. Because it also identifies feedback and feed-forward loops between customers, suppliers, and the process, it jump-starts the team to begin thinking in terms of cause and effect.

---

* Excerpted from Nancy R. Tague, *The Quality Toolbox*, Second Edition, ASQ Quality Press, 2004.

Define, Measure, Analyze, Improve, Control (DMAIC).* DMAIC is a data-driven quality strategy used to improve processes. It is an integral part of a Six Sigma initiative, but in general can be implemented as a stand-alone quality improvement procedure or as part of other process improvement initiatives such as lean. DMAIC is an acronym for the five phases that make up the process:

- *Define* the problem, improvement activity, opportunity for improvement, the project goals, and customer (internal and external) requirements.
- *Measure* process performance.
- *Analyze* the process to determine root causes of variation, poor performance (defects).
- *Improve* process performance by addressing and eliminating the root causes.
- *Control* the improved process and future process performance.

Plan–Do–Check–Act (PDCA) Cycle. PDCA is also called the *plan–do–study–act (PDSA) cycle*. The plan–do–check–act cycle is a four-step model for carrying out change. Just as a circle has no end, the PDCA cycle should be repeated again and again for continuous improvement.

* Excerpted from *The Certified Quality Engineer Handbook*, Third Edition, Connie M. Borror, ed., ASQ Quality Press, 2009.

Chapter 11

Training and Personnel Qualifications

Good manufacturing practices (GMP) require that personnel have the education, training, and experience needed to perform their assigned responsibilities. Inadequate training can cause the United States Food and Drug Administration (FDA) to declare the product to be adulterated under the Food, Drug, and Cosmetic Act (FDCA), potentially preventing the product from being sold or distributed. Training on both specific job tasks and GMP is required.

Why is training important in a pharmaceutical environment? Annual reviews of FDA 483/Warning Letters routinely show 30% or more of observations reference training issues or personnel requirements. From FY2018–2022, 244 Form 483 citations were identified against 21 CFR 211.25(a) regarding the education, training, and experience of those performing tasks within the pharmaceutical manufacturing environment.*

NEEDS ANALYSIS

> Identify the requirements for determining the type of training, qualification, and experience needed by quality staff members, operations personnel, and related functions. (Apply)
>
> **Body of Knowledge II.D.1**

Training activities should commence the day the employee is hired. Human Resources typically starts laying the groundwork for training with an employee handbook that explains company expectations around dress, behavior, vacation/sick time/paid time off, insurance as well as how to deal with grievances with co-workers, etc. Training activities then proceed to departmental or on the job-related activities and can include, but not limited to:

* "FDA Warning Letter & Inspection Observation Trends (Updated 2023)," www.thefdagroup/blog/2019-fda-warning-letter-inspection-observation-trends.

- Safety
- Quality Assurance
- Quality Control
- Manufacturing
- Warehouse Operations
- Inspection

In order to properly organize and standardize this training, it is common to have a job matrix that shows the SOP and training requirements related to each job function. For example, see the difference between the two similar jobs below:

QC Lab Technician (Microbiology)

SOP requirements:	QC-001	Lab Safety
	QC-002	PPE for Microbiology Lab
	QC-003	Gram Staining
	QC-004	Sample Preparation for Testing
On the Job Training	Gram Staining	
	Sample Testing	
	Media Preparation	

QC Lab Technician (Chemistry)

SOP requirements:	QC-001	Lab Safety
	QC-005	PPE for Chemistry Lab
	QC-006	Sample Preparation for FTIR Analysis
	QC-007	Proper Disposal of Reagents
On the Job Training	Lab Safety	
	Sample Testing-FTIR	
	Proper Disposal of Reagents	

It is important to note that all training should be done through consistent and current methodologies, affording each employee the same training to provide standardized processes. Changes to process (See Chapter 12—Change Control) will necessitate updating training processes and ensuring all employees are retrained on the new process.

STAFF DEVELOPMENT REQUIREMENTS

> Determine proof of proficiency based on regulations, guidances, and directives including documented evidence and periodic reassessment. (Apply)
>
> **Body of Knowledge II.D.2**

JOB-SPECIFIC TRAINING

Managers (or supervisors) are responsible for defining the training needed for a specific position in a list of required documents and training courses such as a job curriculum. If retraining needs to be performed periodically, such as for aseptic practices, the curriculum can indicate this timing requirement. The curriculum should indicate what training is needed before specific tasks can be performed by the employee. Training is done and documented before an employee performs the task.

Training should not be considered a one-time exercise that is provided only when an employee is hired. Training related to job tasks is necessary when procedures, batch records, and/or test methods are revised or when a task has not been performed recently. Similarly, when an employee's job changes within an organization, his or her education, experience, and training should be reevaluated to determine what areas need to be addressed to perform the new job. Changes in work experience and education can be documented by maintaining an up-to-date résumé or work history in the employee's training file.

Defect awareness training should make personnel aware of the need to understand that if they do not follow procedures, batch records, or test methods as written, an adverse impact to product quality may occur.

Additionally, a critical part of most pharmaceutical and medical device training involves proper steps to protect controlled areas in a manufacturing suite from the largest source of particulate contamination . . . the workers themselves! Maintaining proper ISO level control (ISO Class 5 is typically maintained for most sterile manufacturing environments for pharmaceuticals, with higher ISO class environments for less controlled spaces) is critical for the purity and sterility of drugs or medical devices manufactured in these areas.

As stated in EU Annex 1 and 21 CFR 211.25, employees working in these areas must undergo an appropriate training protocol focused on proper aseptic gowning educations, understand and explain proper gowning technique, show proficiency in proper gowning, and pass a personnel monitoring (PM) evaluation where multiple locations are tested via RODAC plates by a trained microbiology technician. Typically, this evaluation is performed in triplicate (over three separate gowning events), allowing the employee proper time between each gowning event to avoid sweating, undo stress, etc. Once passing PM results are obtained, an employee is then typically considered "gown qualified" and able to gown and proceed into the controlled areas without the need for an escort. However, the training does not end there. An additional step to aseptic gown qualification typically involves taking part in a media fill event where the employee performs their normal job-related tasks during a media fill, showing that their job related tasks do not negatively impact the results of the media fill. Finally (per FDA Guidance documents*), aseptic gown qualifications must be re-assessed annually at a minimum to maintain proper aseptic gowning techniques.

* Guidance for Industry—Sterile Drug Products Produced by Aseptic Processing—Current Good Manufacturing Practice, US Department of Health and Human Services-Food and Drug Administration, 2004.

GOOD MANUFACTURING PRACTICES TRAINING

Many pharmaceutical companies provide annual GMP training courses to meet the stated requirements in 21 CFR 211.25(a). These courses may include topics such as recent trends in the GMP industry, new or newly revised GMP regulations, summary of deviations and internal audit observations that have occurred in the past year, or a review of GMP regulations that impact the organization. In addition, if a pharmaceutical manufacturer has been distributing its products solely within the United States and gains approval to sell its products in Canada, it would be beneficial to train employees on the Canadian GMP, although they closely resemble the FDA GMP. Pharmaceutical manufacturers also should train employees on the specific GMP related to their jobs (Figure 11.1).

FDA's *Quality Systems Approach to Pharmaceutical CGMP Regulations* guidance document indicates that *quality unit* (QU) personnel should have some understanding of scientific and technical topics, product and process knowledge, and the ability to assess risk. The QU has many responsibilities, the most important of which is releasing or rejecting product. For this responsibility, quality unit personnel should be trained to identify any deviations in manufacturing or testing processes that may not have been observed and addressed during manufacturing and testing, and to conduct effective investigations to determine whether product quality was impacted. If product quality was adversely impacted, or if the impact to product is unknown, the product should not be released.

European GMP regulations include specific requirements for training of personnel whose duties involve manufacturing or *quality control* (QC)–related activities. These are discussed in EudraLex, *Good Manufacturing Practice (GMP) Guidelines: The Rules Governing Medicinal Products in the European Union.* Training also applies to visitors or untrained personnel who must enter production or QC areas. These individuals must first receive information on personal hygiene and use of protective clothing, and should be closely supervised.

European training requirements include basic training on GMP as well as ongoing training. The concept of *quality assurance* (QA) and its implementation should be addressed in GMP training. Specific training should be given if personnel work in areas where contamination may be a hazard. EudraLex specifies that the training programs be approved, preferably by the head of production or quality control, as appropriate, and records of training must be maintained. Effectiveness of training should be assessed as well. World Health Organization (WHO) GMP training requirements are similar to those discussed in EudraLex and also include a requirement for contractors and consultants to receive training.

TRAINING EFFECTIVENESS AND ROLE OF SUPERVISOR

> Apply various methods for testing and evaluating training effectiveness. Identify the role and responsibilities of supervisors including ensuring staff are adequately trained to perform their assigned functions. (Apply)
>
> **Body of Knowledge II.D.3**

According to *Quality Systems Approach to Pharmaceutical GMP Regulations*, managers (supervisors) are responsible for ensuring that training is effective by verifying that the skills are learned and implemented in their employees' daily performance. Management is also responsible for communicating employee roles, responsibilities, and authorities within the quality system and ensuring that interactions between different roles are defined and understood.

Any evidence that indicates the employee's training was not effective, or that the employee does not have the appropriate experience or skills (for example, frequently occurring deviations when executing tasks), should trigger an evaluation of whether the employee is capable and/or whether additional training or retraining is needed. If the original training was not effective, simply retraining an employee may not be sufficient to resolve performance issues. The training may need to be redesigned to be effective.

Training effectiveness can be monitored in a number of ways. Most directly, a post-training assessment (quiz or test, physical skill performance, etc.) can be performed between trainer and trainee to show that the skill is understood and can be performed to the trainer's or companies' expectations/specifications. Additionally, training can also be monitored over time by reviewing trend analysis of deviations related to the specific training. For example, providing skill-based training on proper use of a micrometer should result in a decrease of deviations/investigations resulting from inaccurate micrometer usage. Therefore, it is important to have both an effective training program as well as a monitoring program that show effectiveness for those CAPA items that require improved or repeated training to resolve.

EMPLOYEE QUALIFICATIONS

Managers need to define the qualifications for a position (which is usually defined by a job title). The responsibilities for a specific position are defined in a document such as a job description. In addition to responsibilities, the job description can include minimum education and experience requirements, skills, and required certifications. Certifications provide a good yardstick for knowledge of a topic because a *Body of Knowledge* (BoK) is measured by an examination, but they do not replace the experience and hands-on training provided by performance of a job or task. ASQ certification is peer recognition that an individual has proficiency with, and comprehension of, a specified BoK. A clear job description facilitates hiring the best person for the job.

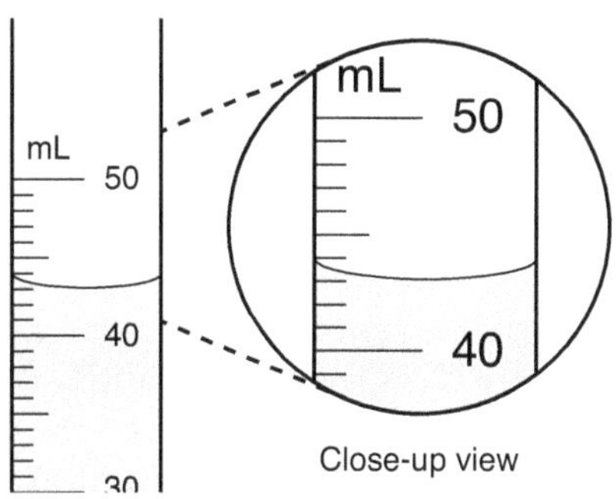

Figure 11.1 Example of graduated cylinder reading training.

Job descriptions for personnel who perform testing and/or calibration tasks have specific minimum requirements as outlined in ISO/IEC 17025:2005, *General requirements for the competence of testing and calibration laboratories.* These requirements include responsibilities for performing and planning tests and/or calibrations and evaluating the results, reporting opinions and interpretations, and modifying, developing, and validating new methods and documented evidence of experience, training, and education.

DOCUMENTATION OF TRAINING

All training should be documented by recording the topic and an overview of the content, the trainer, the date trained, and the trainee's name. Regulatory agencies expect to see documentation that trainers are qualified as well. For example, a GMP trainer should have experience and training in GMP in addition to knowledge of the topic of the training program. It is important that training records be maintained as well, to provide evidence of employee competency.

It is important to note that training records should be maintained in a manner that allows efficient review of the training performed, the employee trained as well as those employees that were not trained (in the case of group training). The training material used at the time of the training should also be maintained to show proof of the specifics of the training at that moment in time, as procedures and best practices will change over time. Training processes should be structured to not only provide training to the employees, but enable management to easily identify an employee that is not available for the training event (that is, vacation, sick, etc.) and ensure that they are updated on the training as soon as possible.

EMPLOYEE PROFICIENCY

Proficiency for a particular task should be documented before the employee performs the task independently. For critical tasks (those that could impact product quality), task training can be performed to ensure that the employee thoroughly understands the task. Task training consists of the trainee reading any related procedures, observing the task being performed by a qualified individual, performing the task under the supervision of a qualified individual with assistance and coaching, if needed, and performing the task independently. The trainer indicates on the training documentation that the trainee satisfactorily completed training, and the employee indicates that he or she understands the task and associated documentation.

Proficiency for gaining knowledge from training can be assessed by administration of a test after the training event. Verification that knowledge was gained during the training session can be accomplished by administering a pretest before the training event occurs and comparing the results to the post-training test. A minimum passing grade is typically defined for GMP tests. Retention of the knowledge gained by attending the training could be assessed by testing a trainee when some time has elapsed after the training event occurs. Additionally, re-evaluating personnel training over time is critical to maintaining a well-trained and proficient workforce. This type of evaluation can catch poor practices before they become routine or shared with other employees. It is important to note that training an employee one time does not mean he or she is trained for life. Periodic retraining and re-evaluation are critical, especially in those roles that impact the SISQP of a pharmaceutical product.

Chapter 12

Change Control and Management

The United States Food and Drug Administration (FDA) provides very specific requirements for change control and management, requiring written procedures for production and process controls to ensure that drug products have the identity, strength, quality, safety, and purity they purport or are represented to possess, and that medical devices are safe and effective. These requirements extend beyond procedures and work instructions to include changes to specifications, products, labels, methods of testing and production, packaging, equipment, and facilities.

21 CFR 211 and 820 provide the requirements for establishing and maintaining documents, including requirements for reviewing, for adequacy, and approval before issuance. The *quality unit* (QU) is ultimately responsible for the change control and management function. However, other functions usually participate in the process depending on the application and requirements, including engineering, production, manufacturing, distribution, regulatory affairs, and clinical.

Procedures for production and process controls are usually birthed or modified using a predetermined *engineering change order* (ECO) system. The ECO system usually includes a list of requirements and considerations, an evaluation checklist, and required signatures. Impact assessment consideration should be evaluated for procedures, work instructions, forms, regulatory compliance, validations, facilities, inventory, equipment, drawings, specifications, risk management (failure mode and effects analysis [FMEA]), packaging, labeling, storage, handling, and distribution. The complexity of the change usually determines the level of control and authorization required. For example, to update a procedure that contains a grammatical error may be considered an administrative change and therefore allow for an abbreviated approval and signature process, whereas a procedural change that modifies a critical process parameter usually requires the full evaluation and approval process, including risk analysis and change effectiveness. Figure 12.1 depicts a typical change control process.

Changes to procedures for production and process controls include a description of the change, identification of the affected documents, the signature of the approving individual(s), the approval date, and the date the change becomes effective. Obsolete documents are promptly removed from all points of use. This process is necessary to prevent unintended use of obsolete documents.

Written production and process control procedures are followed in the execution of the various production and process control functions, and are documented

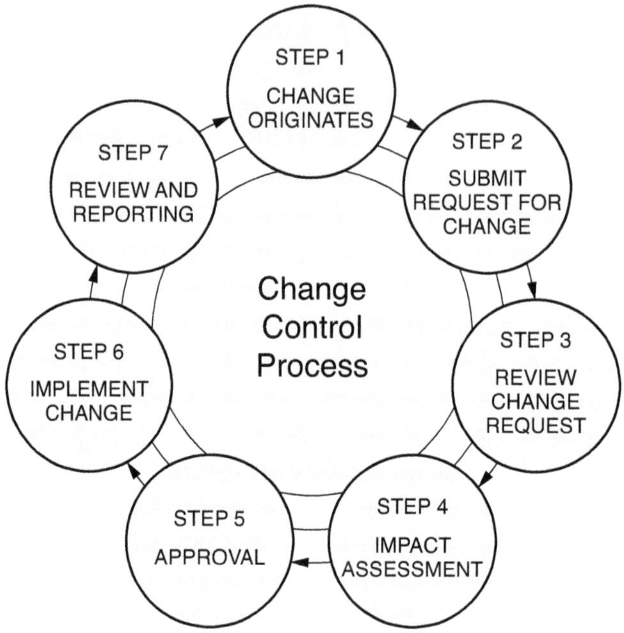

Figure 12.1 Typical change control process.
Source: The Ascent. "A beginner's guide to the Change Control Process," Mark Roy Long (https://www.fool.com/the-ascent/small-business/it-management/articles/change-control-process/), August 2022.

at the time of performance. Any deviation from the written procedures must be recorded and justified. Deviations should be reviewed by the quality unit to ensure no deleterious effects and that the drug products have the identity, strength, quality, safety, and purity they purport or are represented to possess. When deviations occur, it may necessitate the use of the *nonconformance* (NCR) and/or the *out-of-specification* (OOS) process as appropriate.

Another consideration when production and process control functions are birthed or changed is notification and training. It is required that approved changes be communicated to the appropriate personnel in a timely manner, and, where appropriate, the personnel are trained on the changes.

PRE-CHANGE ANALYSIS

> Assess the impact that proposed changes will have on products, processes, facilities, utilities, and systems to minimize risk and ensure regulatory compliance. (Analyze)
>
> **Body of Knowledge II.E.1**

Within the change control process diagram above, there are key steps that must take place to maintain an organization in a state of control. The first of these is Step 3, Review Change Request. Much like ideas dropped into an employee suggestion box, change control ideas are generally easy to generate; however, management must take the necessary steps to evaluate each idea, not only on its merits, but on the impact on the company as a whole. For example, implementing a change to worker walkways within a plant may result in a shorter walk from point A to B, but if that walk now crosses forklift pathways, causing interruptions to the flow of material within the building, that concern should be reviewed prior to implementation. Therefore, all change controls must be reviewed and scoped to the best of management's abilities prior to implementation.

CHANGE IMPLEMENTATION

> Implement and document the change following a change implementation plan. (Create)
>
> **Body of Knowledge II.E.2**

Once the Pre-Change Analysis has been completed and the idea has been approved, implementation can begin. Depending on the complexity of the change, this may range from hours to weeks to complete, during which time employees must be kept aware of any risks or changes to their normal procedures or practices. Continuing the example above, if the new walkway is approved, it would be expected that facilities would need time to order the appropriate signage, walkway barricades, etc. to implement the change. Once facilities begin the work, employees may need to be diverted through different areas to allow the facilities team the ability to safely complete their work. A well-implemented pre-change analysis will capture as many of these as possible, but management needs to remain open to any unforeseen issues that may arise during implementation and be able to react appropriately.

POST-CHANGE ANALYSIS

> Analyze data and other inputs to determine the results of a change and evaluate any new risk factors created by the change. (Analyze)
>
> **Body of Knowledge II.E.3**

Management must also monitor the change post-implementation for any unforeseen issues or complications. This monitoring must be ongoing long enough to assess any environmental impact to the change, change in employee work practices, etc. In our above example, does the new walkway create bottlenecks at different areas, safety concerns, or unforeseen issues that need to be addressed? Do employees now have to travel through a stairway that causes safety concerns? Poor lighting? Any of these concerns must be reviewed and addressed appropriately and timely. If no issues are discovered, then a change control can be closed, but management must remain diligent and may reopen a change control if future considerations warrant a review of the change and a new response.

Chapter 13

Investigations and Corrective and Preventive Action (CAPA)

TRIGGER EVENTS

> Identify events that require investigation, root cause analysis, and impact assessment both directly and indirectly related to the event. (Evaluate)
>
> **Body of Knowledge II.F.1**

According to the United States Food and Drug Administration (FDA) guidance *Quality Systems Approach to Pharmaceutical CGMP Regulations*, "CAPA is the regulatory concept for investigating, understanding, and correcting discrepancies while attempting to prevent their recurrence." *Corrective and preventive action* (CAPA) is an absolute that defines the backbone of the *quality management system* (QMS). The systemic management of the CAPA program is an expectation of most standards and regulations (for example, FDA, European Medicines Agency [EMA], International Organization for Standardization [ISO]). CAPA is essential throughout the life cycle of any product in the pharmaceutical drug and medical device industries; it is not a system used only during the production phase of a pharmaceutical product. ICH Q10 provides a table that allows insight into the various ways in which CAPA is relevant to each stage (Table 13.1). Regulatory agencies have a clear expectation that companies have a defined system in place for the documentation and implementation of corrective actions arising from "investigation of complaints, product rejections, nonconformances, recalls, deviations, audits, regulatory inspections and findings, and trends from process performance and product quality monitoring."

The term *CAPA* is used to define the program in quality manuals as well as in the *standard operating procedures* (SOPs) defining the quality system. Whether there is a single procedure that defines the process for CAPAs or there are separate procedures for CAPA is not important. What is important is that the process be clearly defined, that it is structured, and most critically, that the investigations have the objective of determining the root cause of the issue.

FDA guidance *Quality Systems Approach to Pharmaceutical CGMP Regulations* states that there are four separate concepts used in the quality system model for

Table 13.1 Application of a corrective action and preventive action system throughout the product life cycle.

Pharmaceutical development	Technology transfer	Commercial manufacturing	Product discontinuation
Product or process variability is explored. CAPA methodology is useful where corrective actions and preventive actions are incorporated into the iterative design and development process.	CAPA can be used as an effective system for feedback, feed-forward, and continual improvement.	CAPA should be used, and the effectiveness of the actions should be evaluated.	CAPA should continue after the product is discontinued. The impact on product remaining on the market should be considered, as well as other products that might be impacted.

CAPA (Figure 13.1): remedial containment or corrections to correct the immediate problem identified, root cause analysis to determine and understand the actual cause of the deviation or issue and potentially provide guidance for the prevention of a future occurrence of a similar problem or issue, corrective action, and preventive action, which includes those activities carried out as a result of the root cause analysis to prevent a recurrence of the same or similar issue.

Good manufacturing practices (GMP) requirements dictate that all of the manufacturing, packaging, testing, and holding of drug product must take place according to established and approved procedures. Once the situation is such that this condition is not met, it is necessary to investigate to determine the corrective action necessary. The level of investigation, the corrective action necessary, the formality with which it is documented, and the level of control extended should be based on the risks involved (in accordance with ICH Q9). Many firms have established "alert" and "action" limits on many of their processes to assist in defining the level of formality needed. Regardless of the level of formality involved, specific needs for every investigation include:

- The individuals involved in the investigation, and the decision-making authority, must have sufficient education, experience, and knowledge to carry out their associated tasks, which usually means cross-functional, multidisciplinary teams including Quality Assurance.
- The investigation must be thorough, timely, unbiased, well-documented, and scientifically sound.
- It must include sufficient information to stand on its own and allow an independent reviewer (possibly a regulator) to understand the logic, the rationale for decisions, and who the decision makers were.
- Identify whether immediate containment actions are required.
- It must identify the appropriate corrective action and, to properly take preventive measures, it must identify the root cause of the problem.

Most firms utilize a system to make improvements and fix problems. In the FDA-regulated industry there is a mandate to have such a system; FDA regulations

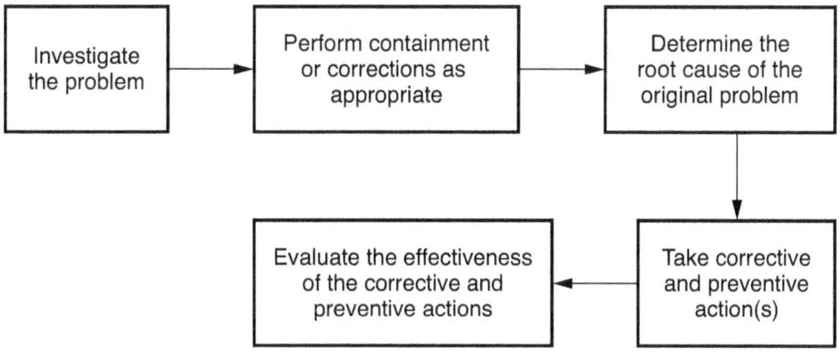

Figure 13.1 Flowchart of interactions of a quality system model.
Source: Figure courtesy of the ASQ Food, Drug, and Cosmetic Division. Used with permission.

21 CFR 211.192 and 820.100 provide the requirements. Most firms call it a CAPA system. It is essential to have a written procedure that indicates how to exercise the system and how to control it. There are a variety of software packages firms may choose from to use in managing and analyzing their CAPA systems. It should be noted that while the pharmaceutical and medical device approaches to CAPA have many parallels; they are not in purest agreement in some respects. The concept of corrective actions in the medical device industry includes any action that helps to resolve and prevent recurrence of a known issue; actions that are titled *preventive actions* are somewhat rare, except for improvements and some management review–based CAPA. In the pharmaceutical industry, however, the idea of preventive actions is more common, as any action that will prevent recurrence is a preventive action. Whatever the differences, the goal is the same: these systems investigate issues to root cause and take actions that stop them from reoccurring. Failure in these systems can cause significant FDA compliance issues, including Warning Letters.

The general flow of a CAPA follows, though it may well be that some of these steps are performed in parallel:

- An issue is identified.
- A CAPA is requested.
- The request is either approved or disapproved.
- If approved, an investigation is conducted to determine the scope, severity, and root cause(s) of the issue.
- A level of risk is determined.
- Based on the investigation, corrections and/or corrective actions and/or preventive actions are identified that will, appropriately to the risk, solve the root cause of the issue.
- These solutions are validated, if necessary.
- Effectiveness of the CAPA is evaluated.
- The case is closed with management approval.

Additionally, the CAPA system should require that the issue and solution are clearly communicated throughout the organization, and ensure that the CAPA record can be retrieved in a timely manner (electronic or paper). Also, CAPA records should be closed in a timely manner in accordance with established SOPs. This eliminates CAPAs being opened and never resolved.

CAPA TRIGGERS

A CAPA can be triggered by any number of incidents within pharmaceutical / medical device manufacturing. A small number of examples could be:

- Customer feedback
- Internal audits
- Safety incidents
- Interventions
- Human error
- Manufacturing deviations
- Batch potency or sterility
- Product recall
- Equipment malfunction

While the above is a common list of CAPA initiation points, it is by no means exhaustive.

CAPA PROCESS

What is the source of the issues to be worked on? There are many systems that feed into the CAPA system, including product or process nonconformance, complaints, audits, management reviews, quality data analysis, and trending (see Figure 13.2). Validation efforts, identified improvement opportunities or suggestions, and sometimes vendor/supplier corrective action requests are other general sources that can indicate or trigger the CAPA process. Product or process nonconformances and complaints tend to be the largest sources of issues and problems. Out-of-specification (OOS) issues are usually grouped with the nonconformances. The implication is that methods for monitoring and escalating from all these sources must be in place.

The investigation stage is conducted to determine the scope, severity, and cause(s) of the issue. There may be a multistage investigation structure first to define scope and risk to health so that immediate containment actions are taken where appropriate, and a further investigation of the root causes is then completed. There is an expectation that when the investigation indicates, the scope of the CAPA may be expanded to all the products or processes that have the same potential issues.

Determining the level of risk is an increasingly global requirement. competent authorities and notified bodies have increasingly explicit requirements concerning including a risk assessment function in your CAPA system; the desire is

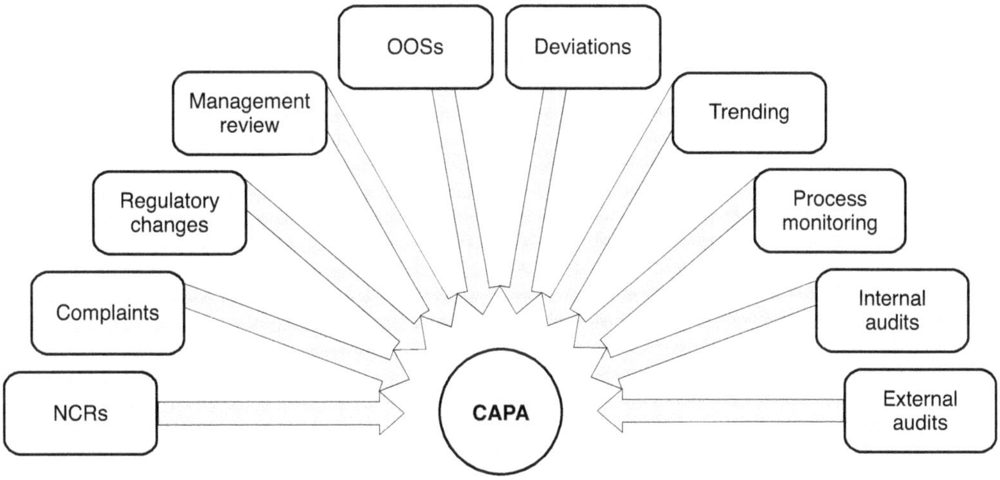

Figure 13.2 Typical CAPA feeder systems.

to ensure that appropriate levels of action, speed, and management attention are brought to any given issue. There is fairly wide latitude regarding the definition of your risk assessment, but a key component must be potential patient harm. If well defined, there is the possibility that many very low-risk issues may be dealt with in a consistent and less strenuous series of steps that can impart an increased level of efficiency.

Corrections are straightforward actions that bring a product into compliance. Corrective actions are those actions that change the process or product so that the issue cannot reoccur. Preventive actions are those actions that in the absence of a known issue will prevent a possible situation from occurring. A few items are often overlooked in these preventive and corrective actions: If a process is changed, the process should require documentation that all the appropriate people are trained, and verification that any change made does not adversely affect something else. For products or processes that are validated, any change that impacts a validation must be revalidated and appropriately documented.

Evaluating the effectiveness of the CAPA usually means that after all the actions are taken, including the validations, the system must designate a period of time (or other criteria) to monitor the performance of the issue to see if it reoccurred, or whether some unintended adverse result occurred. After all this, the CAPA case is closed by documented management approval.

GENERAL CONSIDERATIONS FOR CAPA

Whatever the specifics of the system, a thoroughly successful CAPA system can prevent a host of problems—not only with the process and products but also from FDA. Frequently, demonstrating that an ongoing (that is, not started the day the FDA inspection began), energetic CAPA is in place, a written inspection citation may be avoided. However, FDA most likely will scrutinize the CAPA system during almost any inspection; aside from the documentation of the system, they will likely scrutinize timeliness and efficacy, and investigate specifics on a number of cases. Efficacy can often be demonstrated with a successful quality data trending system that looks for recurring issues. Furthermore, the data from your CAPA system should be one of the items reviewed in your management reviews. Information on CAPA from several FDA sources follows.

CFR 211.192 Production Record Review

All drug product production and control records, including those for packaging and labeling, shall be reviewed and approved by the quality control unit to determine compliance with all established, approved written procedures before a batch is released or distributed. Any unexplained discrepancy (including a percentage of theoretical yield exceeding the maximum or minimum percentages established in master production and control records) or the failure of a batch or any of its components to meet any of its specifications shall be thoroughly investigated, whether or not the batch has already been distributed. The investigation shall extend to other batches of the same drug product and other drug products that may have been associated with the specific failure or discrepancy. A written record of the investigation shall be made and shall include the conclusions and follow-up.

FDA 21 CFR 820.100 Corrective and Preventive Action

(a) Each manufacturer shall establish and maintain procedures for implementing corrective and preventive action. The procedures shall include requirements for:

(1) Analyzing processes, work operations, concessions, quality audit reports, quality records, service records, complaints, returned product, and other sources of quality data to identify existing and potential causes of nonconforming product, or other quality problems.

Appropriate statistical methodology shall be employed where necessary to detect recurring quality problems;

(2) Investigating the cause of nonconformities relating to product, processes, and the quality system;

(3) Identifying the action(s) needed to correct and prevent recurrence of nonconforming product and other quality problems;

(4) Verifying or validating the corrective and preventive action to ensure that such action is effective and does not adversely affect the finished device;

(5) Implementing and recording changes in methods and procedures needed to correct and prevent identified quality problems;

(6) Ensuring that information related to quality problems or nonconforming product is disseminated to those directly responsible for assuring the quality of such product or the prevention of such problems; and

(7) Submitting relevant information on identified quality problems, as well as corrective and preventive actions, for management review.

(b) All activities required under this section, and their results, shall be documented.

FDA Guidance for Industry: Quality Systems Approach to Pharmaceutical CGMP Regulations

(Page 5) CAPA (Corrective and Preventive Action). CAPA is a well-known CGMP regulatory concept that focuses on investigating, understanding, and correcting discrepancies while attempting to prevent their recurrence. Quality system models discuss CAPA as three separate concepts, all of which are used in this guidance.

- Remedial corrections of an identified problem
- Root cause analysis with corrective action to help understand the cause of the deviation and potentially prevent recurrence of a similar problem
- Preventive action to avert recurrence of a similar potential problem

FDA Guide to Inspections of Quality Systems

1. Verify that CAPA system procedure(s) that address the requirements of the quality system regulation have been defined and documented.
2. Determine if appropriate sources of product and quality problems have been identified. Confirm that data from these sources are analyzed to identify existing product and quality problems that may require corrective action.
3. Determine if sources of product and quality information that may show unfavorable trends have been identified. Confirm that data from these sources are analyzed to identify potential product and quality problems that may require preventive action.
4. Challenge the quality data information system. Verify that the data received by the CAPA system are complete, accurate, and timely.
5. Verify that appropriate statistical methods are employed (where necessary) to detect recurring quality problems. Determine if results of analyses are compared across different data sources to identify and develop the extent of product and quality problems.
6. Determine if failure investigation procedures are followed. Determine if the degree to which a quality problem or nonconforming product is investigated is commensurate with the significance and risk of the nonconformity. Determine if failure investigations are conducted to determine root cause (where possible). Verify that there is control for preventing distribution of nonconforming product.

7. Determine if appropriate actions have been taken for significant product and quality problems identified from data sources.
8. Determine if corrective and preventive actions were effective and verified or validated prior to implementation. Confirm that corrective and preventive actions do not adversely affect the finished device.
9. Verify that corrective and preventive actions for product and quality problems were implemented and documented.
10. Determine if information regarding nonconforming product and quality problems and corrective and preventive actions has been properly disseminated, including dissemination for management review.

RESPONSE ACTIONS

> Define immediate action, corrective action, preventive action, management responsibility, and methods of implementing them. (Evaluate)
>
> **Body of Knowledge II.F.2**

A critical, and often overlooked, aspect of the CAPA process is implementing proper response actions once a CAPA has been initiated. Once the error has been identified and a Corrective Action initiated, the immediate response (for example, the response given by the company within the first 24–48 hours post-discovery to contain any at-risk product/material associated with the Corrective Action trigger) must be well defined, encompass all possible related product, and be focused on patient/consumer safety. This immediate response is part of the FDA review during the reporting of the incident via an FDA recall submission.*

Immediate responses are typically broad in scope given that the investigation is likely not completed and all information is not yet known. A typical immediate response could be to stop all product shipments, place holds on production and/or place holds on distributors that have stock of the suspect product. As the investigation into the incident continues and information is developed, holds can be lifted as deemed appropriate.

Failing to implement a proper immediate response, or delaying in initiating an immediate response can lead to difficult conversations with a governing body (that is, FDA, HealthCanada, etc.). They will ask why proper or obvious immediate actions were not implemented in a timely manner and expect well thought out answers to these questions. Lacking clear and scientifically sound answers may lead to a for cause audit from the regulatory agency.

---

* Product Recalls, Including Removals and Corrections—Guidance for Industry, March 2020, FDA.

CAPA FEEDBACK AND TRENDING

> Describe how CAPA trending is used to modify appropriate quality system elements. (Create)
>
> **Body of Knowledge II.F.3**

An equally important aspect of your Quality Management CAPA system will be the effort and details regarding trending of the CAPA data. CAPA trend analysis should be part of the regular Quality Management Review discussions to allow for the management team to review and look at any trends that could adversely impact the business. CAPA trends should be reviewed across any and all aspects of the investigational information. Items such as production line, shift, product, type of incident, and root cause should all be part of the trend analysis to ensure proper corrective or preventative actions are put in place.

Chapter 14
Audits and Self-Inspections

AUDIT PROCESSES AND RESULTS

> Develop audit schedules, differentiate between various audit types (systems, product, and process) conducted either remotely or on-site, document evidence of audit completion, and analyze audit results to assess conformance to requirements. (Evaluate)
>
> **Body of Knowledge II.G.1**

AUDIT TYPES

An *audit* is a "systematic, independent, and documented process for obtaining audit evidence and evaluating it objectively to determine the extent to which audit criteria are fulfilled." Audits can be categorized in different ways (Table 14.1). This chapter focuses on first-party audits in a *good manufacturing practices* (GMP) organization.

FIRST-PARTY AUDITS

Internal audits are performed according to a schedule to evaluate the quality system in order to determine whether it is effectively implemented, covers all requirements, and is maintained. In addition, they are performed to determine whether the organization's processes and products meet established parameters and specifications.

The organization performs *internal audits* to ensure that its systems are compliant by proactively evaluating them and addressing any gaps in compliance. This evaluation process increases the chances that the organization will produce goods and services that meet quality requirements, and help ensure that consumers can trust that their medications and medical devices are safe and effective.

Internal audits can give early warning of a problem before it can impact product quality. Ensuring that quality systems are under control also maximizes the likelihood that a third-party audit of the organization will be successful.

Table 14.1 First-, second-, and third-party audits.

First party	Internal audit (self-inspection) is performed within an organization. A first-party audit could be performed within an organization's site by its own internal auditor, or could be conducted by a corporate auditor of one of the organization's sites. An internal auditor can be a contractor if an organization has no qualified auditors or requires a specialized auditor, however, the audit is still considered a self-inspection.
Second party	Audit is performed by an organization of its supplier to ensure that contractual requirements are being met, or an audit is performed of a potential supplier to determine if it is capable of meeting specific requirements for a product or service. A manufacturer could perform a second-party audit of its raw material suppliers, contract laboratory, sterilization subcontractor, or packaging supplier.
Third party	An audit or inspection to ensure that requirements of an organization are being met such as those performed by a regulator, a QMS certification agency, or a customer and potential customer. In the United States, FDA performs inspections of companies that are subject to GMP to ensure compliance with these regulations.

Table 14.2 System, process, and product audits.

System	A system audit reviews a group of processes and controls, including activities crossing multiple organization, process, and product boundaries.
Process	A process audit examines the input, actions, and output of a single process to ensure that they meet requirements.
Product	A product audit examines a product or service to ensure that it meets requirements or specifications.

System, process, and product audits have different areas of focus that can be used for internal audits (and for second- or third-party audits as well) (Table 14.2). The United States Food and Drug Administration (FDA) has identified six systems that they may review when conducting regulatory inspections: production, materials, laboratory controls, facilities and equipment, packaging and labeling, and quality. *Process audits* can address an aspect of a quality system, such as change control, product disposition, or validation. *Product audits* are usually narrower in scope than system or process audits and focus on a specific product or product family.

AUDIT PROCESS

Internal audits (also referred to as *self-inspections*) are conducted by GMP organizations to evaluate the effectiveness of their quality systems in complying with GMP. The generic steps involved in an audit are:

- Initiation (scope and frequency)
- Preparation (review of documentation, the program, and working documents)

- Execution (opening meeting, examination and evaluation, collecting evidence, observations, closing meeting with the auditee)
- Report (preparation, content, and distribution)
- Completion (report, submission, and retention)

GMP regulations and guidance documents from the United States, Canada, European Union (EU), and Japan have similar requirements for an internal audit program:

- A written procedure to define the audit process
- A defined audit schedule covering all quality systems
- Qualified, independent, trained auditors
- Documentation of the audit report including the findings
- Documentation of the corrective actions
- Management responsibility for timely implementation of effective corrective actions

Written Procedure

All of the steps necessary to conduct the audit should be defined in a written *standard operating procedure* (SOP). The written procedure should include how an auditor will be qualified. An auditor should be knowledgeable about the requirements against which he/she will be auditing (for example, GMP for pharmaceutical products and the specific jurisdictional regulations) and the auditing process. The auditor should have experience in conducting audits.

The procedure should document which requirements are to be evaluated during the audit (for example, US GMP for finished pharmaceuticals, EU requirements for medicinal products for human use) and define how the audit schedule will be established. The frequency of audits and audit topics should be addressed in the written procedure.

The audit SOP includes requirements for the auditor to conduct an opening meeting with the auditee to explain the audit process and scope and to schedule the execution of the audit. The audit methodologies usually include observation of processes, facilities, and equipment, interviews with process participants, and review of procedures and records. At the closing meeting of the audit, the auditor addresses the draft observations with the auditees; in some cases, the auditee may provide missing information that indicates no gap in compliance exists, and the draft observation can be removed.

The procedure indicates that the auditor compiles the observations (gaps in compliance with requirements) in an audit report, which is provided to the auditee. The manager(s) responsible for the areas in which there are observations provide a response for each observation within a defined amount of time (for example, two weeks or one month after the completion of the audit). Finally, the procedure should include the methods used to monitor corrective actions to completion, and mechanisms to determine effectiveness of the corrective actions after they are implemented. The audit procedure may reference a corrective action procedure that contains corrective action monitoring and effectiveness mechanisms in addition to or instead of having that information also being in the audit procedure.

Audit Schedule

The audit schedule is often issued for a one-year period (for example, at the beginning of the calendar year) and indicates the areas and topics to be audited and when the audits will occur. Audits can be prioritized by evaluating the potential risk to product of various quality system elements, but the entire quality system should be evaluated on a periodic basis. Potential risks influencing the audit schedule could include new product launches, significant design changes, facility moves, switching suppliers, and implementation of a new QMS process or software program, etc. The schedule is approved by the management of the audit function. This document can be used to assign auditors to specific audits and to track the execution of audits as they are completed. When a regulatory agency conducts an inspection of an organization, this document can be used to provide evidence that the audits are being conducted as planned. Internal audit reports are not reviewed by FDA during the course of a routine inspection. FDA will only review internal audit reports in the following cases (*FDA Access to Results of Quality Assurance Program Audits and Inspections*):

- In directed or for-cause inspection and investigations of a sponsor or monitor of a clinical investigation
- In litigation (for example, grand jury subpoenas, discovery, or other agency or Department of Justice law enforcement activity, including administrative regulatory actions)
- During inspections made by inspection warrant where access to records is authorized by statute, and when executing any judicial search warrant

Qualified, Independent Auditors

Auditors should be qualified to perform audits based on education, training, and experience. Education and training on GMP and the audit process can be obtained through courses and conferences provided by a variety of GMP consultants and organizations. An auditor's knowledge about current GMP can be increased by reading industry periodicals, regulations, and guidance documents from applicable countries for the type of products being manufactured, inspection results (for example, as summarized in GMP Trends [www.gmptrends.com/] or documented in Warning Letters), and books on GMP topics.

Auditors can become certified through organizations such as the American Society for Quality (ASQ), which has certified quality auditor, certified biomedical auditor, and certified hazard analysis and critical control points (HACCP) auditor designations.

The best way for new auditors to learn about auditing is to work with an experienced (lead) auditor. First, the new auditors need to be trained on the organization's procedure for internal audits. Once they are familiar with the organization's procedures, the new auditors can observe audits in process. Once they are comfortable with the process, the new auditors can assist a lead auditor in performing audits and writing observations and audit reports. As a final step, the new auditors can conduct an audit with oversight by an experienced lead auditor. When the new auditors have proven to the lead auditor that they can conduct an audit and write clear, objective audit reports, they are ready to perform audits independently. It is important to document the new auditors' education, training, and experience to demonstrate their qualifications to conduct audits.

Auditors must be independent of the areas that they audit. An auditor should be able to objectively assess compliance with requirements. Auditors are typically part of Quality Assurance, but it is not uncommon for individuals from a variety of departments to participate on audit teams. Examples of being independent include not auditing one's direct managers' records or processes one is the lead of. For smaller organizations, this may be a challenge; therefore, rationalizing the independence and rationale of which records were chosen for review in the audit report would be important.

Analyze Audit Results to Assess Conformance to Requirements

During the execution of the audit, the auditor needs to determine whether the procedures meet regulatory requirements, and whether actual practices follow the written procedures. Starting with regulatory requirements, the auditor confirms that written procedures exist to meet these requirements. Next, the auditor assesses whether actual practice is consistent with the company procedures by observing the execution of the procedures, interviewing process participants, and/or reviewing records. The auditor asks such questions as:

- Do procedures exist for each regulatory requirement?
- Is there evidence that each procedure is being followed as written?
- Are procedures and records clear and understandable?
- Are records legible?
- Are corrections to records clear and justified?
- Are deviations initiated when procedures are not followed?
- Is the change control process followed to evaluate the potential impact of a change before it is implemented?
- Are training needs identified for each GMP employee for his or her assigned job?
- Is each GMP employee trained to perform a task before the task is performed?
- Do employees receive general GMP training periodically?
- Are employees executing the same task in a repeatable and reproducible manner? Is there a work instruction or a need for a work instruction to ensure consistency?

Documentation of Findings and Corrective Actions

Once the audit has been completed, a report is written by the auditor to serve as a record of the audit. The scope, purpose, and what was observed during the audit, both positive and negative (that is, gaps in compliance), should be documented. The observations of gaps in compliance should be written to clearly indicate which regulation, guidance, or procedural requirement was not met. Many times, auditors will rank the severity of the finding(s) in an effort to help prioritize the nonconformances. Table 14.3 provides some generally accepted nonconformance definitions. The nonconformance rating scales and terms are generally in the company's audit procedure. Responses and actions to audit findings should be prioritized using risk-based methodology.

It is in the internal auditor's and the company's best interest for the internal auditor to maintain a positive relationship with coworkers. Even though a primary goal of auditing is to identify nonconformances, this process should be portrayed as an opportunity for continuous improvement. When an internal auditor conducts an audit, he or she can report on positive results observed (areas observed to be compliant and effective) in addition to the gaps in compliance. It is important to write the observations to address the gaps with GMP and not to place blame on anyone for the gaps. Also, the auditor should not impose his or her solution to an observation; the people performing the process are the experts and owners of the process. The auditor may provide feedback about proposed corrective actions if requested by the auditee. In general, if the internal auditor (or audit team) treats the auditee with respect throughout the audit process, a positive relationship can be maintained.

AUDIT FOLLOW-UP

> Use various methods to evaluate and verify the effectiveness of corrective actions taken. (Evaluate)
>
> **Body of Knowledge II.G.2**

Manager(s) of the areas in which nonconformances were observed are responsible for providing responses to the observations. The responses should identify the corrective actions to be taken to close any gaps in compliance. Each corrective action should indicate the specific action to be taken, who is responsible for implementing the action, and the date when the action will be implemented. The location of the documentation of the observation responses should be documented in the audit procedure. It is often in the audit report or a separate traceable corrective action project.

More than one corrective action may be needed to successfully address the observation. The timeliness of implementation of a corrective action should be based on the criticality of the observation. Nonconforming situations that have a high risk to product quality (for example, potential contamination of product) should be implemented immediately. Observations with a low risk to product

Table 14.3 Typical audit nonconformance definitions.

Nonconformance	Description
Critical	The nonfulfillment of a requirement that represents a high level of risk to the quality, safety, and effectiveness of the product, integrity of the quality management system, or regulatory compliance
Major	The nonfulfillment of a requirement that represents a medium level of risk to the quality, safety, and effectiveness of the product, integrity of the quality management system, or regulatory compliance
Minor	The nonfulfillment of a requirement that needs to be corrected but only represents a low risk to product safety and effectiveness or to the integrity of the quality management system or regulatory compliance
Opportunity for improvement (OFI)	An opportunity to improve the quality system that does not rise to the level of a finding

quality can be implemented as soon as practical. At no time, however, should audit corrective actions languish; they should be given priority by management and implemented in a timely fashion.

After the responses are approved by audit management or similar function per the audit procedure, responsible management ensures that the corrective actions are implemented by the date indicated in the corrective action plan. If for some reason the action cannot be implemented as indicated in the response or by the date in the response, the area management should proactively contact audit management to communicate the change. If the scope of the corrective action or timeliness of implementation need to change, audit management or similar function should approve the change and document it in the audit records. If the nonconformance is critical to product quality, the necessary resources should be found to make it possible to meet the original corrective action plan.

If there are multiple major observations in one process or area, in addition to verification of specific corrective actions per the section below, consider scheduling another audit at an appropriate time to see if there are any other aspects of the process that are not effectively in control or compliance.

Verification of Corrective Actions

The audit department tracks the corrective actions to ensure they are closed as indicated. The audit department may follow up on the dates when the actions were to be completed or check periodically (for example, monthly) to ensure that the actions were completed. The responsible managers provide objective evidence to the audit department to demonstrate that the corrective action was implemented. The objective evidence should be reviewed by the audit department to close the observation. Once the objective evidence is reviewed to confirm that the corrective action was implemented, the audit department can close the observation.

To ensure that the corrective actions effectively resolve the conformance gaps, the next time that an audit is conducted in the area where the observations were made, the auditor will review all previous observations within the scope of the current audit. During the current audit, the auditor reviews the previous nonconformances to make sure that they have been adequately addressed and that the conformance gaps no longer exist.

Another method that could be used to follow up on adequacy of corrective actions is to review the conformance gap after a fixed period of time has elapsed. For example, if an SOP was revised to clarify a process so that it is performed correctly, the audit department could check six months after the SOP is implemented to ensure that no further processing errors have been made as a result of following the revised procedure.

In some cases a noncompliant observation written for one specific compliance gap may be a symptom of a systemwide gap. For example, an observation was listed regarding mechanics not being required to attend annual GMP training. It is possible that other departments who should attend annual GMP training were overlooked, but they do not have this requirement in their training curricula. Finance employees typically do not perform GMP tasks, but if they use the GMP inventory computer system to track product costs, they need to be trained on GMP procedures for this computer system, at a minimum. When determining the corrective actions to be included in an audit observation response, responsible management should evaluate the potential for any systemwide gaps and other preventive actions.

If the responsible area does not implement the corrective actions as indicated in the audit response, the audit department should determine why they were not implemented. If there are resource issues, the audit department should work with management to obtain the necessary resources and adjust the due date for implementation if the risk assessment permits a delay. As a last resort, if the area management does not cooperate with the audit process, the audit department can escalate the issue to upper management or issue a nonconformance for not following the audit procedure.

If a corrective action was implemented but was found not to be effective, the organization needs to find the true root cause and address it with an appropriate corrective action. A follow-up audit could be performed to review more closely the context of the nonconformance, or the responsible area could investigate the root cause of the nonconformance again using more-rigorous problem-solving tools. The second round of corrective actions should be tracked by the audit department, and effectiveness of these corrective actions should be checked after implementation.

Chapter 15

Documents and Records Management

GMP DOCUMENT SYSTEM

> Describe the GMP document system to determine compliance to regulatory requirements including corporate standards, master plans, procedures, manufacturing, and test instructions. (Analyze)
>
> **Body of Knowledge II.H.1**

Documentation is an essential part of a *good manufacturing practices* (GMP) quality system. A well-written document ensures that responsibilities for GMP processes are known and the steps to ensure quality and consistency in the output of the processes are clearly identified and can be followed. A well-documented quality system, combined with the records generated throughout GMP operations, goes a long way toward ensuring that the pharmaceutical products manufactured have the high level of quality that regulators and patients demand.

DOCUMENTATION HIERARCHY

No single right answer exists when it comes to organizing documentation in a GMP environment. Figure 15.1 illustrates an example of how an organization may organize their documents. The external environment (for example, regulations and guidance documents) provide the overarching requirements that shape the documentation system. The regulatory requirements identify the minimum set of documents that must be in place to operate in compliance and to meet the expectations of the regulators. It is not atypical, however, for a firm to expand the scope of their documentation system to include information not required by regulation and to take advantage of the benefits that a document hierarchy brings to an organization.

At the highest level of the documentation hierarchy in most companies sits the *quality manual*, which defines the corporate objectives related to quality and compliance. It establishes a figurative road map including guiding principles

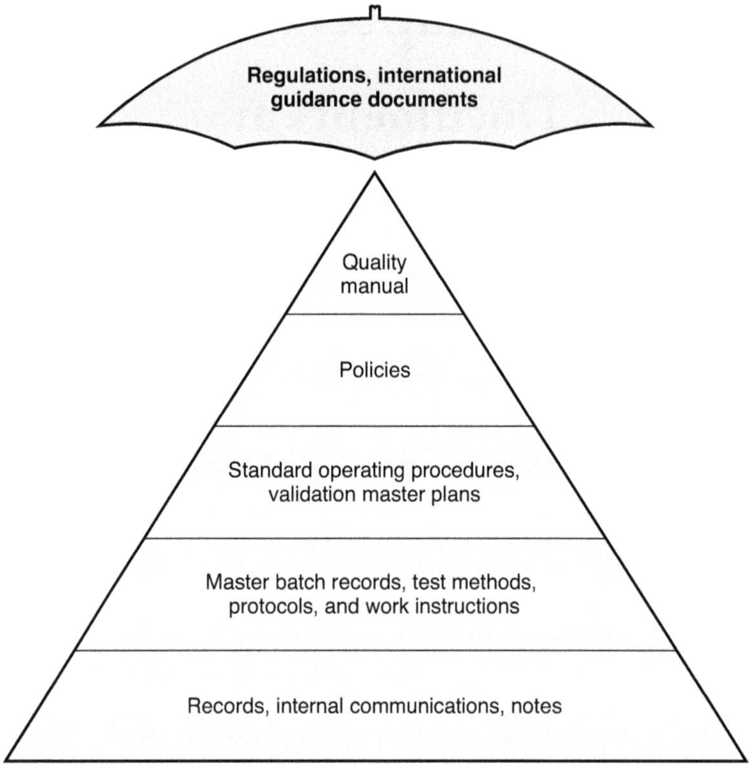

Figure 15.1 Typical documentation hierarchy.
Source: Figure courtesy of the ASQ Food, Drug, and Cosmetic Division. Used with permission.

(regulatory and internal) that employees and other users may refer to in meeting organizational objectives related to the quality system.

Just below the quality manual sit *policies* that describe, in general terms, how the company intends to conduct their business in compliance with the requirements defined in the quality manual. A more detailed discussion of how the company will comply with individual aspects of the requirements in the quality manual is found in documents such as *standard operating procedures* (SOPs) and *validation master plans* (VMPs). SOPs describe, in detail, how firms will conduct operations related to each requirement in the GMP regulations, such as receipt of materials, operation of equipment, and handling of investigations. VMPs provide a detailed description of how a firm will qualify their facility or equipment and the governance program that will ensure it remains in a qualified state. Additional detail, including step-by-step instructions on how to perform a specific task or activity within a GMP environment, can be found in documents such as *master batch records, test methods, protocols,* and *work instructions.* Master manufacturing and packaging batch records (sometimes referred to as *manufacturing formulae, processing, and packaging instructions*) detail each step of the processing and packaging of the drug product and are used by operations personnel to produce the finished drug product. Test methods provide the detailed steps for testing incoming materials and in-process or finished product. Protocols are often generated for special studies, such as validation studies or stability studies. Work instructions may provide even greater detail and are often used when step-by-step instruction is critical to ensure that processes are performed consistently and reliably. *Records, internal communications,* and *notes* represent the data that demonstrate that a company has operated in compliance with their own documentation and the regulations that govern their business.

DOCUMENTS AND RECORDS

It is important to distinguish between the concepts of *documents* and *records.* The term *documents,* in a GMP environment, refers to the how-to instructions that outline what needs to be done to produce the product or operate in compliance with the regulations. *Records,* in a GMP environment, refers to the body of information (history) or data related to the manufacturing, packaging, testing, and holding of each batch of finished product. Once a document has data written on it, it becomes a record.

In 1997, the United States Food and Drug Administration (FDA) issued the Part 11 regulations to ensure the authenticity and integrity of documentation and records and signatures that are maintained in an electronic format. After much discussion with industry and contractors, FDA moderated their position related to Part 11 and narrowed the scope of the requirements. What did not change, however, was FDA's commitment to enforcement of all predicate rule requirements, including those that cover GMP records and record-keeping requirements. If a company chooses to use GMP documents or records in an electronic format in place of a paper format, the requirements of Part 11 apply.

RECORDS

> Review various records (e.g., logbooks, tags, and training evidence) to confirm compliance to requirements such as Attributable, Legible, Contemporaneous, Original, Accurate (ALCOA) and PIC/S guidelines for data integrity. (Analyze)
>
> **Body of Knowledge II.H.2**

Master Documents

GMP regulations require that master documents (for example, SOPs, master batch records, protocols, and specifications) be:

- Designed, prepared, and reviewed in accordance with the requirements in the regulations
- Clearly written and unambiguous
- Approved (signed and dated) by appropriate (that is, qualified by education and/or experience) and authorized individuals

- Identified with a number or code that is unique to the document
- Regularly reviewed and maintained up to date

Executed Documents

GMP regulations specify several practices to be followed in the execution of GMP documents:

- Recording of information and data should be clear, legible, and indelible (typically recorded in ink; pencil and erasable forms of ink are not allowed).
- Information and data should be recorded at the time the activity or step is completed and in a manner that ensures that all of the significant activities associated with the manufacturing and packaging of the drug product are traceable.
- Any alteration or correction to information or data that were previously recorded should be crossed out with a single line, the new information recorded, and signed and dated by the individual making the correction (Figure 15.2).
- Electronic capture of data and information is allowed once the electronic system has been qualified and secured. An audit trail, that is, a record of all data recording, changes, and deletions, must also be in place and secure.
- The executed documents are managed in a manner that allows for ease of retrieval and ensures that the documents are retained in compliance with regulatory requirements.

Globally, GMP regulations mandate the recording of data and information about pharmaceutical products and the materials, equipment, processes, and people involved in their manufacture. It is through this recording of data that GMP records evolve. While some believe that this data recording is an onerous task, there is tremendous value in the data, and they can often be used to benefit the organization.

Training Records

GMP regulations require that personnel have the education, training, and experience needed to perform their assigned responsibilities. Training on both specific

467.4 NLVG 4/12/16

Record the weight of the final blend: ~~464.7~~ kg

Figure 15.2 Example of a correction done according to regulations. Note that the correction is done with a single line through the error, the correct information is added, and the person executing the correction provides their initials and date.

Source: Figure courtesy of the ASQ Food, Drug, and Cosmetic Division. Used with permission.

job tasks and GMP regulations is required. Training related to job tasks is necessary when a task is new to the employee, when procedures, batch records, and/or test methods are revised, or when a task has not been performed recently. Similarly, when an employee's job changes within an organization, his or her education, experience, and training should be reevaluated to determine what areas need to be addressed to perform the new job. Education and experience are typically documented in a résumé or employee work history. Ongoing training is typically documented in separate training records for each training event. These training records can be used to:

- Demonstrate compliance with the personnel requirements in the GMP regulations
- Demonstrate that employees know how their work impacts product and production safety and performance
- Identify knowledge gaps within a functional area and support the justification for additional training
- Assess the needs for additional training after an investigation or product issue
- Assess the effectiveness of various training programs and methodologies

Facility Records

A considerable number of records need to be generated and maintained regarding the qualification and ongoing maintenance of the facilities used to manufacture pharmaceutical products. Some of the records (for example, qualification of utilities or equipment) are generated when the facilities or equipment are first commissioned for use. They may be amended or updated when changes to the facilities or equipment take place, but are relatively static in nature. They provide a snapshot in time of the particular utility or piece of equipment in question. Other records (for example, unscheduled maintenance, preventive maintenance, or calibration documentation) are updated regularly and provide a historical record of what has happened to a piece of equipment over time. These equipment records can be used to demonstrate compliance with the regulations covering the qualification, maintenance, and calibration of facilities and equipment, troubleshoot equipment failures and determine the impact of the failure on previously produced product, and investigate *out-of-specification* (OOS) or other nonconformity test results to determine whether the equipment might have played a role in the product failure.

Material Receipts

GMP regulations require that materials are purchased from approved suppliers and that they are received, sampled, tested, and stored in such a manner as to prevent deterioration or contamination. When a material is first received into the pharmaceutical company's warehouse, a record of its receipt is generated. This initial step can verify that the material received matches the material requested (for example, on a company purchase order) and that it is from the approved supplier. The quantity of the material is typically recorded along with the number of containers and any supplier-specific batch or control number. The material typically moves on to a quarantine area (either physical or virtual) and waits to be sampled. Sampling is usually performed by the *quality unit* (QU), and the material is placed back into a quarantine status until testing is complete and the material has been released. After testing by the QU, the material may move to a different area of the warehouse or, in the case of an electronic quarantine system, have its status changed to "released," which allows it to be used in manufacturing. All of these steps are typically recorded on the material receipt record. This material receipt record can be used to:

- Demonstrate compliance with the GMP requirements for material receipt (that is, demonstrate to the regulators that the material was received appropriately, its identity was appropriately verified, and it was stored appropriately until use)
- Evaluate vendor performance (for example, determine whether the vendor met the required delivery date and supplied the requested quantity of material)
- Facilitate traceability to a specific batch of material in the event of a recall by the vendor
- Improve internal operations by evaluating metrics related to material processing time, number of material movements, testing time, and inventory turns

Batch Records

Pharmaceutical companies are required to develop *batch processing records* (*master batch records* [MBR]) that outline the steps required to manufacture and package pharmaceutical products. Clearly written, detailed batch records are critical to ensuring product quality. The executed batch records provide a record of how the product was manufactured or packaged (for example, the materials and equipment used, the personnel involved). These records can be used to:

- Demonstrate compliance with the requirements for written batch processing records
- Demonstrate compliance with the processes outlined in approved marketing authorizations
- Facilitate investigations into product problems
- Demonstrate employee proficiency through the lack of process problems or deviations
- Demonstrate that the product remains in a state of control (validated state)
- Identify areas for improvement

Log Books

Log books are typically used throughout a GMP facility to document the work that has been done in a particular area or to record the activities performed on individual pieces of equipment. Log books are generally assigned to an area or piece of equipment and provide a chronological history of the products that were produced in the area or on specific equipment, and any other work that has been conducted (for example, maintenance, calibration). Log books can be used to demonstrate compliance with the requirements for capturing equipment or area usage, facilitate investigations into product problems, demonstrate that appropriate cleaning procedures have been followed, and identify opportunities for improvements to equipment based on historical calibration or maintenance activities.

Some companies have electronic quality management documentation and record control systems. In electronic systems, log books may take the form of a generated report or query form which summarize data across individual records.

Distribution Records

Batch records are internally focused on the production operations; and distribution records are externally focused. Distribution records identify key identification and traceability information on where and when product was shipped. This information is often required by regulation and is useful for root cause investigations, adversary notices, recalls and other field corrective actions.

Complaint Records

GMP regulations, specifically in 21 CFR 211.198, require companies to maintain written records of each drug product complaint and the investigation conducted to determine the root cause of the complaint. Complaint records often also contain the evaluation assessment to report serious and unexpected adverse drug experiences to FDA and other regulatory agencies. These records demonstrate a company's commitment to its customers and compliance with the requirements for complaint handling, and identify product or process issues that require correction or improvement.

RECORDS MANAGEMENT

The ISO 15489-1:2001 standard defines *records management* as "the field of management responsible for the efficient and systematic control of the creation, receipt, maintenance, use and disposition of records, including the processes for capturing and maintaining evidence for and information about business activities and transactions in the form of records." Records management is critical, not only from a GMP perspective, but from a company business perspective as well. Company records provide the evidence that an organization has operated in a compliant and lawful manner. In addition, they provide the historical context in which business decisions were made and allow personnel to better understand the reasoning behind those decisions. Records support root cause investigation activities for corrective actions. Without good records management, a company runs the risk of repeating past mistakes.

RECORD RETENTION

> Identify regulatory requirements for record retention. (Understand)
>
> **Body of Knowledge II.H.3**

A key component of any records management program is a documented *record retention* process. Records, and the information contained within those records, have a useful life during which they are an asset to the company. Beyond that useful life, records become an unnecessary expense and can be a corporate liability. It is important to define the retention requirements (or useful life) of each type of record and have a process in place for disposing of records once they have moved beyond their established retention period.

Throughout the GMP regulations, a number of record types have specified retention periods. These retention periods become the minimum requirements for record retention. A company may maintain these records for a period of time beyond the minimum requirements outlined in the regulations, but should do so only if business needs or litigation requirements dictate. Specific retention periods are identified in the GMP regulations (Figure 15.3).

Document and Record Status

Documentation and records can be considered to have a status. Documents are in draft, active (or live), and expired (or superseded). Records are in draft (or

- Any production, control, or distribution record associated with a batch of a drug product must be retained for at least one year after the expiration date of the batch or, in the case of an over-the-counter (OTC) drug product that lacks an expiration date, for three years after the distribution of the batch.
- Records associated with all components, drug product containers or closures, and labeling must be retained for at least one year after the expiration date or, in the case of certain OTC drug products, for three years after distribution of the last lot of drug product incorporating the component or using the container/closure or labeling.
- Complaint records involving a drug product must be retained for at least one year after the expiration date of the drug product, or one year after the date that the complaint was received, whichever is longer. In the case of certain OTC products that lack an expiration date, complaint records must be retained for at least three years after the distribution of the drug product. While not specific to the GMP regulations, it is important to note that adverse event complaint records have a much longer retention requirement. Adverse event records must be retained for a minimum of 10 years from receipt.

Figure 15.3 Specific retention periods identified in the GMP regulations.

work-in-progress), closed or expired (or superseded) because records are often snapshots in time and not maintained. The document revision and status should be apparent to the user. Documents that have been superseded should be labeled and the QMS system should limit who has access.

Litigation Concerns

Properly maintained records can provide significant value to an organization. Unorganized, inaccessible records, or records that are retained after their useful life has passed can become a corporate liability. While not a GMP concern per se, it is important for companies to have processes in place to notify employees regarding new and ongoing litigation matters. Many of these matters will require employees to maintain or produce, for the attorneys, the applicable documents that are the subject of the litigation. If a document is covered by a litigation matter, it cannot be destroyed even if it has reached the end of its established retention period.

Chapter 16

Product Quality Complaints and Adverse Event Reports

The United States Food and Drug Administration (FDA) places a significant amount of emphasis on the need for pharmaceutical companies to develop an effective approach to complaint management, including the establishment of well-written procedures. In fact, the Agency believes that complaint management is such an important element for drug manufacturers that 21 CFR 211 has a dedicated section for the management and retention of complaint files—21 CFR 211.198. Additionally, 21 CFR 314 expands on the post-market surveillance process under 21 CFR 314.80. Equally important is the reporting of adverse events through the use of the *FDA Adverse Event Reporting System* (FAERS). Although the management of complaint files and subsequent investigative process is distinctly different from the voluntary, and in some cases mandatory, reporting of adverse events through the use of FAERS, the two processes are linked.

Although FDA citation for failure to comply with 211.198 in a Warning Letter is a rare event, the Agency does occasionally cite such violations. More common is the citing of a failure to complete and/or document investigations pursued as a result of a complaint.

IMPORTANCE OF POST-MARKET SURVEILLANCE AND REPORTING

As part of the drug application process, FDA wants to reinforce the need for review and reporting of adverse drug experiences (that is, adverse drug event), post-market studies, periodic reporting requirements, and annual reporting requirements, and the importance of record keeping. Similarly to medical devices, drug manufacturers are required to report adverse drug experiences through the use of form FDA 3500A. Once a drug manufacturer becomes aware of an adverse drug experience, the company is required to perform a thorough investigation followed by the completion and submission of form FDA 3500A; however, an exemption is noted for post-market studies. A 15-calendar-day window must be adhered to when reporting an adverse drug experience. Within 15 calendar days of initial receipt of information, the drug applicant is required to report every serious adverse drug experience and every unexpected adverse drug experience. The best practice is to notify FDA has soon as possible when such adverse drug experiences occur. Additionally, drug manufacturers are required to have procedures that delineate their processes for managing surveillance, receipt, evaluation, and reporting of post-market adverse drug experiences to FDA, including record retention.

Additional reporting requirements delineated within 21 CFR 314 include periodic adverse drug experience reports and annual reports. Periodic adverse drug experience reports are required to be filed quarterly with FDA for a period of three years from the date the initial drug application was approved. These quarterly reports are due within 30 days after close of the quarter. At the discretion of the Agency, the quarterly reporting requirement can be extended. Annual reports are due to the Agency within 60 days of the anniversary date of the application approval.

PRODUCT COMPLAINTS

> Describe and distinguish between product complaints and adverse events. Evaluate complaint-handling processes. (Evaluate)
>
> **Body of Knowledge II.I.1**

BUILDING AN EFFECTIVE COMPLAINT MANAGEMENT SYSTEM

The building of an effective complaint management system compliant with 21 CFR 211.198 commences with the establishment of written procedures. From a compliance perspective, FDA requires specific steps to be included in written procedures needed to support an effective complaint management system. When scripting procedures, a single well-written procedure delineating an organization's approach to complaint management should be the goal. That being said, as a minimum, the following components of complaint management should be included in the procedure:

- An applicability statement for oral and written complaints.
- A mandatory review of all complaints by the quality function.
- A discussion on the complaint investigation process; all decisions made in regard to not performing an investigation need to be documented and supported by written rationale.
- A provision for reporting a complaint as an adverse event when deemed necessary. (It is recommended to create a decision tree to support the reporting process.)
- The need to create and retain a complaint file for each complaint.
- Complaint files (record) retention time periods (one year after the expiration of the drug product or one year after the complaint was received, whichever is longer). OTC drug products without expiration dates must maintain records for a minimum of three years.

Additionally, the complaint record must include the name and strength of the drug product, the lot number in question, the name of the individual and/or organization filing the complaint, the nature of the complaint, and the pharmaceutical company's response to the complainant. Furthermore, if an investigation is performed, it is imperative that the result of the investigation be included in the complaint file. As previously stated, if the organization determines that a complaint investigation is not warranted, the decision not to investigate—including supporting rationale and the responsible person making the decision will need to be documented and retained in the complaint file. Finally, ensure that personnel tasked with complaint management duties are trained on the procedure and that the training is documented.

ADVERSE EVENTS

Describe regulatory requirements for the reporting of adverse events (e.g., counterfeit product, fraud). (Analyze)

Body of Knowledge II.I.2

ADVERSE EVENT REPORTING—FAERS

FAERS is simply a database that captures adverse events and medication error reports submitted to the Agency. The primary purpose of FAERS is as a tool to support the Agency's post-market safety and surveillance program. FDA uses the data to assess product safety concerns, and can use the data to drive regulatory actions necessary to protect the public health.

It should be noted that the reporting of adverse events and medication errors is considered voluntary for drug consumers and healthcare professionals; however, if a manufacturer of a drug product receives an adverse event report, the manufacturer is required by law to send the adverse event report to the agency.

DIFFERENCES BETWEEN COMPLAINTS AND ADVERSE EVENTS

The first difference between complaints and adverse events is that not all complaints will result in the reporting of adverse events. Complaint files, however, must be opened and retained by drug manufacturers regardless of whether they report an adverse event. A complaint that is identified as the cause of a serious and unexpected adverse drug experience is required to be reported to the Agency in accordance with 21 CFR 310.305.

The second difference between complaints and adverse events is that complaint files are managed by the drug manufacturer, and adverse events are entered into FAERS and tracked by FDA. The expectation is that the pharmaceutical companies perform a thorough review of all received complaints. FDA is continuously monitoring FAERS for trends in adverse events as part of its post-market safety surveillance program. If the Agency determines that there are issues associated with a drug, it will take appropriate regulatory action. It is an immensely important task to keep accurate, current, and complete complaint file records.

The third difference between complaints and adverse events is that the reporting of adverse events is voluntary for healthcare providers and consumers; however, when complaints are reported to a drug manufacturer (verbal or written), entering the complaint into the complaint management system is not optional. It is strongly recommended that pharmaceutical manufacturers visit FAERS frequently to ensure that a negative trend pertaining to their or competitors' drug performance is not occurring. If a manufacturer observes a FAERS entry on their drug submitted by a consumer or other entity that they were not previously aware of, then the manufacturer shall open a complaint for that incident at that time

KEYS TO COMPLIANCE

Drug manufacturers can pursue several steps to ensure that compliance issues never occur. To start, compliance is always rooted in a well-scripted procedure that complies with all 21 CFR 211.198 requirements. It is strongly recommended that complete and thorough complaint investigations occur for all reported complaints. If a decision is made not to investigate a complaint, the decision always needs to be documented and supported by a written rationale. Build an adverse event reporting decision tree into the procedure. It is always prudent to side with caution when it comes to reporting an adverse event. Adverse event reports are not considered an admission of fault by the FDA. The data can be updated when the complaint investigation has been completed. Always assume that FDA will review complaint files as part of any Agency inspection. Records must be complete, accurate, exhibit *good documentation practices* (GDP), and be available for review. Do not overlook the training piece. All individuals handling complaint calls, performing complaint investigations, or managing the complaint files must be adequately trained on the established procedure(s).

EVENT RESPONSE

> Evaluate the level of action that needs to be taken in response to adverse events such as corrections, and product removal. (Evaluate)
>
> **Body of Knowledge II.I.3**

TYPES OF EVENT RESPONSES

The manufacturer must assess if an adverse event requires a field corrective action. Each country has its own definitions of the types of field corrections, communication, documentation, and registration requirements of those. Not all adverse event response types may be required to be registered with the regulatory authorities or communicated to the public. Common types of event responses include:

- Correction: highest-level description of any type of event response. This includes product removals and safety alerts or similar communication-based corrections.

- Product removal: A subset type of correction in which defective product is removed and cannot be further sold or consumed.
- Recall: A type of product removal. An action taken to remove a product from the market.
- Market Withdrawal: A type of product removal. This is for a minor violation that would not be subject to FDA legal action.
- Stock Recovery: A type of product removal where product that is still in manufacturer's control is returned to be corrected or discarded. For example, if a drug was mislabeled and still in their storage warehouse. If an event cannot be classified as a stock recovery if it has impacted product in the market outside of the manufacturer's control.
- Safety Alerts: A type of correction. It is a communication informing the attended audience (patients, pharmacists, distributors, etc.) of a safety risk. It may recommend an action but is not required to. These may be considered recalls as well.

RECALL LEVELS

A drug recall is classified as Class I, Class II or Class III by FDA based on the severity of the event. These classifications are regardless of if the recall was voluntary or after a regulatory sanction, inspection finding, etc. The actions required are related to the recall's classification.

- Class 1: A dangerous or defective product that could cause serious health problems or death
- Class II: A product that might cause a temporary health problem or pose a slight threat of a serious nature
- Class III: A product that is unlikely to cause any adverse health reaction but violates FDA labeling or manufacturing laws

Market withdrawals, stock recoveries, and safety alerts do not have FDA levels. Manufacturers may designate levels or requirements for various types of these activities within their own quality management systems.

EVENT RESPONSE DOCUMENTATION & COMMUNICATION CONSIDERATIONS

All documentation must be stored and maintained. This includes corrective action investigation, risk and problem scope assessments, recall communications to impacted distributors, clinicians, pharmacists, patients, the public (such as through news media) and government and regulatory agencies. FDA recalls are required to be logged in the FDA.gov MedWatch database by the manufacturer.

Recall plans shall contain the communication plans including the forums and distribution pathways of communication, the communication content, the communication audience, how to handle non-responses, how to handle communications that were returned, and effectiveness check requirements to consider the recall completed.

FDA will communicate recalls through vehicles such as the FDA enforcement report. Maintain copies of communications that were externally sent. This includes the names, addresses, contact information, and dates or where the communications were distributed to. Typically, a recall communication (especially for a Class I or II recall) require multiple attempts and/or multiple methods of attempting communication to an entity with known defective product that are required to perform actions related to the recall. For example, it is not acceptable to leave only one voicemail or email with a distributor or pharmacists with known defective product and consider that sufficient.

Documented evidence of the defective product being reworked, destroyed, or other correction per the recall plan shall be maintained as well. This includes dates and names of employees that conducted these events.

Chapter 17

Product Trend Requirements

> Describe and distinguish between components of periodic product assessment, such as the U.S. annual product review (APR) and the European product quality review (PQR), with regard to data trends and other required elements. (Understand)
>
> **Body of Knowledge II.J**

Organizations, no matter what industry they're in, generate a significant amount of data as a result of running their business. This situation is certainly true of the pharmaceutical industry. *Good manufacturing practices* (GMP) regulations worldwide require pharmaceutical companies to generate a significant body of data related to all their activities, from the receipt of starting materials through the distribution (and return) of finished products. A large amount of data is generated regarding the quality of the finished product, both at release for distribution and throughout the established shelf life of the product.

Regulatory agencies around the world recognize that there is value in the data that are generated or obtained (for example, through receipt of returned goods or product complaints) as a result of complying with GMP regulations. It is this acknowledgment, as well as the recognition that it is important to critically assess the "health" of each pharmaceutical product, that resulted in the regulatory requirement for some form of annual assessment. The goals of the annual assessments are to:

- Confirm that the products and associated processes remain in a state of control
- Identify the need for changes to the products or processes
- Recognize opportunities for continuous improvement

Companies that undertake these reviews simply to meet their regulatory obligations may miss out on the opportunities for continuous improvement that could also positively impact the business and their customers.

REGULATIONS

The United States Food and Drug Administration (FDA) requirement to conduct an *annual product review* (APR) is found in 21 CFR 211.180(e). As the name suggests, FDA GMP regulations require that pharmaceutical companies, on an annual basis (or, minimally, one time per year), complete a thorough review of data associated with each pharmaceutical product.

The European requirement to conduct a *product quality review* (PQR) is found in Chapter 1 of Volume 4, *European Union Guidelines to GMP* (as well as Chapter 1 of the *PIC/S* [Pharmaceutical Inspection Convention and the Pharmaceutical Inspection Co-operation Scheme] PE-009-11 PIC/S GMP Guide—Part I, *Guide for Medicinal Products*). Similarly to the FDA requirement, pharmaceutical companies must, on an annual basis, complete a thorough review of specified data elements. Unlike the FDA GMP requirements, however, this annual assessment must take into account data from previous review periods (previous years). The data elements that must be reviewed also differ from the FDA requirements.

General Requirements

FDA GMP regulations specify that pharmaceutical companies maintain records associated with the production, control, and distribution of each batch of drug product and retain those records for at least one year after the expiration date of the batch (or, in the case of certain *over-the-counter* [OTC] drug products, three years after distribution of the batch). Additional records related to raw materials, packaging containers, closures, and labeling must be retained according to a similar timeline. The data associated with these records are used for evaluating, at least annually, the quality standards of each drug product.

Under FDA requirements, pharmaceutical companies are required to establish a written procedure for conducting annual evaluations and for ensuring that their procedure is followed as written. Additionally, while not specifically identified in the regulations, firms should document the results of each APR in a written report and submit the reports to senior management for review. The reports typically include summary conclusions on the quality of the product and any corrective/preventive actions initiated and/or completed during the review period. Metric targets, specification limits and/or control limits should be apparent in the report where applicable. The APR has a strong link to a firm's *corrective and preventive action* (CAPA) program, and any identified CAPA items should be tracked appropriately and completed in a timely fashion. The APR may be used to support the need for process re-validation frequency and scope.

Data Elements

Required. An evaluation of specific data elements, as specified in the FDA GMP regulations, must be included in the APR:

- Representative number of batches, whether approved or rejected
- Product complaints
- Recalls

- Returned or salvaged drug products
- Investigations

Expected. In addition to the required data elements, the FDA has identified, through the issuance of form FDA 483s and other guidance documents, other data elements that they typically expect to see as part of the APR:

- Deviations or nonconformances
- In-process and finished product testing results and a discussion of any adverse trending
- Other relevant data sources such as those described in the FDA guidance document s (ICH Q7 Good Manufacturing Practice Guidance for Active Pharmaceutical Ingredients)
- Product stability results and a discussion of any adverse trending
- Changes to processes that occurred during the review period

In the spirit of continuous improvement, additional data elements may need to be included in the review process. Firms should assess all of the data that are gathered throughout their operations and determine what makes the most sense for their particular products and processes.

ANNUAL PRODUCT QUALITY REVIEWS

General Requirements

European Union (EU) regulations call for regular, periodic, or rolling quality reviews of all licensed medicinal products, including those that are designated as "export only." The reviews, similarly to FDA requirements, should be conducted and documented annually.

Despite the similarities, there are a number of aspects of the EU regulations that differ from FDA regulations. The primary differences between the EU and FDA requirements related to the quality reviews include:

- EU regulations explicitly require an assessment of the results of the review to determine whether corrective or preventive actions are needed, or whether revalidation is required. FDA likely expects firms to perform this type of analysis, but it is not explicit within the regulations.
- EU regulations require the assessment of current data against data from previous reviews in an attempt to identify adverse trends. FDA does not require firms to look at data from previous review periods. Depending on the number of batches manufactured, however, this type of "prior period" data may be necessary to adequately analyze the data for trends.
- According to EU regulations, quality reviews may be grouped by product type (for example, solid dosage forms, liquid dosage forms, sterile products) where scientifically justified. FDA regulations do not allow products to be grouped; APR must be conducted on individual products.

- EU regulations do not specify a requirement for a written procedure that outlines the PQR process. FDA requirements do explicitly specify that firms adopt a written procedure for the APR process. It is likely best to adopt a written operating procedure that outlines the process for completing these annual product assessments to ensure compliance with expectations of the regulatory agencies involved.
- EU regulations include a number of data elements not required by FDA regulations. It is important to be aware of the differences to ensure compliance with appropriate regulations.

Data Elements Mandated by EU Regulations

Required. An evaluation of the following data elements, as specified in the EU GMP regulations, must be included in the PQR:

- Starting materials, including packaging materials used in the product, with emphasis on those from new sources;
- Critical in-process controls and finished product results;
- All batches that failed to meet established specification(s) and their respective investigation(s);
- Significant deviations or nonconformances, their respective investigation(s), and the effectiveness of resultant corrective or preventive actions taken;
- Changes to processes or analytical methods;
- Marketing authorization variations submitted, granted, or refused, including those for export only;
- Results of stability programs, and any adverse trends;
- Quality-related returns, complaints, and recalls, and their respective investigation(s);
- Adequacy of any other previous product, process, or equipment corrective actions;
- Post-market commitments for new marketing authorizations or variations to marketing authorizations;
- Qualification status of relevant equipment and utilities;
- Contractual arrangements.

Expected. In addition to the required data elements, salvaged products should be included as part of the PQR process. The EU also recognizes the ICH Q7 Good Manufacturing Practice Guidance similar to FDA; therefore, those data sources should also be considered.

ADDITIONAL CONSIDERATIONS

It is important to understand that the use of contractors (for example, contract manufacturers, contract laboratories) does not negate the requirement for the marketing application holder to gather the required data. The application holder is responsible for either ensuring that the contractor completes the APR on their behalf or obtaining the required data from the contractor and completing the annual review on their own. It is critical that the responsibilities for completing the annual reviews be specified in any contracts and/or quality agreements between the contracting parties.

Chapter 18

Supplier and Contractor Quality Management

Suppliers (also referred to as *vendors*) provide goods or services to another organization. Pharmaceutical products are created to improve the health of patients; the pharmaceutical manufacturer is responsible for ensuring that all of its suppliers contribute to product quality.

In Figure 18.1, the pharmaceutical manufacturer has five suppliers: the raw material manufacturer provides chemicals or biologics that become part of the batches of drugs; the filter manufacturer provides filters that are used in the manufacturing process; the vial manufacturer makes the containers for the product; the sterilization contractor sterilizes the vials before they are received at the pharmaceutical manufacturer; and the contract laboratory performs some of the release tests that determine whether the product meets its predetermined specifications. The pharmaceutical manufacturer is a supplier to the distributor, and the distributor is a supplier to the pharmacies that supply drug products to the patients who need them.

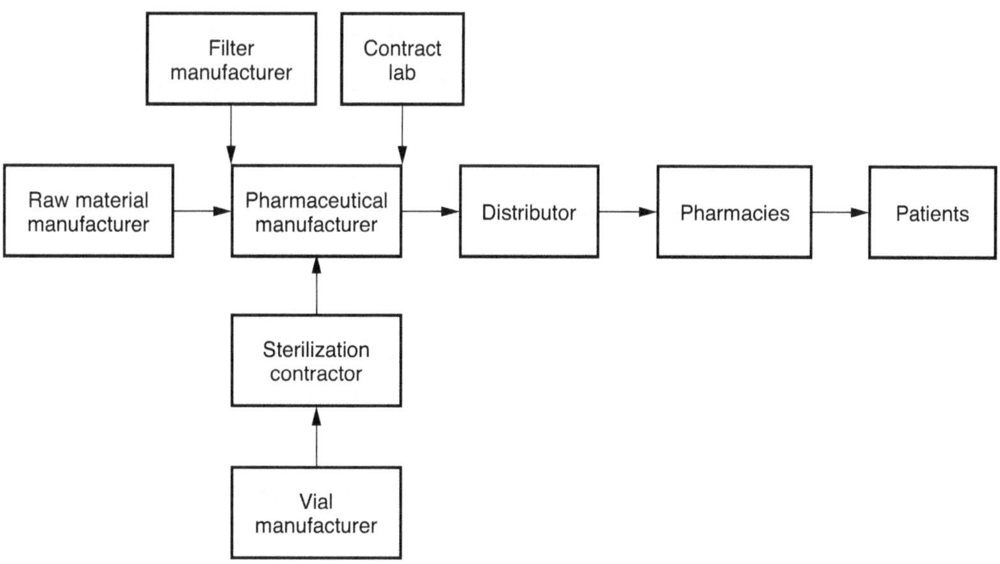

Figure 18.1 Example of suppliers to a pharmaceutical company.

Source: Figure courtesy of the ASQ Food, Drug, and Cosmetic Division. Used with permission.

SUPPLIER QUALITY SYSTEMS

> Identify and interpret standards and regulations related to monitoring supplier, vendor, and contractor quality management systems. (Understand)
>
> **Body of Knowledge II.K.1**

Pharmaceutical *good manufacturing practices* (GMP) vary in the level of detail used to describe the ways that pharmaceutical manufacturers should ensure that their suppliers contribute to the quality of the products they create, but they also have some similarities. In Chapter 7 of the European Union (EU) GMP, *Contract Manufacture and Analysis,* the responsibilities of the contract giver and contract acceptor are defined for subcontracted manufacturing and testing. ICH Q7 addresses controls for contract manufacturers in Section 16, communication of supplier information to customers in Section 17.6, and complaint handling in Section 17.7. Q7 indicates in Section 7.11 that manufacturers should evaluate the suppliers of critical materials, where "critical" is a parameter that "must be controlled within predetermined criteria to ensure that the active pharmaceutical ingredients (API) meet its specification." US Food and Drug Administration (FDA) GMP require that the *quality unit* (QU) be "responsible for approving or rejecting drug products manufactured, processed, packed, or held under contract by another company" [21 CFR 211.22(a)].

ISO/IEC 17025:2005, *General requirements for the competence of testing and calibration laboratories,* provides requirements for the competence of these suppliers to perform tests and/or calibrations. If testing and calibration laboratories comply with ISO/IEC 17025, their *quality management system* (QMS) will also meet the requirements of ISO 9001, *Quality management systems—Requirements.* The unique aspects of this ISO standard regarding supplier competence include:

- Providing appropriate supervision when using staff to perform tasks while training
- Qualifying personnel based on their education, training, and experience, including demonstration of skills required for the position
- Ensuring that personnel have knowledge of the technology used for manufacturing the items tested, and the defects or degradations that may occur during service

In general, pharmaceutical manufacturers develop requirements for their finished product and for the materials they use to manufacture the product. They select suppliers who meet those requirements after evaluating them by conducting on-site audits or surveys. For contract manufacturing and testing, manufacturers develop a written contract to document the agreements between the manufacturer and subcontractor. They monitor the quality of the products and services received and ensure that the quality requirements are fulfilled over multiple receipts of the products and services. Manufacturers may certify a raw material supplier that has consistently met requirements and whose own quality systems have proven to be in control, and thus be able to reduce the testing of the incoming raw materials. They ensure that the supply chain is reliable as close to the original source of the material as possible.

SUPPLIER CONTROLS

> Assess the adequacy of controls over supplier, vendor, and contractor selection and procurement and receipt of raw materials, components, and contract services. Determine the need for formal contracts/quality agreements. (Evaluate)
>
> **Body of Knowledge II.K.2**

It is the responsibility of the pharmaceutical manufacturer to ensure suppliers are adequately controlled so as to not negatively impact the drug's safety or effectiveness. Controls occur during supplier evaluation as the supplier is being selected and the relationship defined. Controls also continue during design and development, production or other post-market surveillance activities depending on the supplier's services provided. Examples of supplier controls include establishing raw material or component quality requirements, purchasing requirements, quality management system requirements, quality agreements that clearly define roles and responsibilities, supplier audits, and supplier performance trending.

REQUIREMENTS FOR MATERIALS AND SERVICES

Suppliers of raw materials, components, and contract services need to meet requirements for the product or service they provide to the pharmaceutical organization. When a pharmaceutical product is being developed, specifications are written to define the quality characteristics that must be met for raw materials and components that are used in the manufacture of the product. Contracted services such as manufacturing, testing, and sterilization have defined requirements to ensure that their activities support product quality. The specifications can come from many development data sources including safety risk assessments, regulatory requirements, clinical performance needs, and manufacturing risk assessments.

When written specifications or requirements are available for a purchased material or contracted service, the supplier must be knowledgeable of and meet these requirements. The specifications are developed to ensure that the final product will have the safety, identity, strength, quality, and purity as designed. Two different aspects of contracts are purchasing requirements and quality requirements.

Purchasing requirements, often defined in a purchase order or purchase agreement, that need to be considered include:

- Minimum order quantities
- Lead times for filling orders after they are placed
- Batch sizes

- Price and payment terms
- Liability and insurance
- Confidentiality requirements
- Warranty

Quality requirements and change notifications may include:

- Quality attributes as defined in written specifications
- Permission to periodically audit the supplier
- Preapproval of subcontracting to other suppliers
- Preapproval of changes to the product or service
- Notification of audit/inspection findings from regulatory, accreditation, and certification bodies
- Shipment methods (if they could impact quality, such as refrigeration, overnight delivery)
- Packaging materials (for example, tamper-proof, temperature maintenance and monitoring, humidity control through desiccant packs, prevention of contamination)
- Notification of changes in licensure, certification, and registrations
- Traceability requirements of materials and components to their manufacturing processes and sub-suppliers such as the use of SKU numbers, batch numbers, etc.
- Length of time to retain documentation related to manufacturing or provision of services
- Communication and investigation of nonconformances that may impact the product or services provided
- Identification of material (labeling)
- Changes in organizational structure, leadership, and management
- Manufacturing validation requirements (such as sterilization) Quality management system requirements such as training requirements of supplier's manufacturing employees, certifications, etc.
- Documentation of testing results
- Changes to the facility, including relocation

SUPPLIER EVALUATION

> Assess the quality systems of suppliers, vendors, and contractors using various methodologies including supplier qualification, certification, evaluation, audit, and supplied product or service performance trending. (Evaluate)
>
> **Body of Knowledge II.K.3**

Regulatory authorities expect manufacturers to ensure appropriate standards are met for suppliers of materials and services. Suppliers may provide materials such as excipients, components, containers, and closures; suppliers may also provide services such as outsourced manufacturing, contract laboratory testing, warehousing, transportation and distribution.

Pharmaceutical manufacturers must evaluate the adequacy of their suppliers to meet their requirements. This evaluation is a continual process. Typically, manufacturer's assign suppliers a criticality level based on risk assessments and are assigned a supplier evaluation status. A supplier may be considered "under evaluation" or "under probation" while being initially evaluated per the pharmaceutical manufacturer's internal QMS supplier evaluation process requirements.

Examples of activities performed during a supplier evaluation include the use of an on-site audit, a quality system survey, trending, supply chain continuity assessment, regulatory compliance assessments and materials/component quality testing. Regulatory assessments include the supplier's ability to meet requirements around the user of hazardous materials such as ones described in this section below.

Materials and components should be assessed not only for the product's requirements but also where they are sourced from. Pharmaceutical supplier's may have their own sub-suppliers; each with their own level of quality. It is important to understand how many and which sub-suppliers they may use for a critical material or component and to ensure no counterfeits are being used.

After a supplier has successfully completed their evaluation, their status may change to "active" or similar terminology. If the supplier is not meeting expectations they may be changed to "on probation," "re-evaluation," or even "removed from service." This supplier status notifies the other employees at a pharmaceutical manufacturer of their performance, which could impact the likelihood of selecting the supplier for a future drug launch.

USP <1083> Supplier Qualification defines supplier qualification as the process in the supply chain for systematically ensuring a material or service is compliant and does not increase risk to the manufacturing of a product. USP <1083> describes high-level supplier qualification practices applicable to all organizations regardless of the material or service provided. Systematic evaluation of suppliers based on:

- The risk to the quality of the product or service supplied
- Compliance of their quality systems to applicable regulations and requirements of the supply contract or quality agreement

- The reliability of the supplier to avoid quality deviations and shortages of components and/or products

The USP <1083> Supplier Qualification Life Cycle is summarized below:

1) Preparation
    a. Supplier Qualification process owner is defined.
    b. Cross-functional subject matter experts to support the supplier qualification program are identified. Cross functional teams typically include Quality, EH&S, Manufacturing, and Purchasing. Other participants may include Logistics, R&D, or Regulatory Affairs.
    c. Material specifications are defined.
    d. Service requirements are defined.
    e. Material and service criticality is defined.
2) Identification and selection—Supplier requirements are defined, and potential suppliers identified for evaluation
3) Evaluation and acceptance—Perform evaluation based on risk and criticality of the material and supplier, which may include any or all of the following:
    a. documentation review
    b. audit
    c. sample request for analysis or testing
4) Performance monitoring
5) Disqualification
6) Conditional approval of an existing supplier

ON-SITE AUDIT

Suppliers must be evaluated periodically, and the primary way of performing the evaluation is an on-site audit. Some suppliers who have many customers charge a fee to permit an audit to be conducted at their facility, or they offer one day when multiple customers can perform audits at the same time to minimize the time the supplier spends hosting audits. Suppliers such as contract laboratories may be audited once a week because they have so many customers, which requires substantial resources to support the audits and respond to audit observations. An audit of a supplier should be performed by a qualified auditor. The auditor performs the audit against specific requirements, such as specifications for the product, the quality agreement, and any applicable regulatory requirements for the type of supplier being audited. Suppliers of pharmaceutical manufacturers may not be subject to pharmaceutical GMP regulations themselves. Some suppliers may be ISO 9001 certified, which provides some assurance of understanding of quality; others may have few controls in place to assure quality. The auditor should work with the supplier to ensure that the supplier has sufficient quality management systems in place so that their product or services will meet requirements.

After an audit is performed of a supplier, the pharmaceutical manufacturer provides the supplier with an audit report. The audit report is generally signed by both the manufacturer and the supplier. The audit report summarizes the results of the audit, including any nonconformances (observations) from the requirements against which the supplier was audited. Typically, a response to the audit observations is requested within a month from the report issuance. The supplier is requested to respond with corrective action plans for the observations. When the audit response is received, the pharmaceutical manufacturer reviews the response to determine whether the corrective action plans adequately address the observations. If the supplier refuses to respond or to take action on critical observations, the pharmaceutical manufacturer has to determine a path forward. The supplier may need to be replaced with one who will cooperate with efforts to ensure that the items or services supplied to the pharmaceutical manufacturer meet the specifications.

QUALITY SYSTEM SURVEY

An alternative method to an on-site audit for evaluation of a supplier is a written *quality system survey*. Surveys may be used to evaluate suppliers whose products are a low risk to pharmaceutical product quality, or to monitor suppliers' quality systems in years when on-site audits are not performed. The survey contains questions about the quality systems that the supplier has implemented to ensure that their products or services meet requirements. Surveys can be customized based on the type of supplier: calibration suppliers, raw material manufacturers, or packaging materials manufacturers. The manufacturer sends the survey to the supplier and requests completion in a timely fashion. The manufacturer may include some of that supplier's performance or product metrics on in the survey too. The manufacturer reviews the response to ensure that quality systems are implemented. If the supplier is not FDA regulated or ISO certified, the manufacturer will need to carefully evaluate the survey response to ensure that the minimum quality system elements are implemented.

In some cases, the survey is sent to the supplier as part of the initial approval of the supplier with the purchase agreement and quality agreement, before the supplier is approved for use. Many suppliers who are frequently requested to complete surveys for their customers have prepared documents that describe their quality systems to minimize their time spent completing surveys. These documents are sent to the pharmaceutical manufacturer in place of a survey response. The pharmaceutical manufacturer can accept the quality system document in place of a response to the quality systems survey if it addresses the required information in the survey. Surveys are generally approved by both the manufacturer and supplier.

CHANGE CONTROL

In addition to evaluating suppliers periodically, the pharmaceutical manufacturer needs to ensure that any changes made by the supplier are evaluated before implementation, if possible, to prevent potential impact to the drug product. The quality agreement between the supplier and manufacturer should include a section that makes it clear to the supplier that any changes that could potentially impact the supplied item or service should be preapproved by the pharmaceutical manufacturer.

For example, if a rubber stopper supplier changes the formulation of the silicone coating applied to the stoppers to prevent them from sticking together, the pharmaceutical manufacturer should be provided with some samples to test with the drug product to ensure that the silicone coating does not interact with the drug. The pharmaceutical manufacturer should be provided with adequate time to test the stoppers before the change is implemented at the stopper supplier's facility. If a contract laboratory changes to a new instrument using a state-of-the art technology to perform a test, the contract laboratory should work with the pharmaceutical manufacturer to ensure that the test result is the same as before the instrument was upgraded.

If a change is made by the supplier, but the pharmaceutical manufacturer is not given advance notice, an unanticipated product impact may result in deviations, investigations, and retroactive analysis of product implications. If product was released before the impact was known, the change could lead to a product recall and possible adverse impact to patient health. The supplier should clearly understand that changes need to be preapproved by the pharmaceutical manufacturer. Unapproved changes without advanced notices are typically addressed through the issuance of a supplier corrective action. The supplier corrective action request duration is related to the risk. The supplier corrective action request should contain improvements addressing the lack of an advanced notice in addition to any product quality issues that arose from the change.

TRENDING

FDA guidance *Quality Systems Approach to Pharmaceutical CGMP Regulations* recommends that data for acceptance and rejection of materials be trended to evaluate supplier performance. If 0.1% of a supplier's batches received by the pharmaceutical manufacturer were rejected over the past year, and the frequency of rejection is increasing, it is time to address the issue with the supplier. If occasional noncritical issues arise between audits of the supplier, they can be addressed at the time of the next audit; and the issue should still be communicated prior to the audit giving the supplier time to investigate. When an auditor is preparing for a periodic audit of a supplier, the auditor should obtain the trended acceptance and rejection data to determine whether any issues should be addressed during the audit. If a quality issue such as contamination is observed in incoming material, a for-cause audit should be scheduled at the supplier as soon as possible to investigate the root cause of the issue, the QMS gaps that lead to the escape of the issue, supplier's corrective actinon plan, and/or effectiveness of the supplier's corrective actions of the issue. In addition, the supplier should work with the pharmaceutical manufacturer to communicate how the investigation is progressing toward the root cause of the contamination and what corrective actions will be implemented to ensure that future receipts of the material are not contaminated.

MATERIAL OR SERVICE REQUIREMENTS AND CRITICALITY

Once the final product attributes and desired material performance are defined, the material criticality needed to meet the final product specification is defined. Material criticality should be aligned with the following International Council for Harmonisation of Technical Requirements for Pharmaceuticals for Human Use (ICH) guidelines:

- Q8 (Pharmaceutical Development)
- Q9 (Quality Risk Management)
- Q10 (Pharmaceutical Quality System)
- Q12 (Technical and Regulatory Considerations for Pharmaceutical Product Lifecycle Management)

Service requirements specifying what is required from the service suppliers should also be defined. For example, for storage and transportation service providers, the storage conditions, category of product, licensing needed for handling a specific product, transportation modes, transportation lanes, duration of trips, and types of shipping containers should be defined.

SUPPLIER CERTIFICATION

Pharmaceutical manufacturers have several options for categorizing a supplier after the supplier meets the requirements for approval. The pharmaceutical manufacturer can maintain a list of approved suppliers and those that have been evaluated as not meeting requirements for approval. The pharmaceutical manufacturer can also choose to take extra steps to certify a high-performing supplier in order to document that the supplier understands the pharmaceutical manufacturer's requirements and has the quality systems in place to ensure that they will meet or exceed the pharmaceutical manufacturer's requirements.

Certification usually entails building a partnership between the supplier and pharmaceutical manufacturer. In addition to normal approval steps, according to the *Supplier Management Handbook* (Bossert 2004), certification requires the supplier to document the process control details to show how critical parameters are controlled to deliver the desired result. Statistical process control, process capability, and process reliability analysis can contribute to control of critical parameters. The review for certification should include cost, quality, delivery, technical support, and management attitude over an extended period of time. After the certification is approved by both the supplier and the pharmaceutical manufacturer, the supplier is recognized to celebrate the new status of the supplier. Finally, the supplier's performance is monitored to ensure that the certification can be maintained.

SUPPLY CHAIN CONSIDERATIONS

One aspect of supplier quality management that has received more attention in recent years is the identification of the supply chain and the pharmaceutical manufacturer's responsibility for ensuring the quality of materials used in drug products. The heparin contamination that originated with pig intestine processing in

China demonstrated the complexity in tracing the supply chain and ensuring quality at each hand-off.

The pharmaceutical manufacturer should do its best to determine the original source of the materials used in its drug products and should evaluate the quality systems of the manufacturers in the supply chain as far back as they can. The pharmaceutical manufacturer is responsible for ensuring the safety, identity, strength, quality, and purity of its products as designed, and evaluation of its suppliers plays a critical role in fulfilling this responsibility. There are several concerns of which pharmaceutical and device manufacturers must be aware when evaluating their suppliers, including the use of restricted chemicals, restricted minerals, animal tissues, and counterfeit medicine and medicinal components.

Supply chain risk assessments may highlight areas of focus within the manufacturing lines. Business risks can also be considered such as critical materials that are purchased from a single source, supplier's ability to scale with increased demands, etc.

REGISTRATION, EVALUATION, AUTHORIZATION, AND RESTRICTION OF CHEMICALS (REACH)

REACH Regulation (EC) No. 1907/2006 aims to improve the protection of human health and the environment through better and earlier identification of the intrinsic properties of chemical substances. This is accomplished with four processes: registration, evaluation, authorization, and restriction of chemicals. In 2018, REACH was updated to include nanomaterials, the specific clarifications and requirements of which were not called out in the original REACH revision.

One of the major elements of REACH is the requirement to communicate information on chemicals up and down the supply chain. This ensures that manufacturers, importers, and also their customers are aware of information relating to the health effects and safety of the products supplied.

RESTRICTION OF HAZARDOUS SUBSTANCES (RoHS)

RoHS Directive 2015/86 originated in the EU and restricts the use of 10 hazardous materials found in electrical and electronic products. It is referred to as RoHS 3, as there was originally a 2002 (RoHS 1) and 2011 (RoHS 2) version. All applicable products in the EU market pass RoHS compliance. RoHS impacts the entire electronics industry and many electrical products as well.

RoHS specifies maximum levels for the following ten restricted materials:

- Lead (Pb)
- Cadmium (Cd)
- Mercury (Hg)
- Hexavalent chromium (Cr6+)
- Polybrominated biphenyls (PBB)
- Polybrominated diphenyl ethers (PBDE)

- Bis(2-Ethylhexyl) phthalate (DEHP) (new to RoHS 3)
- Dibutyl phthalate (DPB) (new to RoHS 3)
- Benzyl butyl phthalate (BBP) (new to RoHS 3)
- Diisobutyl phthalate (DIBP) (new to RoHS 3)

CONFLICT MINERALS

Conflict resources are natural resources extracted in a conflict zone and sold to perpetuate the fighting. The Conflict-Free Sourcing Initiative was founded in 2008 by members of the Electronic Industry Citizenship Coalition (EICC) and the Global e-Sustainability Initiative (GeSI) to ensure that all products or materials supplied do not contain any conflict minerals, defined as columbite–tantalite (tantalite), cassiterite (tin), wolframite (tungsten), and gold from countries of concern, including the Democratic Republic of Congo (DRC) or an adjoining country, presently Angola, Burundi, Central African Republic, the Republic of the Congo, Rwanda, South Sudan, Tanzania, Uganda, or Zambia.

ANIMAL TISSUE

FDA has issued *Medical Devices Containing Materials Derived from Animal Sources (Except for In Vitro Diagnostic Devices); Draft Guidance for Industry and Food and Drug Administration Staff* to update the policy regarding the use of animal-derived material in medical device manufacturing. The role of animal-derived material in medical devices is well established. However, these materials may carry a risk of transmitting infectious disease when improperly collected, stored, or manufactured.

This draft guidance updates the 1998 guidance by addressing aspects of potential risk from the use of animal tissues (for example, viruses and bacteria) and now includes recommendations related to all *transmissible spongiform encephalopathies* (TSEs). The 1998 guidance addressed ways to reduce the potential for exposure to *bovine spongiform encephalopathy* (BSE).

The draft guidance recommends that information be collected and documented when animal tissue–derived materials are used as either device components (for example, pericardium, viscera, bone, hyaluronic acid, collagen) or manufacturing reagents (for example, tissue culture media, enzymes). Information that is helpful during the review of animal-derived device materials includes certificates of analysis (COA) and material safety data sheets (MSDS), when available.

COUNTERFEIT MEDICINE AND MEDICINAL COMPONENTS

Counterfeit medicine and medicinal components are a growing problem in the global economy. These products may be contaminated, contain the wrong active ingredient, contain no active ingredient, or contain the right active ingredient but at the wrong dose. Counterfeit drugs are illegal and may be harmful.

FDA takes all reports of suspected counterfeits seriously and, in order to combat counterfeit medicines, is working with other agencies and the private sector to help protect the nation's drug supply from the threat of counterfeits.

QUALITY AGREEMENTS

> Explain the purpose and describe the main elements of a quality agreement. (Understand)
>
> **Body of Knowledge II.K.4**

Quality agreements may also be known as *technical agreements* or *supplier quality agreements*, or another term may be used. These agreements are a vital form of supplier control and establishing requirements, expectations, and communication pathways early on between the manufacturer and the supplier. Some pharmaceutical companies may require quality agreements for all of their suppliers; others may use them only for contracted services, for raw materials that are critical to product quality (for example, excipients), for single-sourced materials, for custom materials, or for materials for which they would like to reduce testing upon receipt.

According to ISO/IEC 17025, when laboratories review contracts when preparing to test or calibrate for a potential customer, the review should include:

- Clearly defined and understood requirements (including the methods to be used)
- Determination of whether the laboratory has the capability and resources needed (including personnel skills and expertise)
- Selection of the appropriate test and/or calibration method

Once the supplier has been identified and the quality agreement has been approved by the supplier and manufacturer, the manufacturer starts receiving the supplier's products or services. The supplier should be evaluated by audit or survey and by trending information about receipt of product or services provided. In addition, changes to the supplied product or service should be monitored to ensure that the changes do not impact product quality. To improve the likelihood that a supplier will provide products or services that meet requirements, the supplier can be qualified or certified.

In the FDA guidance *Quality Systems Approach to Pharmaceutical CGMP Regulations*, issued in September 2006, FDA recommends periodic auditing of suppliers based on a risk assessment. Suppliers who provide *active pharmaceutical ingredients* (APIs), which are an important component of product quality, would be audited more frequently than a manufacturer of sodium chloride. The API manufacturer's quality systems should be thoroughly examined to ensure that they can reliably provide an API that meets the pharmaceutical manufacturer's specifications.

If the pharmaceutical manufacturer accepts the test results on the *certificate of analysis* (COA) from the API manufacturer and does not perform full analytical tests itself, the pharmaceutical manufacturer should evaluate the API release testing processes during the audit of the API manufacturer. In addition, if the pharmaceutical manufacturer accepts the supplier's COA for test results, the pharmaceutical manufacturer should, at a minimum, perform an identity test for each receipt of material.

Additional assurance that a critical supplier, such as a contract manufacturer for an API, will be able to meet requirements can be achieved by the pharmaceutical manufacturer placing a person-in-the-plant at the contract manufacturer to observe manufacturing and/or to review batch records and test results on site. A person-in-the-plant will be able to work in a timely manner with the contract manufacturer on any issues that may arise during manufacturing or testing.

OUTSOURCE PROCESSES

> Use various methods for the management of contract manufacturing, drug development, testing laboratories, and other outsourcing activities including defining the roles and responsibilities of each party. (Apply)
>
> **Body of Knowledge II.K.5**

Pharmaceutical manufacturer's quality management system documentation should clearly define what QMS functions or regulatory requirements are being outsourced. Outsourced processes are when an entire process is being executed by a supplier even if the legal responsibilities are still ultimately with the pharmaceutical manufacturer. A supplier of printing nonsterile packaging component for a pharmaceutical manufacturer's own internal manufacturing lines may not considered an outsourced process because the manufacturer still can perform quality checks and other verification activities on the component.

If a supplier is performing sterilization activities, then that would be considered outsourced because the pharmaceutical manufacturer cannot easily verify the output on their own manufacturing line.

The same supplier evaluation and supplier control activities as described earlier in the chapter apply to suppliers of outsource processes. Given the higher risk nature of outsource processes, more of these supplier control activities may be required. For example, an on-site audit and verification of the supplier's independent QMS or laboratory certifications would be required beyond a QMS survey.

Outsourced process suppliers may meet the definition of a pharmaceutical legal manufacturer, designer, labeler, or distributor. This required regulatory registration of the outsource supplier as a manufacturer so that they may be periodically inspected by regulatory government agencies or even the pharmaceutical manufacturers' own QMS certification agency.

Given the complexities and higher risk nature of outsourced processes, a well-documented quality agreement with clear roles and responsibilities is especially important. In addition to the typical quality agreement content already described earlier in this chapter, regulatory responsibility such as facility registrations, recalls, safety adverse event reporting, handling of external audits, etc. should also be in the quality agreement.

Part III

Laboratory Systems

Chapter 19

Compendia (US, Europe, and Japan)

This chapter covers understanding, navigating, and applying information found within the three major pharmaceutical compendia to the selection and implementation of qualitative or quantitative release tests for drug substances (*active pharmaceutical ingredients* [APIs]), drug products, or excipients). While regulatory and consensus standard references cited and annotated herein are important, proper, real-time lookup and learning from the current or effective-in-future General Chapters (plus specific monographs) of the three major compendia are required for full understanding and knowledge. Detailed interpretation of General Chapter content is confined to only those mandatory and informational General Chapters considered to be a minimum for *good manufacturing practices* (GMP) literacy.

REQUIRED VS. INFORMATIONAL COMPENDIA

> Describe and distinguish between required and informational ("general") compendial chapters. (Apply)
>
> **Body of Knowledge III.A.1**

A compendium is simply a collection or compilation of distinct pieces of writing or chapters. Vis-à-vis drug substances, drug products, and excipients (collectively referred to as official articles contained within the Unites States Pharmacopeia [USP]), the term "compendium" refers to the USP, the European Pharmacopeia (PhEur), or the Japanese Pharmacopeia (JP). The USP, PhEur, and JP comprise General Notices, General Chapters, and official article-specific chapters called *monographs* that contain the official article's definition, packaging, storage, and other requirements and specifications. The USP General Chapters provide frequently cited procedures common to multiple official articles. For drug substances and drug products, the USP General Chapters <1> through <999> are mandatory, whereas General Chapters <1000> through <1999> are intended to be informational only by the USP. In reality, the United States Food and Drug Administration (FDA) considers the latter General Chapters to represent good science, which translates into FDA often enforcing informational General Chapters as extensions of *current good manufacturing practices* (cGMP).

MARKETING REQUIREMENTS VS. COMPENDIA

> Distinguish among the U.S. Pharmacopoeia (USP), European Pharmacopoeia (PhEur or EP), and Japanese Pharmacopoeia (JP) in terms of requirements for marketing authorization. (Understand)
>
> **Body of Knowledge III.A.2**

Drug substances, drug products, and excipients marketed in the United States must comply with the USP compendial standards. Prior to FDA approval, manufacturers must prove that their products are high quality, safe, effective, and free of contamination and defects. Analytical test methods used to evaluate these critical quality attributes are developed and validated as part of the drug development process. In the USP Pending Monograph Program new monographs or monograph revisions may be proposed and developed concurrently with the drug application process. The monograph will not become official until after FDA approval of the drug article.

To market products in the signatory states of the European Medicine Convention, the quality standards from the PhEur must be applied. European drug marketing authorization dossiers must include identification of the features of the formulation and manufacturing process which are essential for the quality of the product with specifications. Validation studies for these tests are performed during the development process and the resulting data is included in the drug application. These validation studies are utilized during the work of the European Pharmacopoeia commission in evaluating submission and revision requests.

The Japanese Ministry of Health, Labor and Welfare publishes the Japanese Pharmacopeia (JP) to standardize and control properties and qualities of drugs. Drugs sold or supplied in Japan must meet the standards of the JP. The Japanese Pharmaceutical and Medical Device Agency (PMDA) conducts evaluations of pharmaceuticals and medical devices for the Ministry of Health, Labor and Welfare. Revisions and updates to the JP are submitted by industry and reviewed by experts prior to inclusion in the JP.

All major pharmacopeias incorporate a period of public comment prior to adding a new monograph or revision of an existing monograph.

COMPENDIAL METHODS REVIEW

> Review compendial methods to ensure they are verified as suitable for use in the testing lab. (Analyze)
>
> **Body of Knowledge III.A.3**

Verification is the process to demonstrate that a compendia procedure is suitable for use for a drug product or matrix when performed for the first time by a laboratory to ensure acceptable results may be obtained with the personnel, reagents, and equipment available. Laboratories using compendial test procedures are not required to validate the procedures, but documented evidence of suitability is required. Verification consists of assessing selected analytical performance characteristics. The extent of the verification should be risk based. Consideration should be given to the experience of the user, the type of procedure, instrumentation, and the drug matrix, and method of manufacture. Per USP <1226> verification is not required for basic compendial test procedures routinely performed such as loss on drying or pH. Verification studies are performed for validated methods to show their suitability for a particular laboratory and set of conditions. If the formulation, strength of the product or delivery system is different than the monograph scope, full validation studies will be required. Validation is required for articles or the application of a general test chapter to a new article.

COMPENDIAL OR NON-COMPENDIAL REQUIREMENTS REVIEW

> Review test methods, qualifications, validation, and verification against required compendial chapters (general and informational, as needed). (Analyze)
>
> **Body of Knowledge III.A.4**

Proposed new or revised analytical procedures not found in the compendial must be validated to ensure the accuracy and reliability of the method [21 CFR 211.194(a)]. New analytical procedures as part of drug registration applications must be validated. During validation laboratory studies are performed to establish that the performance characteristics of the procedure meet the requirements of the analytical application. Typical characteristic studies are accuracy, precision, specificity, detection limit, quantitation limit, linearity, range, and robustness. ICH Validation of Analytical Procedures: Text and Methodology Q2 (R1) has recommended procedures and specifications.

Revalidation is required if there is a change in product synthesis, composition or procedure.

BIOLOGICAL, MICROBIOLOGICAL, CHEMICAL, AND PHYSICAL TEST METHODS

> Identify and interpret results from compendia identification tests, quantitative analysis, qualitative analysis, and other tests or studies for biological, microbiological, chemical, and physical tests. (Apply)
>
> **Body of Knowledge III.A.5**

ANALYTICAL TESTS

Analytical tests measure a physical, chemical, microbiological property or characteristic that should be within an appropriate limit, range, or distribution to ensure the desired product quality. A risk-based approach is recommended to identify these critical quality attributes (CQA) and control these CQAs throughout the drug development lifecycle.

There are two main types of chemical analysis: qualitative (identification) and quantitative. *Qualitative analysis* is performed to establish the composition of natural or synthetic substances. These tests are performed to indicate whether the substance or compound is present in the sample or not. Common identification tests utilize wet chemistry reactions with reagents to indicate the presence of a functional group by observation of a color change or precipitate. Other more specific identification tests will compare a spectra or chromatogram of a sample to a known standard. *Quantitative* analytical techniques are mainly used to identify the amount or concentration of an analyte and quantify any compound or substance in the sample. Many tests for impurities are limit tests. If the specification is "not more than x," the actual value may not be precisely determined but rather sample response is compared to standard response, and the result is reported as conforms or does not conform to the specification.

An assay is a quantitative test to determine the amount or concentration of the article and evaluates purity for an article. An assay can also be used to evaluate the amount of impurity. Most monographs include an assay test. Physical tests may be performed to identify a physical characteristic of the article, such as appearance, viscosity, tablet hardness, and tablet/capsule size.

Microbiological tests evaluate the presence of microorganisms in non-sterile products or the resistance to growth of microorganisms by a sterile product (sterility). The most performed microbiological tests are Total Aerobic Microbial count (TAMC) and Total Yeast and Mold Count (TYMC). Additional tests are performed for specific microorganisms. The ability of the test to detect microorganisms in the presence of product must be established. Results are generally reported as Number of Colony Forming Units (CFU).

In sterility tests, the articles are inoculated with microorganisms and growth is evaluated over time. If no growth is observed, the product complies with the sterility test.

Biotechnology derived products developed through the manipulation of living organisms usually consist of large molecule proteins. Specialized techniques are utilized to assess the strength and potency of the product. Bioassays may be quantitative or qualitative.

PHARMACEUTICAL WATER QUALITY CONTROLS

Bulk pharmaceutical water serves as a pharmaceutical component (a manufacturing material that is later completely evaporated or sublimated) or a material that decontaminates, cleans, and/or sanitizes product-contact equipment. It is produced and distributed by a system that undergoes initial qualification and then must be monitored. Thus, bulk pharmaceutical water is a utility, a process material, an excipient, or all three items.

Total organic carbon (TOC) and water conductivity make up the common tests prescribed by the USP, PhEur, and JP. *Water for injection* (WFI) additionally requires the *bacterial endotoxin test* (BET) with a specification of not more than 0.25 endotoxin units/mL (international units/mL in Europe). The Japanese aseptic processing guidance (Appendix A2, "Pharmaceutical Waters") states, "Whenever water quality is monitored and controlled by conductivity and TOC testing, it is not usually necessary to monitor individual metals or inorganic ions." Therefore, BET, TOC, and water conductivity may be considered sufficient process control and monitoring tests for maintaining a pharmaceutical water storage and delivery system in control.

For *quality control* (QC) monitoring tests of pharmaceutical water, neither USP nor JP require heavy metals or other specific ion testing of the pharmaceutical waters because the Environmental Protection Agency's (EPA) National Primary Drinking Water Regulations have sufficiently tight specifications to preclude their retesting in the more purified water. Unfortunately, the PhEur is not in harmony with USP or JP by the requirement of several tests (Table 19.1).

Though not a compendial specification, the PhEur suggests a *total aerobic microbial count* (TAMC) action limit of 10 CFU/100 mL in its WFI monograph, which agrees with the 1993 FDA *Guide to Inspections of High Purity Water Systems*. The PhEur and the 1993 FDA guide give a TAMC action limit of 100 CFU/mL for purified water. The USP, in its informational General Chapter <1231>, "Water for Pharmaceutical Purposes," explains that microbial specifications would be inappropriate for a continuously produced material such as bulk pharmaceutical water and for such a validated process. The microbial incubation time lag inherently defeats the real-time production and usage of bulk pharmaceutical water.

The PhEur TAMC action limit statement, in both the purified water and WFI monographs, specifies the use of agar medium S, which is R2A media. Neither the USP nor JP specifies an agar for TAMC of bulk pharmaceutical water, permitting different process validation results to dictate the appropriate medium.

Two guides exist for the frequency and location of bulk pharmaceutical water sampling for monitoring or testing off-line. The 1993 FDA guide also created the generally accepted sampling frequency and sampling locations within a bulk pharmaceutical water distribution system, namely, daily sampling (except during shutdown) from a minimum of one point of use, with all points of use tested weekly. USP General Chapter <1231> states that "sampling frequencies should be based on system validation data and should cover critical areas including unit operation sites samples should be collected from use points using the same delivery devices,

Table 19.1 Tests and specifications for the principal pharmaceutical waters as per European Pharmacopeia.

Pharmaceutical water → / Quality control release test ↓	Purified water	Water for injection in bulk
Nitrates	0.2 ppm	0.2 ppm
Heavy metals	0.1 ppm	0.1 ppm

ppm = parts per million

such as hoses, and procedures, such as preliminary hose or outlet flushing, as are employed by production from those use points."

Pure steam is manufactured such that, when condensed, it satisfies WFI specifications. Guidelines for monitoring pure steam are less well defined to nonexistent compared with those for bulk pharmaceutical waters. Providing it is manufactured within a system that already passes TOC and water conductivity, it is presumed not to deviate significantly from its source pharmaceutical water. In addition, pure steam is inherently self-sterilizing such that TAMC is highly unlikely to provide trends on the steam but only on incorrectly sampled condensate. BET results on pure steam use points or sampling points provides enough assurance against grossly contaminated (by Gram-negative bacteria) draw-off valves.

Water conductivity monitoring/testing may be performed off-line or in-line. In-line conductivity monitoring inherently prevents the progression to stage 2 or stage 3 of the USP or PhEur water conductivity test; however, the distribution system must be designed such that an *out-of-specification* (OOS) result (based on conductivity versus water temperature) automatically results in either diversion to drain or recirculation into the water generation system. If using the off-line pharmacopeial water conductivity test, stages 2 or 3 must be performed if the previous stage fails. No OOS investigation (per FDA out-of-specification guidance) or stoppage of the conductivity or any multistage compendial test may occur between stages. USP General Chapter <645> requires an electronic calibration of the conductivity meter by replacing the conductivity sensor with a nationally traceable resistance device, for example, Wheatstone Bridge.

The TOC compendial test for bulk pharmaceutical water is harmonized among the USP, PhEur, and JP. The apparatus for measuring TOC, like water conductivity, is either in-line or off-line. The periodic suitability testing of the apparatus is specified as employing both a standard solution of compendial sucrose and a system suitability standard solution of 1,4-Benzoquinone. The diluent/solvent is reagent water having a TOC of not more than 0.10 mg/L.

A monitoring test required for, and applicable only to, WFI and pure steam is the BET. The test, being highly dependent on the particular *Limulus amebocyte lysate* (LAL) and corresponding test equipment and instruments, is not a candidate for compendial test suitability verification, but must undergo complete method

- Gel-clot method
- Semiquantitative test
- Turbidmetric endpoint method
- Turbidmetric kinetic method
- Chromogenic endpoint method
- Chromogenic kinetic method

Figure 19.1 Six methods for testing for bacterial endotoxins.

validation for parenteral products. For WFI and pure steam, pharmacopeial test verification of suitability may suffice. The LAL reagent used in the BET has been classified as a biologic, and is licensed by the Center for Biologics Evaluation and Research (CBER). FDA requires that a finished parenteral pharmaceutical manufacturer operating in the US or shipping to the US use an LAL reagent licensed by CBER in all validation, in-process, and end-product LAL tests.

Portions of the BET General Chapter have been harmonized among the three compendia. USP, PhEur, and JP permit and describe six methods (Figure 19.1). All three compendia contain similar wording that "in the event of doubt or dispute, the final decision is made based upon method A [Gel-clot method: limit test] unless otherwise indicated in the monograph [for a specific finished parenteral pharmaceutical]."

Chapter 20

Laboratory Investigations of Atypical Results

The International Conference on Harmonization (ICH) defines *specifications* as a "list of tests, references to analytical procedures, and appropriate acceptance criteria that are numerical limits, ranges, or other criteria for the test described." Specifications establish the set of criteria to which a material should conform to be considered acceptable for its intended use. "Conformance to specification" means that the material (for example, raw material, finished drug product), when tested according to the listed analytical procedures, will meet the listed acceptance criteria. The term *"out-of-specification* (OOS) results" includes all test results that fall outside the specifications or acceptance criteria established in drug applications, *drug master files* (DMF), or official compendia, or by the manufacturer. The term also applies to all in-process laboratory tests that are outside of established specifications.

An OOS result requires a concerted effort to determine whether it is due to a product failure or a laboratory error. In either case, an investigation is required to determine the source of the OOS result, the disposition of the product, and the impact on associated products and processes.

REGULATORY GUIDANCE

ICH Q7 *Guidance for Industry: Good Manufacturing Practice Guidance for Active Pharmaceutical Ingredients* states that OOS results must be immediately reported to laboratory supervisors and/or managers and must be fully investigated. The guidance further states that any OOS result obtained should be investigated and documented according to a procedure. This procedure should include analysis of the data, assessment of whether a significant problem exists, allocation of the tasks for corrective actions, and conclusions. Any resampling and/or retesting after OOS results should be performed according to a documented procedure.

The United States Food and Drug Administration (FDA) OOS guidance further elaborates how to conduct an OOS investigation: To be meaningful, the investigation should be thorough, timely, unbiased, well documented, and scientifically sound. The first phase of such an investigation should include an initial assessment of the accuracy of the laboratory's data. Whenever possible, this should be done before test preparations (including the composite or the homogenous source of the aliquot tested) are discarded. If this initial assessment indicates that no meaningful errors were made in the analytical method used to arrive at the data, a full-scale OOS investigation should be conducted. For contract laboratories, the laboratory should convey its data, findings, and supporting documentation to the manufacturing firm's *quality control unit* (QU), who should then initiate the full-scale OOS investigation.

A microbiological OOS, referred to as a *microbial data deviation* (MDD), is often associated with the aforementioned guidance documents, as well as with United States Pharmacopeia (USP) <1117>, but these references either exclude microbiological tests and biological assays or intimate the difficulties associated with resolving a microbiological OOS. A better guideline to use when conducting an MDD investigation is the FDA aseptic processing guidance, *Sterile Drug Products Produced by Aseptic Processing—Current Good Manufacturing Practice.* Although this guideline addresses positive sterility results, it can be judiciously applied to investigate an MDD for a nonsterile raw material, product, and assay.

TEST DATA

> Describe and develop procedures for investigating each type of test data including biological, microbiological, chemical test, and unknowns. (Analyze)
>
> **Body of Knowledge III.B.1**

Investigation of aberrant or out-of-specification test data varies with the type of test. Some items to consider include:

- *Timeline.* Verify the time of the experiment and check for any unusual delays. Compare to successful results. For example, was the extraction time shorter or longer than expected?
- *Equipment.* Was the equipment used within its calibration period? Were all system suitability tests performed and within specification?
- *Reagents and Standards.* Were reagents used appropriate for the procedure? Consult COAs for purity or impurities that could have adversely affected the procedure? Were reagents and standards within their expiration? Were they stored appropriately prior to use? Were all dilutions and preparations correct?
- *Chromatograms and Spectra.* Carefully observe any graphs for anomalies. Are there unexpected peaks? Is the baseline stable and appropriate? If a blank correction was performed, verify that it was correctly performed.
- *Linearity.* If a result is based on linearity, check that the linearity has a suitable correlation coefficient and standards bracketed the result. Verify that one standard reading or preparation did not adversely affect the curve.
- *Calculations.* Verify all calculations for correctness and that they are scientifically appropriate.

- *Environmental Conditions.* Check for unusual or out-of-specification testing environments such as laboratory temperature or humidity, oven temperatures, storage conditions.
- *Media Preparation and Growth Promotion.* For microbiological tests, review media preparation and growth promotion records for anomalies.
- *Aseptic Conditions.* If required, check records for anomalies.

ATYPICAL RESULTS

> Identify, analyze, and interpret data on processes or products that are out-of-specification or out-of-trend. Determine the outcome of the laboratory portion of the investigation and the criteria for further investigation. (Evaluate)
>
> **Body of Knowledge III.B.2**

ANALYTICAL ERROR

All qualitative and quantitative analytical testing is subject to error. *Error* is the numerical difference between the observed value and the true value. The true value and observed value in chemical analysis can be related with each other (error) by the equation

$$E = O - T$$

where

E = Error

O = Observed value

T = True value

The smaller the error, the greater the accuracy of the measurement. Errors can be expressed as

$$E/T \times 100 = \%\text{ error}$$

$$E/T \times 100 = \text{Per hundred error}$$

CONDUCTING AN OOS INVESTIGATION

The FDA guidance for conducting an OOS investigation applies to chemistry-based laboratory testing of drugs regulated by the Center for Drug Evaluation and Research (CDER); it is directed toward traditional drug testing and release methods. These laboratory tests are performed on *active pharmaceutical ingredients* (APIs), excipients, and other components, in-process materials, and finished drug products to the extent that *current good manufacturing practices* (cGMP) regulations (21 CFR 210 and 211) and the Federal Food, Drug, and Cosmetic Act (FDCA) [Section 501(a)(2)(B)] apply. The principles in this guidance also apply to in-house testing of drug product components that are purchased by a firm. This guidance can be used by contract firms performing production and/or laboratory testing activities on behalf of the marketing authorization holder. This guidance does not apply to biological assays (for example, in vivo, immunoassays) or microbiological testing.

Specifically, the FDA OOS guidance discusses how to investigate OOS test results, including the responsibilities of laboratory personnel, the laboratory phase of the investigation, additional testing that may be necessary, when to expand the investigation outside the laboratory, and the final evaluation of test results. It is important to remember, however, that FDA guidance documents, including the OOS guidance, do not establish legally enforceable responsibilities. Instead, guidance documents describe the Agency's current thinking on a topic, and should be viewed only as recommendations unless specific regulatory or statutory requirements are cited. The use of the word "should" in Agency guidance documents means that something is only suggested or recommended but establishes the current thinking of the agency.

Analytical OOS Investigations

FDA regulations require that an investigation be conducted whenever an OOS test result is obtained (21 CFR 211.192). The purpose of the investigation is to determine the cause of the OOS result. The source of the OOS result should be identified either as an aberration of the measurement process or an aberration of the manufacturing process. Even if a batch is rejected based on an OOS result, an investigation is necessary to determine whether the result is associated with other batches of the same drug product or other products. Batch rejection does not negate the need to perform the investigation. The regulations require that a written record of the investigation be made, including the conclusions and follow-up.

Responsibility of the Analyst. In accordance with the cGMP regulations in 21 CFR 211.160(b)(4), the analyst should ensure that only those instruments meeting established performance specifications are used, and that all instruments are properly calibrated. Systems not meeting suitability requirements should not be used. If reference standard responses indicate that the system is not functioning properly, all data collected during the suspect time period should be properly identified and should not be used. The cause of the malfunction should be identified and, if possible, corrected before a decision is made whether to use any data from before the suspect period.

Analysts should check the data for compliance with test specifications before discarding test preparations or standard preparations. When unexpected results are obtained and no obvious explanation exists, test preparations should be retained, if stable, and the analyst should inform the supervisor. An assessment of the accuracy of the results should be started immediately.

If errors are obvious, such as the spilling of a sample solution or the incomplete transfer of a sample composite, the analyst should immediately document what happened. Analysts should not knowingly continue an analysis they expect to invalidate at a later time for an assignable cause (that is, analyses should not be completed for the sole purpose of seeing what results can be obtained when obvious errors are known).

Responsibility of the Laboratory Supervisor. The supervisor's assessment should entail:

- Discussing the test method with the analyst; confirming analyst knowledge of, and performance of, the correct procedure;
- Examining the raw data obtained in the analysis, including chromatograms and spectra, and identifying anomalous or suspect information;
- Verifying that the calculations used to convert raw data values into a final test result are scientifically sound, appropriate, and correct; determining whether unauthorized or nonvalidated changes have been made to automated calculation methods;
- Confirming the performance of the instruments;
- Determining that appropriate reference standards, solvents, reagents, and other solutions were used and that they met quality control specifications;
- Evaluating the performance of the test method to ensure that it is performing according to the standard expected, based on method validation data and historical data;
- Fully documenting and preserving records of this laboratory assessment. The assignment of a cause for OOS results will be greatly facilitated if the retained sample preparations are examined promptly. Hypotheses regarding what might have happened (for example, dilution error, instrument malfunction) should be tested. Examination of the retained solutions should be performed as part of the laboratory investigation.

Laboratory management should ascertain not only the reliability of the individual values obtained but also the significance these OOS results represent to the laboratory quality assurance program. Laboratory management should be especially alert to developing trends, and upper management should appropriately monitor these trends and ensure that any problem areas are addressed.

Laboratory error should be relatively rare. Frequent errors suggest a problem that might be due to inadequate training of analysts, poorly maintained or improperly calibrated equipment, or careless work. Whenever laboratory error is identified, the firm should determine the source of that error and take corrective action to prevent recurrence. To ensure full compliance with the cGMP regulations, the manufacturer also should maintain adequate documentation of the corrective action. When clear evidence of laboratory error exists, laboratory testing results should be invalidated. When evidence of laboratory error remains unclear, a full-scale OOS investigation should be conducted by the manufacturing firm to determine what caused the unexpected results. It should not be assumed that OOS test results are attributable to analytical error without performing and documenting an investigation. Both the initial laboratory assessment and the following OOS investigation should be documented fully.

Full-Scale OOS Investigation

When the initial assessment does not determine that laboratory error caused the OOS result, and testing results appear to be accurate, a full-scale OOS investigation using a predefined procedure should be conducted. This investigation may consist of a production process review and/or additional laboratory work. The objective of such an investigation should be to identify the root cause of the OOS result and take appropriate corrective and preventive action. A full-scale investigation should include a review of production and sampling procedures and will often include additional laboratory testing. Such investigations should be given the highest priority. Among the elements of this phase is evaluation of the impact of OOS result(s) on already distributed batches.

Production Review

The investigation should be conducted by the *quality unit* (QU) and should involve manufacturing, process development, maintenance, and engineering. In cases where manufacturing occurs off-site (for example, performed by a contract manufacturer), all sites potentially involved should be included in the investigation. Other potential problems should be identified and investigated.

The records and documentation of the manufacturing process should be fully reviewed to determine the possible cause of the OOS result(s). A full-scale OOS investigation should consist of a timely, thorough, and well-documented review. A written record of the review should include, at a minimum:

- A clear statement of the reason for the investigation
- A summary of the aspects of the manufacturing process that may have caused the problem
- Results of a documentation review, with the assignment of actual or probable cause
- Results of a review made to determine whether the problem has occurred previously
- Description of corrective actions taken

If this part of the OOS investigation confirms the OOS result and is successful in identifying its root cause, the OOS investigation may be terminated and the product rejected. A failure investigation that extends to other batches or products that may have been associated with the specific failure must also be completed (21 CFR 211.192). If any material was reprocessed after additional testing, the investigation should include the comments and signatures of appropriate production and QC personnel.

OOS results may indicate a flaw in product or process design. In such cases, it is essential that redesign of the product or process be undertaken to ensure reproducible product quality.

Retesting. The sample used for the retesting should be taken from the same homogeneous material that was originally collected, tested, and yielded the OOS results. For a liquid, it may be from the original unit of liquid product or composite of the liquid product; for a solid, it may be an additional weighing from the same sample composite prepared for the original test.

Retesting is indicated when instrument malfunctions or sample handling problems (for example, a dilution error) are suspected. Decisions to retest should be based on the objectives of the testing and sound scientific judgment. Predefined retesting plans should include retests performed by an analyst other than the one who performed the original test. A second analyst performing a retest should be at least as experienced and qualified in the method as the original analyst.

The practice of testing into compliance is unscientific and objectionable under cGMP. The maximum number of retests to be performed on a sample should be specified in advance in a written *standard operating procedure* (SOP). The number should be based on scientifically sound principles and should not be adjusted depending on the results obtained. The firm's predetermined retesting procedures should contain a point at which the additional testing ends and the batch is evaluated. If the results are unsatisfactory at this point, the batch should be considered suspect and rejected or held pending further investigation [21 CFR 211.165(f)]. Any deviation from this SOP should be rare and done in accordance with 21 CFR 211.160(a), which states that any deviations from written specifications, sampling plans, test procedures, or other laboratory control mechanisms shall be recorded and justified. In such cases, before starting additional retesting, a protocol should be prepared (subject to approval by the QU) that describes the additional testing to be performed and specifies the scientific and/or technical handling of the data. In the case of a clearly identified laboratory error, the retest results would substitute for the original test result. All original data should be retained, however, and an explanation recorded. This record should be initialed and dated by the involved persons and include a discussion of the error, and supervisory comments. If no laboratory or calculation errors are identified in the first test, there is no scientific basis for invalidating initial OOS results in favor of passing retest results. All test results, both passing and suspect, should be reported and considered in batch release decisions.

Resampling. *Resampling* involves analyzing a specimen from any additional units collected as part of the original sampling procedure or from a new sample collected from the batch, should that be necessary. The original sample from a batch should be sufficiently large to accommodate additional testing in the event an OOS result is obtained. It may be appropriate to collect a new sample from the batch. Control mechanisms for examination of additional specimens should be in accordance with predetermined procedures and sampling strategies [21 CFR 211.165(c)].

When all data have been evaluated, an investigation might conclude that the original sample was prepared improperly and was therefore not representative of the batch quality [21 CFR 211.160(b)(3)]. Improper sample preparation might be indicated, for example, by widely varied results obtained from several aliquots of an original composite (after determining there was no error in the performance of the analysis).

Resampling should be performed using the same qualified, validated methods that were used for the initial sample. If, however, the investigation determines that the initial sampling method was inherently inadequate, a new, accurate sampling method must be developed, documented, and reviewed, and approved by the QU [21 CFR 211.160 and 211.165(c)].

Reporting Testing Results

Averaging—Appropriate Uses. If the sample can be assumed to be homogeneous (that is, an individual sample preparation designed to be homogeneous), using averages can provide a more accurate result. In the case of microbiological assays, the USP prefers the use of averages because of the innate variability of the biological test system.

If a test requires a specific number of replicates to arrive at a result (for example, a *high-performance liquid chromatography* [HPLC] assay result is determined by averaging the peak responses from a number of consecutive, replicate injections from the same preparation), then the assay result would be calculated using the peak response average. This determination is considered one test and one result, distinctly different from the analysis of different portions from a lot—intended to determine variability within the lot—and from multiple full analyses of the same homogeneous sample.

The use of replicates to arrive at a single reportable result, and the specific number of replicates used, should be specified in the written, approved test method. Acceptance limits for variability between the replicates should also be specified in the method. Unexpected variation in replicate determinations should trigger remedial action as required by 21 CFR 211.160(b)(4). If acceptance limits for replicate variability are not met, the test results should not be used.

If a series of complete tests (full run-throughs of the test procedure), such as assays, are part of the test method, it may be appropriate to specify in the test method that the average of these multiple assays is considered one test and represents one reportable result. In this case, limits on acceptable variability between the individual assay results should be based on the known variability of the method and should be specified in the test methodology. A set of assay results not meeting these limits should not be used. These appropriate uses of averaging test data should be used during an OOS investigation only if they were used during the original testing that produced the OOS result.

Averaging—Inappropriate Uses. All individual test results should normally be reported as separate values. Where averaging of separate tests is appropriately specified by the test method, a single averaged result can be reported as the final test result. In some cases a statistical treatment of the variability of results is reported (for example, in a test for dosage form content uniformity), and the standard deviation (or relative standard deviation) is reported with the individual unit dose test results.

Averaging can conceal variations in different portions of a batch, or within a sample, for example, when performing powder blend/mixture uniformity or dosage form content uniformity determinations. In these cases, testing is intended to measure variability within the product, and individual results provide the information for such an evaluation.

For additional testing performed during an OOS investigation, averaging the result(s) of the original test that prompted the investigation and additional retest or resample results obtained during the OOS investigation may hide variability among the individual results. These data can be misleading when some of the results are OOS and others are within specifications. The laboratory must provide all individual results for evaluation and consideration by the QU, which is responsible for approving or rejecting, for example, drug products or in-process materials (21 CFR 211.22).

Outlier Tests. The cGMP regulations require that statistically valid QC criteria include appropriate acceptance and/or rejection levels [21 CFR 211.165(d)]. A value that is markedly different from the others in a series obtained using a validated method may qualify as a statistical outlier. An outlier may result from a deviation from prescribed test methods, or it may be the result of variability in the sample. It should never be assumed that the reason for an outlier is error in the testing procedure rather than inherent variability in the sample being tested.

Outlier testing is a statistical procedure for identifying from an array those data that are extreme, and the possible use of outlier tests should be determined in advance. It should be written into an SOP for data interpretation and be well documented. The SOP should include the specific outlier test to be applied, with relevant parameters specified in advance, and should specify the minimum number of results required to obtain a statistically significant assessment from the specified outlier test.

For biologic assays having a high variability, an outlier test may be an appropriate statistical analysis to identify those results that are statistically extreme observations. The USP describes outlier tests in the General Chapter "Design and Analysis of Biological Assays." In these cases, the outlier observation is omitted from calculations. The USP states that "arbitrary rejection or retention of an apparently aberrant response can be a serious source of bias . . . the rejection of observations solely on the basis of their relative magnitudes is a procedure to be used sparingly."

An outlier test is only a statistical analysis of the data obtained from testing and retesting validated chemical tests with relatively small variance, and when the sample being tested can be considered homogeneous. It will not identify the cause of an extreme observation and should not be used to invalidate the suspect result. Outlier tests have no applicability in cases where the variability in the product is what is being assessed, such as for content uniformity, dissolution, or release rate determinations. In these applications, a value perceived to be an outlier may in fact be an accurate result of a nonuniform product.

When using these practices during the additional testing performed in an OOS investigation, the laboratory will obtain multiple results. It is critical for the laboratory to provide all test results for evaluation and consideration by the QU in its final disposition decision. In addition, when investigation by a contract laboratory does not determine an assignable cause, all test results should be reported to the customer on the *certificate of analysis* (COA).

Concluding the Investigation

To conclude the investigation, the results should be evaluated, the batch quality should be determined, and a release decision should be made by the QU. The SOP should be followed in arriving at this point. Once a batch has been rejected, there is no limit to further testing to determine the cause of the failure so that a corrective action can be taken.

Interpretation of Investigation Results

The QU is responsible for interpreting the results of the investigation. An initial OOS result does not necessarily mean the subject batch fails and must be rejected. The OOS result should be investigated, and the findings of the investigation, including retest results, should be interpreted to evaluate the batch and reach a decision regarding release or rejection (21 CFR 211.165).

When an investigation has revealed a cause, and the suspect result is invalidated, the result should not be used to evaluate the quality of the batch or lot. Invalidation of a discrete test result may be done only upon the observation and documentation of a test event that can reasonably be determined to have caused the OOS result.

When an investigation indicates that an OOS result is caused by a factor affecting the batch quality (that is, an OOS result is confirmed), the result should be used in evaluating the quality of the batch or lot. A confirmed OOS result indicates that the batch does not meet established standards or specifications, and should result in the batch's rejection, in accordance with 21 CFR 211.165(f), and proper disposition.

For inconclusive investigations that do not reveal a cause for the OOS test result, the investigation changes from an OOS investigation into a batch failure investigation, which must be extended to other batches or products that may have been associated with the specific failure (21 CFR 211.192). For inconclusive investigations that do not confirm the OOS result, the QU might still ultimately decide to release the batch. Any decision to release a batch despite an initial OOS result that has not been invalidated should come only after a full investigation has shown that the OOS result does not reflect the quality of the batch. In making such a decision, the QU should always err on the side of caution.

An assay result that is low, but within specifications, should raise a concern. One cause of the result could be that the batch was not formulated properly. Batches must be formulated with the intent to provide not less than 100% of the labeled or established amount of active ingredient [21 CFR 211.101(a)]. This includes situations where the analytical result meets specifications, but caution should be used in the release or reject decision.

All records pertaining to the OOS test result should be retained. Records must be kept of complete data derived from all tests performed to ensure compliance with established specifications and standards (21 CFR 211.194).

MICROBIOLOGY OOS INVESTIGATIONS

No definitive guideline exists on how to conduct a microbiological OOS investigation. The FDA OOS guidance essentially excludes microbiological and biologic assays; the ICH Q7A OOS guidance states that OOS investigations are not normally needed for in-process tests that are performed for the purpose of monitoring and/or adjusting the process; and USP <1117> intimates the difficulties associated with

- Identification (speciation) of the organism
- Records of laboratory tests and deviations
- Monitoring of production area environment
- Monitoring of personnel (laboratory and production)
- Product presterilization bioburden
- Production record review
- Manufacturing history (including media fills)

Figure 20.1 The aseptic processing guidance identifies seven areas that should be scrutinized.

resolving a microbiological OOS. In fact, the term "microbiological OOS" is better served by the term "MDD."

One guideline, the FDA aseptic processing guidance, addresses an MDD, albeit it is focused on sterile products. As sterility is the most critical attribute of a product, however, application of these stringent investigational principles can be applied to other products or raw materials to ensure that the decision to retest or fail a product has been made with the greatest scientific scrutiny and the lowest risk. When applied judiciously, the guidance is an effective tool that can be used to perform an MDD investigation.

The aseptic processing guidance identifies seven areas that should be scrutinized when conducting an MDD (Figure 20.1). Four additional areas of investigation not included in the guideline complete the cycle for investigating an MDD, and include API source, excipient source, satellite facilities (if used), and laboratory (in-house or contract).

The principles of conducting an analytical OOS investigation heretofore described should be invoked throughout an MDD investigation—including proper documentation, collection of data, and interpretation of data by the QU—and final disposition by the QU and company management. In both instances the investigation must be conducted using sound scientific principles, and the conclusion must be based on empirical data that are elucidated during the process. The underlying assumption of an MDD is that it evaluates the microbiological disposition of a product or material, and the product must be placed on hold until it is proven that the MDD is a result of laboratory or operator error that does not affect product integrity.

Standard Operating Procedures and Protocol

The firm should have an SOP that describes how an MDD investigation will be conducted, and should include a section for preparing a protocol, subject to approval by the QU, that describes the additional testing to be performed and specifies the scientific and/or technical handling of the data. The protocol should be based on scientifically sound principles and should clearly state when retesting is appropriate. It should indicate that any unsatisfactory results incriminate the batch, which must be rejected or held pending further investigation [21 CFR 211.165(f)]. Any deviation from this SOP should be rare and done in accordance with 21 CFR 211.160(a), which states that any deviations from written specifications, sampling plans, test procedures, or other laboratory control mechanisms shall be recorded and justified. In such cases, before starting additional retesting, a protocol should be prepared (subject to approval by the QU) that describes the additional testing to be performed and specifies the scientific and/or technical handling of the data.

Identification (Speciation) of the Organism

Microorganism isolates should be identified to the species level. Microbiological monitoring data should be reviewed to determine whether the organism is found in laboratory and production environments, personnel, or product bioburden. Advanced identification methods (for example, nucleic acid based) are valuable for investigational purposes. When comparing results from environmental monitoring and product/production positives, both identifications should be performed using the same methodology. Identifying the same organism from environmental and product/production is helpful in determining the source of the contamination but by itself does not prove it. Additional data must support the findings.

Based on the species, an initial determination of the source can be made. For example, isolation of *Staphylococcus hominis* would indicate the source to be human skin, while *Bacillus sphaericus* would implicate an environmental source. Interpretation of data by a qualified microbiologist is an essential element of the MDD investigational process.

Record of Laboratory Tests and Deviations

Review of laboratory deviations and investigations can help to eliminate or to implicate the laboratory as the source of contamination. If the organism is seldom found in the laboratory environment, product contamination is more likely than laboratory error. If the organism is found mainly in the laboratory and seldom in the production environments, product contamination may still be more likely than laboratory error. Review of laboratory test and deviation data from at least one year before the MDD is suggested to develop a comprehensive understanding of the dynamics associated with the laboratory testing procedures and environment.

Proper handling of deviations is an essential aspect of laboratory control. When a deviation occurs, it should be documented, investigated, and remedied. If any deviation is considered to have compromised the integrity of the microbial testing, the test should be invalidated immediately without incubation.

An MDD result can be viewed as indicative of production or laboratory problems, and the entire manufacturing process should be comprehensively investigated as such problems often can extend beyond a single batch. To more accurately monitor potential contamination sources, maintaining separate trends by appropriate categories such as product, container type, filling line, sampling, and testing personnel is recommended. Where the degree of test sample manipulation is similar to that during production and an MDD is found, the production process may be implicated more than the laboratory.

Microbial monitoring of the testing area of the laboratory and personnel can reveal trends that are informative. Upward trends in the microbial load in the testing area of the laboratory should be promptly investigated as to cause, and corrected. In some instances, such trends can appear to be more indicative of laboratory error as a possible source of an MDD.

Where a laboratory has a good track record with respect to errors, this history can lower suspicion of the laboratory as a source of contamination as chances are greater that the contamination arose from production; however, the converse is not true. Specifically, where a laboratory has a poor track record, firms should not assume that the contamination is automatically more attributable to the laboratory and consequently overlook a genuine production problem. Accordingly, it is essential that all MDDs be thoroughly investigated.

Monitoring of Production Area Environment

Trend analysis of microorganisms in the critical and immediately adjacent areas is especially helpful in determining the source of contamination in an MDD investigation. Consideration of environmental microbial data should not be limited to the results of monitoring the production environment for the lot, day, or shift associated with the suspect lot. Results showing little or no recovery of microorganisms can be misleading, especially when preceded or followed by a finding of an adverse trend or atypically high microbial counts. It is therefore important to look at both short- and long-term environmental trend analyses. A review of at least the past year of environmental monitoring data from the production area is useful in determining trends and overall performance of the area. It can help identify areas that are more susceptible to microbial contamination, facilitating *corrective and preventive action* (CAPA).

Monitoring Personnel (Laboratory and Production)

The review of data and associated trends from daily monitoring of personnel can provide important information indicating a route of contamination. The adequacy of personnel practices and training also merits significant review and consideration. Trending these data can help pinpoint specific technicians and operators who may be the source of the contamination, and appropriate remedial actions can be taken, including decertifying them to perform their jobs, if necessary. In certain instances, the technician or operator may be removed and asked to undergo comprehensive training and evaluation to ensure he or she meets standards established in corresponding SOPs.

Product Bioburden

A review of product bioburden is required to determine if adverse bioburden trends have occurred in the past. One-year data analysis will help ascertain whether the bioburden levels contributed to the MDD. Excipients and APIs that show increased levels of microbial activity may be implicated as the source of the MDD, and remedial actions should be taken, including bioburden reduction procedures or a change in the supplier. High-quality starting materials are necessary to produce high-quality finished products.

Production Record Review

Complete batch and production control records should be reviewed to detect any signs of failures or anomalies that could have a bearing on product integrity. The investigation should include:

- Events that could have impacted a critical zone;
- Batch and trending data that indicate whether utility and/or support systems are functioning properly. For example, records of air quality monitoring for filling lines could show a time at which there was improper air balance or an unusually high particle count;
- Review of information on construction or maintenance activities that could have had an adverse impact;
- Review of production records from at least the past year to determine whether trends were apparent.

Manufacturing History (Including Media Fills Where Applicable)

The manufacturing history of a product or similar products should be reviewed as part of the investigation. Past deviations, problems, or changes (for example, process, components, equipment) are among the factors that can provide an indication of the origin of the problem. Depending on the product(s) produced, sterility, bioburden, personnel, and environmental data from at least the past year should be reviewed to determine whether trends were apparent.

API Source

As part of the total quality system, microbiological testing should be performed on every batch of API received into the firm's plant(s) to confirm the microbiological acceptability of the material and to generate data for trending purposes. If a monograph exists for the API and includes microbial testing, those tests should be performed. For APIs with no microbial monographs, the firm should determine what microbiological tests are critical in assuring the microbial integrity of the material and perform the appropriate tests, including bioburden, microbial limits, endotoxin, and sterility testing. As part of the MDD investigation, a microbiological review of at least the past 15 batches should be conducted to ascertain both the species of organisms that are inherent in the material and any trends that may exist.

Confirmation that the indigenous microbes in the API are the same as those found in the affected product implicate the API as a probable root cause of the organism, and appropriate preventive measures in the manufacture of the API should be conducted:

- Suspend delivery of new API lots to the firm
- Place all existing in-house lots of the material on *quality assurance* (QA) hold

- Place all in-process finished product lots containing the API on QA hold
- Require a written investigation from the manufacturer, including CAPA
- Reinspect the API facility
- Increase incoming testing and inspection of the material
- Consider elimination of the manufacturer as a provider of the API
- Conduct a formal investigation, including risk analysis, of the MDD to determine the disposition of the product and any affected batches of products from any of the firm's production plants using the API

Excipients

It is highly recommended to perform incoming microbiological testing on excipients to ensure their integrity. Although it may not be feasible to test every lot or batch of excipient, they should all be tested on a regular basis as governed by the firm's supplier certification SOP. The same process and criteria for investigating an API MDD should be applied to the excipient manufacturer, including a risk analysis. Microbial testing for certain products known not to harbor microorganisms (for example, concentrated acids and bases) may be waived. As part of the MDD investigation, a microbiological review of at least the past 15 batches should be conducted to ascertain both the species of organisms that are inherent in the material and any trends that may exist. Purified water or water for injection should be treated as an excipient, and a review of microbial water quality from the past year should be included in the investigation.

Satellite Facilities

Some firms use multiple production facilities to formulate products. (For example, they may weigh the API and/or excipients in one facility and transport them to another facility, either off site or at the same site.) Each facility must be treated in the same manner as the main production facility, and the areas to be investigated should be the same as described earlier:

- Speciation of the organism from the API/excipient/environment
- Production area environment
- Personnel (laboratory and production)
- Production record review
- Manufacturing history (including media fills)

Trend analysis is critical in ascertaining whether the organism has been associated with a satellite site, and the satellite should be under continuous microbiological surveillance to ensure it supports the microbial integrity of the process.

Contract Laboratories

Many firms use contract laboratories to test APIs, excipients, and bulk and finished products, and ask the laboratories to conduct an investigation of an MDD. Whenever possible, the firm should request that they be allowed to conduct the investigation of the laboratory; this request is not to indict the laboratory, but only serves the best interest of the firm. Generally, a contract laboratory has many clients, and serving one client over the others is not in their best interest. This philosophy is applicable not only to microbiological testing but also to analytical testing and to contract manufacturers. Allowing a firm access to the inner workings of the laboratory may be the best method to uncover the source of an MDD, or OOS in the case of analytical deviations.

If the laboratory does not allow outside personnel to conduct an investigation, the firm should require the laboratory to provide written documentation of these items:

- *Receiving area.* The physical condition of the receiving area may be indicative of the manner in which the laboratory operates. Good laboratories have well-defined areas that are organized, and materials are quickly logged in and sent to the appropriate departments. Sample logging, whether manually recorded or electronically recorded, is a critical component of the area, and an SOP should cover sample receipt and tracking.
- *Sample staging area.* Similarly to the receiving area, the sample staging area should be well organized and clean, and this should be documented.
- *Autoclave cycles.* Autoclave cycles should be reviewed for the past year to determine whether any excursions occurred.
- *Cleaning agents and processes.* Cleaning agents and frequencies should be reviewed for the past year, and trends should be documented.
- *Environmental isolates.* One-year environmental monitoring data should be reviewed and the organisms found compared to those found in the product. Additionally, the isolated organisms should be reviewed in relation to the cleaning agents being used. If *Bacillus* are frequently found and no sporicidal agent is used, there is a possibility that *Bacillus* spores may be a potential problem.
- *Personnel.* Personnel may contribute to an MDD, and the records from the past year of technicians conducting the testing should be thoroughly reviewed. Organisms associated with the technicians should be compared with those found in the MDD. If retraining is required, the training records should be reviewed.
- *Incubators.* A review of the cleaning and environmental monitoring of and organisms found in the incubators should be conducted. Incubators are a major source of environmental excursions and may contribute to an MDD.

- *Filters* (including *high-efficiency particulate air* [HEPA] filters, if used). Cleaning, maintenance, and environmental monitoring of filters from the past year should be reviewed and the organisms found compared with those found in the MDD.
- *Purchased media*. The manufacturer of the purchased media should be asked to provide a one-year review of the batches used to test the product in question. Media producers occasionally issue a recall of contaminated media, which can be a source of the MDD.
- *Equipment*. A one-year review of the microbiological history of biohazard hoods, work top spaces, pipettes, hockey sticks, and other areas and tools should be conducted to determine whether they contributed to the MDD.
- *Sterility testing*. A one-year review of the anteroom, sterility suite, HEPA filters, and failure rate should be conducted, and any organisms should be compared with those in the MDD.
- *General supplies*. Many supplies are shipped in cardboard boxes, which are a major source of contamination. Sampling of the cardboard may reveal the source of the MDD.
- *Media preparation and growth promotion*. A one-year review of media preparation and growth promotion should be conducted and correlated with the autoclave cycles used and test organisms used for the media preparation and growth promotion, respectively. Although not a common source of contamination, these areas should be well controlled as an overall investigational assessment.
- *Training*. Properly trained technicians are the most important component of well-functioning laboratories. Their records should be examined to ensure they received appropriate training for the tests they are performing. Aseptic training may be the most critical area to examine.
- *Observation of testing*. If possible, the firm should be allowed to observe the technician(s) who originally performed the testing perform a mock test from start to finish, including receipt, preparation, handling, dispensing, diluting, and incubating samples. Minor inconsistencies may be revealed, which could be the source of the MDD.

CASE STUDY

A contract testing laboratory encountered an MDD in a sterility test. After conducting an MDD investigation, the firm was allowed to observe the laboratory's technicians perform a mock test. As part of the test, 20 ampules were pooled into sterile containers before filtration. The sterile containers were urine sample containers that were sterilized by gamma radiation; they were shipped to the laboratory in a large cardboard box containing a polyethylene bag filled with

1000 containers. The containers were not individually wrapped, as it was cost-efficient to purchase them in bulk. The laboratory had recently made the change from individually wrapped containers to the bulk package containers. The laboratory was asked to sample the inside, outside, and flaps of the cardboard box, as well as the polyethylene bag. Two samples were found to harbor the offending microorganism, which allowed the original test to be invalidated and retesting to be performed. Had the firm not been within the core observing the testing, the source of the contamination might not have been found, and the product would have been rejected. The laboratory discontinued purchasing in bulk and returned to purchasing individually wrapped sample containers. Witnessing the operation enabled appropriate CAPA to be conducted: the corrective action was to sample the cardboard and invalidate the original testing and retest according to USP <71>; the preventive action was to purchase individually wrapped sample containers.

OBJECTIONABLE MICROORGANISMS

Many companies believe that if their nonsterile product meets the USP requirements, it will be safe from FDA dispute. This belief is correct, and the manufacturer is responsible for all contents of its drug product. Should question arise over the appropriateness of a particular organism, the manufacturer is expected to have a justification for the presence of that organism, preferably as part of the batch release document. 21 CFR 211.113 requires that medicines should be "free of objectionable microorganisms."

USP <1111> provides little specific guidance other than the following: "The significance of microorganisms in nonsterile pharmaceutical products should be evaluated in terms of the use of the product, the nature of the product, and the potential hazard to the user." The USP recommends that certain categories be routinely tested for total counts and specified indicator microbial contaminants, including natural plant, animal, and some mineral products for *Salmonella*, oral liquids for *E. coli*, topicals for *P. aeruginosa* and *S. aureus*, and articles intended for rectal, urethral, or vaginal administration for yeasts and molds.

Other than these listed microorganisms, no others are mentioned as objectionable in the USP. FDA has a publication of objectionable microorganisms that is used for the food industry but can also be applied to pharmaceutical sciences: the *Bad Bug Book*. This publication details numerous microorganisms that may present a health hazard, and whose presence may invoke a rejection or recall of a product or excipient; however, the difficulty facing the industry is determining how objectionable a microorganism actually is. For products that are not labeled as "sterile," circumstances exist where an objectionable microorganism could be acceptable for product release based on a risk assessment analysis.

Similarly to an MDD, little guidance exists on conducting a risk assessment for objectionable microorganisms. FDA suggests that once all organisms grown in the total count studies (total aerobic as well as total yeast and mold) are identified, a qualified microbiologist would conduct a risk analysis on the presence of those organisms in the medication. This risk analysis should incorporate a minimum of five separate analyses (Figure 20.2).

- Absolute numbers of organisms seen
- Microorganism's characteristics
- Product characteristics
- Route of administration
- Potential impact on patients

Figure 20.2 Risk analysis should incorporate a minimum of five separate analyses.

Absolute Number of Organisms Seen

Most products have microbial specifications for total bacterial counts. When the counts exceed these limits, even if the organisms identified are nonpathogenic or nonobjectionable, an MDD should be conducted, most likely resulting in the rejection of the batch and possibly other batches. Although high numbers of nonpathogenic organisms may not pose a health hazard, they may affect product integrity. An unusually high number of organisms seen in the product may indicate a problem during the manufacturing process, or an issue with an excipient or API. The high bacterial counts may indicate that the microorganisms are thriving in the product. For a preserved product, this indication could suggest that the product's preservative system is not functioning, is missing, or incorrectly formulated.

THE CHARACTERISTICS OF THE MICROORGANISM

The characteristics of the microorganism can be determined by textbook, library, or Internet searches, and should include synonyms of the organism. The *Bad Bug Book* is an excellent reference for determining whether the organism is a known pathogen or an objectionable. The potential for the organism to cause spoilage of the product is another important factor. If the substances used by the microorganism for growth are found in the product, the risk is great that it will degrade the product, supporting the rejection of the batch and associated batches.

A microorganism is *objectionable* if it has the potential to degrade the product's stability. Evaluate the microorganism's tolerance to unusual conditions, such as low or high pH, high salt concentration, high sugar concentration (osmotic conditions), low water activity, and growth temperatures. Organisms with a proclivity for harboring plasmid-mediated antibiotic resistance may be considered objectionable, particularly if the product is to be used in vulnerable patients with compromised immune systems.

Product Characteristics

The characteristics of the product affect the ability of the microorganism to grow within it. Some of these include anhydrous versus water based (that is, enough free water to support microbial growth) and container design and closure (that is, the container should be designed to minimize contamination and subsequent spoilage and should also retard access to the environment and prevent contamination from the environment). Special consideration should be given to an anhydrous medication's exposure to water, providing the potential for microbial proliferation. And a review of the production records, environmental monitoring trends, and product complaints is also warranted, and comprises part of an MDD investigation.

Route of Administration

Some microorganisms that could be tolerated in a topical medicine may cause severe distress to a patient if taken orally or inhaled. Similarly, a medication orally administered can tolerate some microorganisms that would be devastating in a medication meant to be applied topically to abraded skin or to rashes. Inhalants, although not required to be sterile, are a particularly sensitive area, and great care should be taken in classifying any contaminate as "nonobjectionable."

Patient Population

The targeted patient plays a critical role in assessing the risk of a potentially objectionable microorganism. Microorganisms isolated from products intended for pediatric or immune-compromised patients present a greater risk to these patients than microorganisms in those products intended for use in relatively healthy patients. *Burkholderia cepacia* in cough medicine poses a greater risk than if it was found in a topical cream. Although the manufacturer cannot control patient abuse or off-label use of the product, the reasonable use of it should be considered as part of the risk analysis. Certain patient populations may be exposed to increased risk if they use the medication harboring the particular microorganism.

Chapter 21

Instrument Management

INSTRUMENT CONTROLS

> Apply operating procedures for instrument identification, classification, qualification, calibration, and preventive maintenance. (Apply)
>
> **Body of Knowledge II.C.1**

All laboratory instrumentation must be maintained to ensure proper functioning. Instrumentation is numbered so it can be easily identified during testing. If an instrument is used in a *good manufacturing practices* (GMP) environment, it will have the appropriate qualification and calibration performed. In a research and development environment, instrumentation does not require routine calibration, although a minimum standard for verification should be established to ensure data accuracy.

As an instrument is acquired in a GMP environment, it should be assigned a unique identification number. Some companies will use an abbreviation of the instrument name with a set of numbers for the unique identification (for example, balances will be numbered as Bal 22, Bal 23, Bal 24). The associated documentation links this unique identification to the instrument type, model, and serial number. Documentation includes the date an instrument was put into service and when it was retired.

Instrumentation can be classified into one of three categories: minor, intermediate, or major. USP <1058> *Analytical Instrument Qualification* uses the designations A, B, and C. Appropriate categorization of instruments depends on the intended use of the end user. Minor instruments that do not perform any measurements, such as stirring plates, water baths, and sonicators, do not require any calibration or qualification. Intermediate instruments such as thermometers and timers have little or no configurability, and typically perform a single task. Qualification for intermediate instruments is typically limited to *installation qualification* (IQ) and *operational qualification* (OQ), with the OQ consisting of calibration. Major instruments are those that are multifunctional and configured with multiple components, such as *high-performance liquid chromatography* (HPLC), *gas chromatography* (GC), *infrared* (IR), and *ultraviolet-visible* (UV-Vis) spectrophotometers. These instruments require full qualification (IQ, OQ, and *performance qualification* [PQ]).

Based on the instrument category, IQ, OQ, and PQ may be required when a new instrument is introduced into the laboratory, a critical part is replaced, or instrumentation is moved from one location to another. A protocol defines the purpose or objective, test procedures to be followed, parameters to be evaluated, and the criteria necessary to demonstrate acceptability. This may be done through the change control process.

The IQ is conducted to ensure that the instrument and software, if applicable, were installed in the proper environment and manner as intended by the manufacturer. The OQ is conducted to ensure that the instrument and software, if applicable, operate as designed and generate accurate data. The PQ is conducted to ensure that the instrument and software, if applicable, meet the end user–specific requirements and function as intended in the production environment. A report summarizes the test results for the qualifications and any deviations. Upon completion of the qualification, the instrument can be put into service.

For instruments in the GMP environment, routine maintenance and calibrations are required. The maintenance and calibration schedules for laboratory instrumentation can be established based on severity of environment, severity of use, frequency of use, durability of the instrument, accuracy of measurement required, precision of measurement required, and history of maintenance and calibration. Each type of instrumentation should have a procedure that specifies its operation, maintenance, and calibration. The procedure must specify what activity should be conducted, what the acceptance criterion is, and the frequency. An instrument log book may be used to track the usage, repair, and calibration of the instrument. All maintenance and calibration information, which includes reagents or standards used, must be documented. When an instrument requires repair or calibration, it should be flagged as out of service in some manner, such as with an "out of service" sticker. If an instrument fails the calibration, an impact assessment must be performed. During the impact assessment, all data generated since the last successful calibration must be evaluated. Upon completion of the repair or calibration, the out-of-service indicator is removed and the instrument is placed back into service. If the instrument was being calibrated, a new calibration sticker, which designates the next calibration due date, is applied.

INSTRUMENT CALIBRATION

> Determine that instruments are calibrated within the specified range of operation and that they are accurate and precise. (Apply)
>
> **Body of Knowledge II.C.2**

Calibration is the act of adjusting the accuracy of a measuring instrument by comparison with a standard. Prior to the purchase of an instrument, user requirement specifications should be determined for the laboratory defining the needed range, accuracy, and precision of the instrument. Instruments should be qualified to show that they can meet the user requirement specifications. For example, if a temperature must be measured to accuracy +/– 0.1°C, then a thermometer calibrated to +/− 0.5°C would not be appropriate. Accuracy is determined by comparing the measurement of a standard by the unit under test (UUT) to a standard (NIST traceable standards are ideal). Precision is determined by examining the variability of multiple measurements. Laboratories may elect to calibrate the instrument to a subset of its designed operating range if the entire range is not needed for testing.

Most laboratories calibrate instruments to be three or more times as accurate as the testing requirement. This allows for additional uncertainty due to testing variation. Instruments should be labeled with the range and accuracy verified during the performance qualification. Instrument qualification testing should verify that measurements meet predetermined accuracy and precision specifications throughout the intended range of use. Instrument users should verify that the range and accuracy of the instrument meets the testing requirements prior to use.

Chapter 22
Specifications

Specifications are an important element of the overall quality strategy of ensuring that the materials used to produce the final drug product, and the drug product itself, are acceptable for their intended use. It is important to understand what constitutes a specification, how specifications are developed, the different types of specifications, and what is required to change a specification once it is approved (either by a regulatory agency in the case of submitted drug products, or as part of the initial validation effort for non-submitted drug products).

TYPES OF SPECIFICATIONS

> Determine whether approved specifications exist for raw materials, intermediates, packaging components, labels, and finished products. (Analyze)
>
> **Body of Knowledge III.D.1**

The International Conference on Harmonization (ICH) defines *specifications* as "a list of tests, references to analytical procedures, and appropriate acceptance criteria, which are numerical limits, ranges, or other criteria for the tests described." Specifications are established (proposed and justified) by the manufacturer and serve as the basis for quality evaluation. "Conformance to specifications" means that the materials and/or drug product, when tested according to the listed analytical procedures, will meet the established acceptance criteria. It is important to note, however, that specifications are established to confirm the quality of these items and are not intended to provide complete characterization. As a result, the specifications should focus on those characteristics that are most critical to ensuring the safety and efficacy of the particular material or drug product.

DEVELOPMENT OF SPECIFICATIONS

Specifications for raw materials, intermediates, packaging components, and the finished product are first established during the drug development process. Like the drug product itself, specifications evolve over time as more information about the product and manufacturing process is generated. After drug product approval and/or launch, the experience and data generated during commercial manufacture of the drug product may result in changes to the established specifications.

At the early stages of drug product development, pharmaceutical companies have limited knowledge of the product. Different raw materials are being assessed, different manufacturers of the materials may be involved, and the manufacturing process has not been established. This limited knowledge of the product and manufacturing process is further compounded by limited experience in the clinic with the investigational drug product. As a result, the specifications that are established for the drug product during the early stages of development should be focused on product safety.

As drug product development progresses, and more information is gathered about the product, its components, and the processes used to manufacture the product, additional quality characteristics are identified and incorporated into the specifications. The final specifications that are ultimately developed are the quality standards (that is, tests, analytical procedures, and acceptance criteria) that are used to assess the quality and performance of the drug substance, excipients, container closure systems, intermediates, and drug product to ensure safety and efficacy. Final specifications are approved as part of the regulatory approval process or must be approved before validation of the drug product manufacturing process for non-submission drug products. Once they are approved, they are subject to change control processes to ensure that proposed changes to the specifications are adequately justified and that no negative impact on product quality, safety, and efficacy will occur.

CRITERIA FOR ESTABLISHING SPECIFICATIONS

It is important to consider three categories of requirements when establishing specifications for a particular material, component, or product:

- *Compendial requirements.* Quality attributes, test methodology, and acceptance criteria that have been established by national or international compendia. Pharmacopoeial requirements establish the base or the minimum quality standards for evaluating a material. Additional quality attributes and tests may be necessary to ensure that the material is fit for its intended use.
- *Regulatory requirements.* Quality attributes and acceptance criteria that are well defined by regulatory agencies (for example, specifically identified in regulatory guidance).
- *Product-specific requirements.* Quality attributes and acceptance criteria that are unique and critical to the material or product itself.

RAW MATERIAL SPECIFICATIONS

Drug Substances

According to ICH guidelines, tests and acceptance criteria for drug substances typically fall into one of two categories: *universal tests* (tests applicable to all substances)

Table 22.1 Typical universal and product-specific tests for chemical drug substances.

Universal tests	Description	Qualitative statement about the state (for example, solid, liquid) and color of the drug substance
	Identification	Test(s) to discriminate between compounds of closely related structure that are likely to be present; specific for the new drug substance; if the new drug substance is a salt, identification testing should be specific for the individual ions
	Assay	A specific, stability-indicating test to determine the content of the new drug substance
	Impurities	Test(s) for organic and inorganic impurities and residual solvents
Specific tests	Physiochemical properties	Properties, such as pH of an aqueous solution, melting point/range, and refractive index; test(s) determined by the physical nature of the drug substance
	Particle size	Test to determine the particle size distribution
	Polymorphic forms	Test(s) to determine the crystalline form(s) of the new drug substance
	Water content	Test (for example, loss on drying, Karl Fischer titration) for residual moisture in the drug substance; may be important for hygroscopic substances or substances known to degrade due to moisture
	Inorganic impurities	Inclusion of tests (for example, residue on ignition) is based on the knowledge of the manufacturing process and the types of impurities likely to be present
	Microbial limits	The type of microbial test(s) and acceptance criteria are based on the nature of the drug substance, method of manufacture, and the intended use of the finished dosage form manufactured with the drug substance
Additional considerations	Chiral drug substances	New drug substances that are optically active may need specific identification testing

and *product-specific tests* (tests unique to the substance itself or the specific form of the drug substance). Table 22.1 identifies the typical universal and product-specific tests for a chemical drug substance. A similar outline of tests for biologic drug substances can be found in ICH Q6B. Additional tests may be critical, and the manufacturer of the drug substance should have fully characterized the material and is in the best position to identify the tests that are needed to establish the quality of the material.

Inactive Ingredients (Excipients)

Excipients are defined as ingredients added to the drug formulation for purposes other than the therapeutic or medicinally active effect. They are largely inert substances that are added to the formulation to aid in delivery of the active ingredient(s).

Well-established and well-characterized excipients are typically covered by a national or international compendium, and most pharmaceutical companies will adopt the compendial specifications and test methodology. When establishing specifications for compendial items, however, it is important to note that there may be specific attributes of the inactive ingredient that are important for the company's product and/or manufacturing process but that are not identified as part of the compendial specifications. A good understanding of the manufacturing process, as well as knowledge of the inactive ingredient's function in the finished dosage form, is necessary to ensure that the compendial specification captures all of the tests necessary to ensure that the inactive ingredient will perform as expected in the final drug formulation.

INTERMEDIATES

Specifications for intermediates are necessary to ensure that the material is acceptable for continued processing. Full characterization of the intermediate may not be necessary, however, and the specifications (appropriate tests and acceptance criteria) identified should be those that provide additional assurance that the final product will conform to established specifications.

PACKAGING COMPONENTS

Packaging components are initially qualified during drug development. It is through the development process that it is shown that the selected packaging components adequately protect the product, that they do not interact with the product, and that they are constructed of materials that will not leach harmful or undesirable substances into the product during routine storage.

Once the packaging components are qualified (that is, shown to be suitable for their intended use), specifications must be established for routine quality control. Primary packaging components—those that are intended to have direct product contact—are most critical, and specifications need to be established to ensure batch-to-batch consistency of these components.

According to the United States Food and Drug Administration (FDA) guidance *Container Closure Systems for Packaging Human Drugs and Biologics, quality control* (QC) specifications for packaging components typically include physical characteristics (dimensional criteria, physical parameters, performance characteristics) and chemical composition (colorants, material of construction). For most drug products, a pharmaceutical company may accept a packaging component lot based on acceptable results documented on a *certificate of analysis* (COA) or *certificate of compliance* (COC) from the packaging component vendor, and the performance of an appropriate identification test. This type of reduced testing is acceptable provided that the vendor's test data are periodically validated/revalidated (21 CFR 211.84). Additional specifications may be appropriate for components used in the packaging of injectable or inhalation products.

FINISHED DRUG PRODUCT

Similarly to the requirements for drug substances, tests and acceptance criteria for drug products typically fall into one of two categories: universal tests (tests applicable to all drug products) and product-specific tests (tests unique to the drug product itself and/or the dosage form). Table 22.2 describes the analytical tests that are typical to chemical drug products.

Other dosage forms (for example, oral liquids, parenteral products) will have tests specific to the type of product. Other tests may be critical, in which case the manufacturer of the drug product should have fully characterized the product during drug development and thus will be in the best position to identify the tests that are needed to ensure the quality of the finished drug product.

TEST DATA AND SPECIFICATIONS

Compare test data with specifications to determine whether raw materials, intermediates, packaging, labels, and finished products meet requirements. (Analyze)

Body of Knowledge III.D.2

Acceptance Limits. The acceptance limits for specifications of the different quality characteristics are established by taking into account all significant elements related to the quality of the drug (medicinal) product (constancy of its characteristics), its activity (level of active constituent), and, if necessary, its safety (risk of microbial contamination, breakdown products).

Acceptance Limits of Pharmacotechnical Parameters. For most pharmacotechnical specifications, the European Pharmacopoeia, or failing this, the national pharmacopoeias of the member states, describes general test procedures with, in some cases, standards or maximal limits. In the context of these specifications, it is necessary to establish minimal and/or maximal limits specific and adapted to the medicinal product that is the object of the marketing authorization, to guarantee the reproducibility of the finished product at manufacture.

Maximum Acceptable Deviation in Content of Active Substances. Directive 75/318/EEC states that the maximum acceptable deviation in the active substance content of the finished products shall not exceed 5% at the time of manufacture. On the basis of the stability tests, the manufacturer must propose and justify maximum acceptable tolerance limits in the active substance content of the finished product up to the end of the proposed shelf life:

Release limits of 5% are acceptable without further justification.

- Release limits wider than 5% would need to be justified in the development pharmaceutics documentation with experimental results, which are normally based on a confidence level of 95%. The wider limits also include both the variation of the production and of the test procedure for the assay.

Table 22.2 Typical analytic tests for chemical drug products (for tablets and capsules).

Universal tests		Description	Qualitative statement about the state (for example, solid, liquid) and color of the drug substance
		Identification	Test(s) to discriminate between compounds of closely related structure that are likely to be present; specific for the new drug substance; if the new drug substance is a salt, identification testing should be specific for the individual ions
		Assay	A specific, stability-indicating test to determine the content of the new drug substance
		Impurities	Test(s) for organic and inorganic impurities and residual solvents
Specific tests	Tablets/capsules	Dissolution	Test to measure the release of drug substance from the drug product
		Disintegration	Testing is most appropriate when a relationship to dissolution has been established or when disintegration is shown to be more discriminating than dissolution
		Hardness/friability	If the characteristics of hardness and friability have a critical impact on drug product quality, acceptance criteria should be included in the specification
		Uniformity of dosage units	This term includes both the mass of the dosage form and the content of the active substance in the dosage form
		Water content	Test for water content should be included and justified with data on the effects of hydration or water absorption on the drug product
		Microbial limits	Tests for the total count of aerobic microorganisms, the total count of yeasts and molds, and the absence of specific objectionable bacteria should be conducted except when its components are tested before manufacture and the manufacturing process is known, through validation studies, not to carry a significant risk of microbial contamination or proliferation
Additional considerations		Oral liquids	Tests may be required for: • Uniformity of dosage units • pH • Microbial limits • Antimicrobial preservative content • Antioxidant preservative content • Extractables • Alcohol content • Dissolution • Particle size distribution • Redispersibility • Rheological properties • Reconstitution time • Water content

Table 22.2 *Continued.*

Additional considerations	Parenteral drug products	Tests may be required for: • Uniformity of dosage units • pH • Sterility • Endotoxins/pyrogens • Particulate matter • Water content • Antimicrobial preservative content • Antioxidant preservative content • Extractables • Functionality testing of delivery systems • Osmolarity • Particle size distribution • Redispersibility • Reconstitution time

- Use of inadequate manufacturing procedures or inadequate test procedures (low precision) will not be accepted as a justification for wider release limits.
- To satisfy the 5% requirements, it is left to the responsibility of the manufacturer to apply an adjustment of the amount of active substance in the production of the finished product. In such a case, the overage must be clearly stated. The release limit will remain within 5% of the stated content.
- References to pharmacopoeias cannot normally be accepted as a justification for wider limits, as monographs do not describe a defined composition of the medicinal product and do not pertain to release, but to recontrol by official laboratories over the whole shelf life of the product.
- Exceptionally, for certain products with a well-known degradation process and that pose no safety problems (for example, vitamins), an overage at release can be tolerated. This overage, the aim of which is to guarantee a sufficient level at the end of shelf life, must not cause an excessive level at release, and release limits must be adapted accordingly.
- At recontrol by official laboratories, the release limits of the manufacturer will be taken into consideration despite the fact that the limits used at recontrol may not be identical to the release limits of the manufacturer (due to interlaboratory variability). Where recontrol tests are carried out by another laboratory, the test procedures must be validated in their hands.
- Overfilling to guarantee the delivery of the theoretical unit dose must be justified. The acceptance limits for determination of unit content are adapted in consequence.

Acceptance Limits for Excipients

Excipients that affect the bioavailability of an active substance must be the object of a quantitative determination in each batch unless bioavailability is guaranteed by other appropriate tests, established on a case-by-case basis as a function of development studies. In the case of preservatives, content limits of 90% to 110% at release should be acceptable without further justification, except in special cases. On expiry, the lower limit of antimicrobial preservatives may be reduced subject to the results of satisfactory preservative efficacy testing. For chemical preservatives (antioxidants), the lower limit may be considerably lower than 90% during shelf life because of the preferential degradation of these agents.

Conformance to Specifications

Analytical results observed in the laboratory are compared to the acceptance criteria in the specifications to determine conformance. Numerical values should not be rounded until the final calculation. Acceptance criteria are significant to the last digit shown. Printed labels should be examined for identity and conformance to the specifications of the master production record. Packaging and labeling materials should be examined to determine conformance to specification prior to use.

Certificate of Analysis issued for raw materials, intermediates, APIs, or drug products should include each test method performed, the acceptance criteria, and the actual numerical result if applicable. Additionally, the COA should include a reference to any associated OOS investigations with the collected data.

ADDITIONAL CONSIDERATIONS

For all incoming materials (those purchased from third-party suppliers), the specifications should clearly identify the approved suppliers for the raw material in question. Documentation specific to the approved manufacturing sites is important, as many suppliers have multiple manufacturing sites, and all of them may not be approved to supply the material. If the material is provided by a broker or agent for the manufacturer, care should be taken to ensure that the materials received are coming from the approved supplier and the approved manufacturing sites.

In addition to documenting the approved supplier(s), the specifications should include a description of the approved shipping containers (these are typically the containers that the supplier has used to establish the raw material shelf life) and any special environmental conditions for maintaining the quality of the material during storage. Specifications should include the name and internal (company-specific) code for the material(s); the name also should include any reference to a pharmacopoeial monograph (for example, United States Pharmacopeia [USP]).

Finally, information should be provided that is related to the sampling plan to be used to sample the raw materials, and any special instructions necessary to protect the individual sampling the material, as well as any precautions to protect the material itself. Directions for sampling may be included in the specification (or reference provided to procedures that describe the sampling process).

SPECIFICATION REVISION

> Review and update specifications when methods are revised, or compendia are changed. (Evaluate)
>
> **Body of Knowledge III.D.3**

Post-approval changes (or post-launch changes in the case of non-submission products) in the laboratory are not uncommon. As the manufacturer gains experience with the drug product and the analytical methodology, changes may be necessary to ensure that the methods are robust and are capable of producing consistent and reliable analytical results. Changes made in the manufacturing process, or in the pharmaceutical product itself, may have an impact on the analytical methodology; revisions to the methodology may be necessary to accommodate these changes.

Compendial changes are not uncommon. Pharmaceutical manufacturers should have mechanisms in place to track proposed compendial changes and to evaluate the impact of the changes on applicable drug products. Pharmaceutical manufacturers need to have a robust change control process in place to capture and incorporate necessary changes.

Changes to a packaging component (for example, changes to the materials of construction or the manufacturing process used to produce the component) must be assessed and, if appropriate, reported to the applicable regulatory agency. The filing requirements vary depending on the type and impact of the change. The finished dosage form manufacturer is responsible for conducting the appropriate studies to demonstrate that the change in question does not have an adverse impact on the drug product. Specifications are updated to incorporate the change(s) once the change has been shown to be acceptable and the change has been approved by the regulatory agency, if required.

Chapter 23

Laboratory Record-Keeping and Data Requirements

RECORD-KEEPING REQUIREMENTS

> Identify and review record-keeping requirements for data acquisition systems to ensure data integrity. (Apply)
>
> **Body of Knowledge III.E.1**

The laboratory is responsible for testing incoming material and finished products. The laboratory uses their approved specifications and test methods, which have been validated or verified for suitability, to perform the required testing. When testing is conducted, all the pertinent information must be documented in real time (as the test is conducted). The analyst must use permanent ink, blue or black, when making data entries. The analyst is responsible for accurately recording the pertinent information on a controlled document, which can be a standard laboratory notebook or controlled analytical data forms or a validated software package. Analytical data forms can be blank, so the analyst is required to document everything, or prepopulated with pertinent information, so the analyst only has to fill in identification numbers, reagent batch numbers, calibration dates, and so on.

The analyst records the date the testing is being started. The batch number, and sample identification number are entered on the controlled document, and the method number and revision number recorded. As a sample is weighed, the weighing information is documented. If a printer is attached to the balance, the analyst ensures that the sample identification, balance number, and calibration date are recorded on the weigh tape (Figure 23.1). If the weigh tape is thermo paper, the analyst must make a copy of the tape, which will ensure that the information captured on the tape is not lost as the thermo paper fades over time. Both the original weigh tape and the copy, if applicable, are securely adhered to the controlled document. One can securely attach raw data to the controlled document by initialing and dating across the edge of the raw data in a manner such that the initials/date are partly on the attachment and partly on the controlled document, then all four sides must be taped down. Clear tape that is not easily removed must be used, and the attachment must not cover any information on the page.

ADF09192803
TM-450-02

Bal 25	Cal Due: 4-12-16
05/11/2009	12:42
ID	09-1928
1	1.7288 g
2	1.7653 g

Figure 23.1 Raw data attachment example.

Preparation of solutions to be used in the analysis can be documented on the controlled document or in a log book designated for such preparations. Most laboratories have a solution preparation book where all solution preparations are documented. All pertinent information, such as chemical name, manufacturer, grade/purity, lot number/identification number, and expiration date, pertaining to reagents used and the method used to prepare the solution are documented. A unique identification number is assigned to the solution along with the appropriate expiry date. The solution is labeled with the unique numbers, solution description, and expiration date for future use. As the analyst follows the test method, all standards, if applicable, and sample preparations must be documented. If the analyst is making further dilutions, the dilution scheme must be documented. If the analyst uses timers, thermometers, sonicators, or water baths, all associated identification numbers and calibration/service due dates are recorded.

Once the sample preparation is complete, the analyst prepares the instrument used in the analysis. The analyst verifies that the instrument is within calibration; if not calibrated, the analyst ensures that the calibration is completed before use. The analyst captures the instrument identification number and calibration due date on the controlled document. The analyst conducts the analysis. If the instrument requires a sample sequence, the analyst will complete the necessary data entry before starting the analysis. Most computerized systems have a set method of analysis in the system; the analyst will verify that the correct method is used for the analysis. In most data acquisition systems, only system administrators with a certain level of authority are allowed to set up methods and revise them. The analyst conducting routine testing only has access privileges to create run sequences. The analyst either documents or prints out the parameters used in the analysis. All printouts generated must be labeled with notebook and page/controlled document number, sample identification, number of pages, and the analyst name/date. Once the run is complete, the analyst can evaluate the data. In some instances, the data acquisition system used has validated templates in which the data require no further calculations. If there is no associated template, the analyst may need to manually calculate the results or use a nonvalidated template in Excel. A sample calculation should be included to demonstrate how the analyst calculated the result. Once the analyst has completed the calculations, all paperwork can be completed, including reporting the final results on an analytical result form or in a data management system.

~~stoed~~
stored
ABC 4-12-16 spelling

Figure 23.2 Correct documentation of an error example.

The analyst must review his or her work to ensure there are no obvious errors and must initial/date it to signify that the work is complete. If errors were made, the error must be crossed out with a single line and initial/date/explanation included, then the correct information can be added (Figure 23.2). If one digit of a number or one letter of a word is incorrect, the analyst must cross out the entire number or word and make the correction. Any deviations from the method must be documented either as a planned or unplanned deviation and approved by management.

The analyst can begin to clean up the work area, but keeps all glassware in the bench area until it can be verified by a second analyst. All solutions are maintained until the analysis is complete and known to be correct.

Once the analysis is complete and documented, a second individual must review and understand the data and calculations. The reviewer ensures that all documentation is complete and accounted for. The data reviewer must ensure that the correct method and revision were used to perform the analysis. The data reviewer checks that all solutions used were prepared correctly and that all materials used were within expiry dating. The reviewer verifies that all instrumentation used was identified and within calibration. The data reviewer ensures that all the documentation is clear, and that comments are appropriate and easily understood. All laboratory investigations and deviations are reviewed, which will ensure that laboratory management has approved them. Any calculations made that are not part of a validated template must be verified. The data reviewer initials and dates the controlled document and any attachments to signify review. If any information needs to be corrected or added, the data reviewer does not sign off until the analyst has made the correction/addition. When adding information after the analyst has added the final sign-off, he or she must designate the additional information with a late entry, initials, and date to flag the additional information.

If a company has a validated electronic system, data are captured electronically. Such validated electronic systems include instrumentation data acquisition systems or electronic notebooks. These systems have routine backups and built-in security. System administrators are designated to maintain and qualify the system, and change controls are used to document any changes to the validated system. Each user must have a unique user name and a password to access the system. For a review, some validated electronic systems will flag the errors, such as *out of specification* (OOS) or *out of calibration* (OOC). Calculations do not need to be verified each time, since the system was validated. All analyst and administration activities are documented in an audit trail that the reviewer can view. The audit trail review will indicate any suspicious activity that may require additional investigation.

Upon completion of the final data review, laboratory management conducts a final review to approve the data. Management ensures that all results meet the approved acceptance criteria and that all documentation is complete. Upon completion of the management assessment, laboratory management approves the data.

Once the raw data have been reviewed, they will be archived. If raw data are on paper, all raw data will be filed in a document control system. If the raw data have been collected electronically, the electronic information will be saved to a server that has routine backup and scheduled maintenance. Each company establishes its own document retention policy. Typically, raw data are maintained for seven years.

SIGNIFICANT DIGITS AND ROUNDING

Significant digits relate to the precision of a measurement. Accuracy and precision are concepts that are very important in laboratory calculations. Precision and accuracy are related, but they refer to different aspects of a measured or calculated value. *Accuracy* refers to how correct a value is. *Precision* refers to how many digits are used in the number. Figure 23.3 graphically represents the difference between accuracy and precision. For consistency, use of significant digits and rounding should be specified in a procedure.

Rules for Significant Digits/Figures

- Exact numbers have an infinite number of significant digits, but they are generally not reported.
- Defined numbers have an infinite number of significant digits.
- All digits 1 through 9 are significant.

  Example: 5.678 contains four significant digits.
- Zero is significant when it is between two non-zero digits.

  Example: 708, 40.5, 102, and .308 contain three significant digits.
- A terminal zero to the right of a decimal point in a number greater than 1 is significant.

  Example: 325.0, 42.30, and 7.250 contain four significant digits.
- Zeros to the right of the decimal point are significant if they are at the end of the number.

  Example: 1750.0 contains five significant digits.
- A zero used to fix a decimal point in a number less than 1 is not significant.

  Example: 0.123, 0.0753, and 0.00619 contain three significant digits.

Maintaining Significant Digits in Calculations

- For addition or subtraction, the answer must be rounded off to contain only as many decimal points as are in the value with the least number of decimal points.

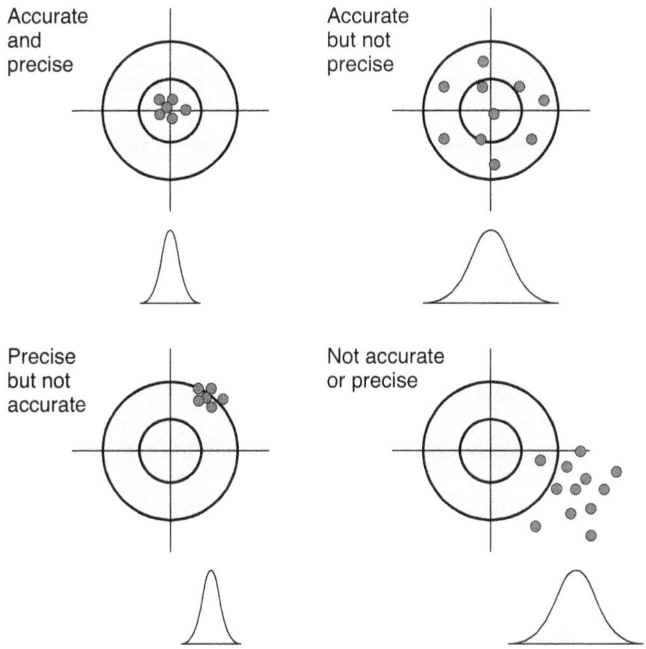

Figure 23.3 Accuracy versus precision.
Source: Mark A. Durivage. 2016. *Practical Design of Experiments (DOE)*. Milwaukee: ASQ Quality Press.

- For multiplication or division, the answer must be rounded off to contain only as many digits as are in the value with the least number of significant digits.
- When calculating the logarithm of a number, retain in the mantissa (the number to the right of the decimal point in the logarithm) the same number of significant figures as there are in the number whose logarithm is being found.
- When calculating the antilogarithm of a number, the resulting value should have the same number of significant figures as the mantissa in the logarithm.
- In combined math operations that consist of multiplication/division and addition/subtraction, the answer must be rounded off to contain only as many significant figures as are in the value with the least number of significant digits.

Rounding with Significant Figures

Once the correct number of significant digits to keep has been determined, identify the next digit. If this digit is > 5, then round the last significant digit up one. If this digit is < 5, then keep the last significant digit the same.

RECORD REVIEW

> Ensure data integrity and prevent loss of data by reviewing laboratory records including audit trails of electronic data to detect errors, falsification, and fraud. (Evaluate)
>
> **Body of Knowledge III.E.2**

Data integrity refers to the completeness, consistency, and accuracy of data. ALCOA is a useful acronym to describe data integrity. All *good manufacturing practices* (GMP) laboratory data should be Attributable, Legible, Contemporaneously recorded, Original, and Accurate. The FDA issued the guidance "Data Integrity and Compliance with Drug GMP," effective December 2018, detailing current recommendations for data integrity, which addresses many data integrity concerns.

Data integrity is critical and must be maintained throughout the GMP data life cycle, from creation, through modification, to archiving. Metadata is data about data. Metadata contains contextual information about how data was collected such as by whom, the time, and conditions. Metadata should be maintained along with the associated data throughout the record's retention period.

Audit trails are secure computer-generated time-stamped records of the events that occurred during the creation, modification, and deletion or retention of an electronic record. Personnel responsible for reviewing records should review the associated audit trail at the same time the record is reviewed. Review of the audit trail must be documented by the reviewer's signature and date, either electronically or handwritten.

Audit trails should capture all modifications to data. All data—including obvious errors and failing, passing, and suspect data—must be in the current good manufacturing practices (CGMP) records that are retained and subject to review and oversight. An investigation with documented, scientifically sound justification is necessary for data to be invalidated and not used in determining conformance to specification. Testing and sampling with the goal of achieving a specific result or overcoming an unacceptable result is not consistent with GMP practices.

Processes should be designed so that a record cannot be created or modified without a record of the modification. Reviewers should always check for the presence of orphan data. Orphan data is data that has been separated from the reported data and results, and it is not GMP compliant. Data may often be filtered by date collected and/or analyst to detect orphan data.

Computer systems used in the collection, processing, or retention of GMP data should be validated for the intended purpose. Changes to records should only be permitted by authorized personnel. Individual unique passwords should be maintained, and the system should record the time, date and user making any modification as part of the audit trail. Administrative control of computer systems with the rights to alter or delete records should be assigned to different personnel than those responsible for the record contents. Blank forms should be controlled by the quality unit and incomplete or erroneously completed forms should be retained as part of the complete record.

An accurate and true copy of data should be maintained throughout the record's retention period. For static unchangeable data this can be in the form of a paper record or an electronic record. For dynamic data that can be reprocessed records must be maintained electronically so that the dynamic nature of the record format is maintained. Electronic records are subject to regulatory inspection in the same manner as paper records.

CERTIFICATES OF ANALYSIS (COAS)

> Review Certificates of Analysis (COAs) to ensure they are complete, internally reviewed, and appropriately retained. (Apply)
>
> **Body of Knowledge III.E.3**

Once the data are approved, a *certificate of analysis* (COA) is generated. If data are stored in a validated electronic system, a COA may be generated automatically. The COA contains descriptions of the material tested including sample or batch number. The COA lists all tests performed, with the method used and the actual result, the specification, and a statement of compliance to the specification. Numerical specifications should be rounded to the same number of significant figures as the specification. In some cases, a GMP compliance statement may be included. The COA should include a reference to any associated OOS investigations with the collected data. The COA may contain an electronic signature or a wet signature that indicates all results have been reviewed or approved. COA information must be reviewed and approved to ensure accuracy. If a mistake is found, and the COA must be reissued a revision history is included in the document. In most cases, the COA is produced upon request from a customer, and the original is sent to them. All records related to drug production should be maintained for one year after the expiration date of the product or five years after release, whichever is longer.

Chapter 24

Laboratory Handling Controls

All samples and reagents coming into the laboratory must be checked and logged in before use. Laboratory procedures should be in place so materials are consistently received.

SAMPLE HANDLING

> Determine whether samples are identified and handled in accordance with requirements including name, sample identification, and chain of custody. (Apply)
>
> **Body of Knowledge III.F.1**

SAMPLES FOR PRODUCTS MANUFACTURED IN-HOUSE

During the manufacturing process for a material, operations personnel remove a representative composite sample from the manufacturing line as designated in their batch record or associated paperwork. Depending on the nature of the material being manufactured, the sample may come from an in-process manufacturing step, completed semifinished good, or final packaged product.

The batch record, or associated paperwork, should detail how and when to pull the sample. The sample container is typically identified in these procedures. The operator should use the appropriate sampling method as defined in the written instructions. The sample container should be labeled with at least the product name, batch number, and sampled-by date. Depending on the nature of the sample, other pertinent identifying information may be included on the label. If the manufacturing of the batch occurs over several shifts, staff performing the sampling should initial and date their activities on the sample label. At the completion of the sampling, the sample is sent to the laboratory. The sample is logged into the laboratory system and is scheduled for testing.

Stability Samples

Depending on regulatory commitments as well as the nature of the product being manufactured, stability testing will vary. Once a packaging batch is identified for a stability study, operations personnel pull an adequate amount of packaged product throughout the packaging run to meet the study requirements. At the completion of the sampling, the samples are transported to the laboratory and placed under appropriate environmental stability testing conditions. At the appropriate intervals, a sample is pulled and scheduled for testing.

Laboratory Receipt of Samples

The laboratory may receive samples for analytical quality testing from within the company or from contract clients. The sample receiving area should be in a controlled access environment. Personnel should first examine the packing list and submission paperwork. Suitable personal protective gear should be donned for the type of sample received. The receiving employee opens the packaging, making a note of any damage or spillage. A description is recorded of the contents received including packaging size, weight or volume, and container description. If damage is noted or the contents received do not match the shipping information, the shipping company or department should be notified. Analytical testing should not proceed if a sample has been exposed to contamination from damaged packaging or if temperature requirements were not maintained during shipping. The receiving laboratory enters the received samples into their sample management system and applies their own sample ID #. The label should also include safety and storage conditions. Samples should be stored under specified conditions (for example, refrigerated or under desiccant). Controlled substances should be logged into the controlled substances log by trained personnel and stored in a safe.

If subsampling is required, this should be performed in a controlled way to appropriate nonreactive containers. A unique laboratory number system should be established for subsamples with the time and date of the subsampling recorded.

REAGENTS, SOLUTIONS, AND STANDARDS IDENTIFICATION

> Determine whether reagents, solutions, and standards are identified and labeled, and quantities are traceable in accordance with requirements including opened-on, expiry, (validated) use-by, or recertify-by dates. (Apply)
>
> **Body of Knowledge III.F.2**

Receiving personnel in the warehouse are responsible for verifing that the material received matches the associated paperwork and is what was ordered. This check is typically done by reviewing what was received against what was ordered on the purchase order. Once material is received into the system, warehouse personnel move the appropriate number of containers (based on the predefined sampling plan) to the inspection area. The containers moved into the inspection area are selected randomly by the warehouse personnel.

The quality inspectors compare the name of the material and vendor/ manufacturer's lot number on the container label against the information on the paperwork, which provides documentation that the material was received from an approved vendor. The quality personnel verify the shipping container to ensure it is appropriate, which is required to ensure that the packaging configuration is appropriate for use in production, as well as provide confirmation that the material is stable in the shipping container. If there are specific storage requirements, the quality inspector will ensure that all documentation is in order to ensure that the material had been stored appropriately during transit.

The quality inspector prepares a sample container with the appropriate identifying information, such as material name, batch number, and vendor batch number, if applicable. The person performing the sampling and the date sampled are recorded on the sample container. Proper labeling is a *good manufacturing practices* (GMP) requirement and helps avoid costly mistakes and rework. Typically, laboratory samples are given a unique sample identification number.

The quality inspector ensures that the container is clean and places the container in an appropriate sampling area. Depending on what is being sampled, the sampling area may be anything from a room or tent with appropriate *high-efficiency particulate air* (HEPA) ventilation in a cleanroom, laminar flow hood, or biosafety cabinet. Once the quality inspector has donned the appropriate personal protective gear, the sampling can occur. The tools used in sampling must be clean to prevent contamination. Disposable sampling tools are often used; the benefit of this approach is that it nearly eliminates the potential of contaminating the sample with an improperly cleaned sampling tool.

The quality inspector opens the appropriate number of containers and begins to sample from each drum. Depending on the testing requirements for the material, individual samples may be pulled from each container, or a smaller amount may be pulled from each container and combined to make a composite sample. A reserve sample is often pulled in conjunction with the laboratory sample. Upon completion of sampling, the sample is delivered to the laboratory for testing. The sample is logged into the laboratory system and is scheduled for testing.

Reagents or Solutions

Proper control and identification of reagents and solutions in the laboratory ensures that the materials used in the analysis of products are appropriate. A *material safety data sheet* (MSDS) should accompany each receipt of a reagent. The MSDS contains much of the information needed to ensure proper handling and use of the material, what to do in the event of a spill, and so on. Before the material is handled, the MSDS should be reviewed to ensure that appropriate precautions are taken. The containers must be inspected to ensure their integrity. If a seal is broken, the solution has been frozen, the container is leaking, or the labeling is inconsistent with what was ordered, the material should not be used. A label containing information (Figure 24.1) should be secured on each of the reagent or solution bottles. Often, reagents inventory is managed though a *laboratory information management*

Received by/Date: ____________

Expiration date: ______________

Open by/Date: ________________

Use-by date: _________________

Figure 24.1 Example label for purchased reagents.

Table 24.1 Typical solution expiry dates.

Solution	Expiry
Water	1 week after opening
Buffers/dissolution media	pH 6–8: 1 week
	Others: 2 weeks
Buffers/dissolution media (purchased)	5 years (unless otherwise stated by manufacturer)
Laboratory solvents < 10% organic	1 month
Laboratory solvents ≥ 10% organic	3 months
Storage solvents (50% organic)	6 months
Test solutions	1 year (unless otherwise specified by USP)
Test solutions (purchased)	5 years (unless otherwise specified by manufacturer)
Volumetric solutions	1 year (unless otherwise specified by USP)
Volumetric solutions (purchased)	Manufacturer certified date

system (LIMS). If this is the case, the information listed on the label may be derived from the database by a bar code instead of being listed on the label.

The expiration date of the reagent should be readily apparent on the container. For commercially procured reagents, the manufacturer-assigned expiration date is often used. If no expiration date is available, and the material is stable, a company may have a procedure in place that would allow it to assign a predefined expiration period based on the date of receipt and type of material. Shorter expiration dates may be required if stability is an issue for the material. For example, for solutions that are prone to microbial growth, such as *high-performance liquid chromatograph* (HPLC) water, an opened date and use-by date may be added to the container upon opening. Table 24.1 contains some typical expiry dating periods for common solutions.

Some reagents and solutions may be used past their expiration date if the material can be recertified or re-qualified. For example, it is common to re-standardize or recertify volumetric solution upon use or at monthly intervals. The current compendia (European Pharmacopeia [PhEur], Japanese Pharmacopeia

[JP], and United States Pharmacopeia [USP]) typically define the process for re-standardizing the most commonly used volumetric solutions.

The preparation of all solutions prepared in the laboratory must be documented. The prepared solution should be identified with, at a minimum, the name/composition of the solution, concentration, storage, re-standardization date, expiration date, reference, and analyst/date.

CALIBRATION AND LABORATORY STANDARDS

Laboratories use primary and secondary standards to determine material results and equipment calibration. A primary standard is purchased from an official source, such as USP, British Pharmacopeia (BP), or National Institute of Standards and Technology (NIST). If no official source for a standard exists, material can be thoroughly characterized to ensure its identity, strength, quality, purity, and potency. Secondary or in-house standards are raw materials that have been assayed against a primary standard.

When standards are received into a laboratory, traceability must be established. Most laboratories will assign a unique standard number and complete corresponding paperwork indicating the standard's name, lot number, purity, expiration date, and drying requirements. USP has a bimonthly publication that reports expiration dates. Analysts must verify they are not using expired standards. Based on the USP publication, most companies will establish a bimonthly review of their standards to ensure no expired ones are in the laboratory.

STORAGE REQUIREMENTS

> Describe and use procedures to store samples, reagents, solutions, solvents, and standards in appropriate environmental conditions (e.g., temperature, humidity, light exposure, and absence of oxygen) to maintain the material's characteristics for testing. (Apply)
>
> **Body of Knowledge III.F.3**

Samples

The storage of samples is extremely important to ensure the integrity of the material. Once sample containers are received into the laboratory, the containers should be stored in a centralized location until testing is started. Samples should have some form of unique identification to differentiate them from the other samples in the laboratory. Once testing has been started, procedures will dictate how samples are prepared and stored. For example, if a sample is light sensitive, amber glassware is used in the sample preparations. During method validation, solution stability is determined. Upon completion of testing, most laboratories require solutions to be retained, even if the solution has expired, until data review is complete. If there is a need for an investigation, the expired solutions may be used for information.

21 CFR 211.170 requires a reserve sample that is representative of the batch to be retained. The reserve sample consists of at least twice the quantity necessary to conduct all required testing. Reserve samples are typically pulled during inspection sampling and/or final packaging. The reserve samples are maintained at warehouse conditions and evaluated once a year for signs of deterioration. The finished product reserve samples are maintained for one year past their expiration date. The raw material reserve samples must be maintained one year past the expiration of the last finished product they were used in. Most pharmaceutical companies have finished products with a maximum of 36-month expiration dating; thus, they can assign a five-year expiration date to the raw material reserve sample, which ensures they will meet the CFR requirement.

REAGENTS OR SOLUTIONS

Chemicals, reagents, and solutions must be stored in the laboratory in a manner that considers safety, chemical integrity, and efficient management. Materials must be stored in tightly sealed and chemically resistant containers. Strong bases should not be stored in glass containers because the hydroxide will slowly react with the silica. Exposure of materials to heat or direct sunlight should be avoided. If a material is easily oxidized, it may be stored with a layer of nitrogen. Some companies store those chemicals in desiccators and flush with nitrogen. Materials requiring refrigeration or freezing should be stored as such. Chemically incompatible materials will be segregated according to their MSDS requirements. Acids and bases are not stored together.

ANALYTICAL STANDARDS

Analytical standards are stored according to the label instructions. Typically, standards are maintained in desiccators until use. If drying is required before use, the analyst should take only the amount needed and dry in an appropriate container. Once the standard is dried, the analyst can store it in a desiccator until it is needed. Unless otherwise specified, standards are typically weighed and diluted within 24 hours of removal from the oven. After weighing, the remaining dried material is discarded. The appropriate storage of the standard solution should be outlined in the company's method. During method validation, standard solution stability should be determined.

Chapter 25
Stability Programs

Laboratory testing, which is required by 21 CFR 211.160 and 211.165, is necessary to confirm that components, containers and closures, in-process materials, and drug products conform to appropriate standards of identity, strength, quality, and purity. Release testing provides evidence that the materials conform to appropriate standards at the time of release. Stability testing provides evidence on how the quality of a drug substance or drug product varies over time. Environmental factors and product-related factors may impact the stability of the drug substance/drug product over time. The impact of environmental factors, such as temperature, humidity, and light, must be evaluated. The impact of product-related factors, such as the chemical and physical properties of the drug substance/drug product, the manufacturing process, and the container-closure system, must also be evaluated.

RELEASE TESTS VS. STABILITY-INDICATING TESTS

Define and distinguish between release and stability-indicating tests. (Analyze)

Body of Knowledge III.G.1

TYPES OF STABILITY STUDIES

Stability programs can be divided into three main types: product development stability, commercial support stability, and change management stability. While there are overlaps in the execution of the stability studies, the objectives of each of the stability types are different.

Product Development Stability

The type of product that is being considered defines the activities of the development stability program. For generic drug products, the stability requirements for submission of the Abbreviated New Drug Application (ANDA) include the comparison of the final drug product stability data to those of the innovator product.

Stability data on the innovator product are of further value, as they provide insight into formulation and packaging requirements.

For new drug products, no innovator studies can be performed; however, clinical studies are frequently conducted using another therapeutic compound as a comparator, as well as a placebo product. Since these clinical studies are often blinded, the identity of the target formulation, the comparator formulation, and the placebo must be hidden by over-encapsulation or other means of obscuring the identity. Stability testing of these formulations is required to evaluate impact to drug release of the obscuration process, and to ensure patient safety throughout the duration of the clinical study.

Other stability studies common to both types of products include photostability testing, excipient compatibility testing, stress testing, and packaging evaluation studies. Further, through well-planned *design of experiments* (DOE) studies, appropriate ranges and tolerances in component ratios and processing parameters can be established. Stability is an integral part of these DOE studies, as the various changes may not result in observable differences in the release results.

Photostability studies are performed as two distinctive studies; one as a form of forced degradation, and the other to demonstrate appropriate protective qualities of container closure systems (also known as *confirmatory photostability studies*). For the forced degradation photostability study, the test article is exposed to high levels of visible light and *ultraviolet* (UV) radiation with the intent of degrading the test article. The results from this testing help to identify potential degradation pathways and provide guidance in formulating and packaging the product. Confirmatory photostability studies are performed as indicated by International Conference on Harmonization (ICH) guidance Q1B *Photostability Stability Testing of New Drug Substances and Products.* Studies are performed in the least protective packaging available (such as open petri dish), progressing to more protective packaging (primary commercial packaging) as required. The data generated from these studies will be used to demonstrate that the packaging provides adequate protection from light.

Excipient compatibility studies are performed to determine whether any intended components contribute to instability of the product. Forced degradation studies are performed as a tool in developing stability-indicating analytical methods, while the packaging evaluation studies help to identify appropriate packaging configurations and material.

Stress testing is performed to help identify likely degradation products and validate stability-indicating methods. Most companies target to degrade a material 5% to 15%, but no more than 20%. Successful degradation is demonstrated when the analysis clearly indicates the presence of degradation peaks and a lowered assay. Companies may use acid stress, base stress, thermal degradation, or oxidative degradation to degrade their material. An appropriate stability-indicating method will have appropriate resolution between the analyte and degradation peaks and will demonstrate mass balance.

A company can acid-stress the analyte and placebo of the drug substance and/or drug product by adding HCl such that the $pH \leq 3$. Samples are maintained in that environment up to a week. The sample is neutralized before analysis. If no degradation occurs, a higher concentration of acid (maximum of 6N) and/or an elevated temperature (60°C to 80°C) can be used. The base stressing is done by stressing the analyte and placebo of the drug substance and/or drug product by adding NaOH such that the pH ≥ 9. Samples are maintained in that environment up to a week. The sample is neutralized before analysis. If no degradation occurs, a higher concentration of base (maximum of 6N) and/or an elevated temperature (60°C to 80°C) can be used.

The thermal degradation is done by stressing the analyte and placebo of the drug substance and/or drug product by heating in an oven or water bath to temperatures of 70°C to 105°C, but at least 10°C below the melting point. Samples are maintained in that environment for up to a week. Lower temperatures can be selected if the sample melts or excessive degradation occurs. If no degradation occurs, a combination of thermal exposure and humidity (for example, 50°C/80% *relative humidity* [RH]) can be used. The oxidative degradation is done by stressing the analyte and placebo of the drug substance and/or drug product by adding up to 3% H_2O_2. Samples are maintained in that environment for up to a week.

Packaging evaluation studies are performed near the end of the development process when a likely formulation and process have been identified. For these studies, a firm would place the same batch of material in various container-closure systems and submit them to the same storage conditions. Results can be compared side by side to determine if one package provides more protection than another. The results obtained from all these studies will be used to help select a final formulation, establish appropriate specifications, and determine product shelf life.

The final stability study from the development program is the *submission study*. This study is conducted following the guidance provided in ICH Q1A (R2) *Stability Testing of New Drug Substances and Products* (referred to as "Q1A guidance"). The batches used in the submission study represent the final formulation, manufacturing processes, and commercial packaging. Sample material is stored under controlled conditions (Table 25.1) for specific time periods and tested. The data generated from these studies are evaluated and submitted to regulatory bodies as part of the drug approval process.

Commercial Support Stability

As part of the submission process, a company will include a commitment to continue stability evaluation over the commercial life of the product. The stability commitment may vary depending on the properties of the product, but generally specifies that one representative batch per year be held at the long-term storage condition (Table 25.1) and tested at the same time points and by the same methodology as were used to generate the submission data.

The results generated for a commitment batch have far-reaching implications. Thus, a batch that fails to meet specification during shelf life would cause all batches of the same packaging configuration to be evaluated. In such a scenario, a firm could be required to reduce expiry or recall product from the market.

Change Management Stability

The components, processing, packaging, and labeling of a drug product are specified in the approved application. A change to any of these requires evaluation to ensure that the change has no deleterious effect on the product. Stability testing

Table 25.1 Definitions of ICH conditions.

Condition set points	Acceptance criteria	Suggested use
25°C/60% RH	± 2°C/±5% RH	• General case long-term stability • Refrigerated product accelerated stability
30°C/65% RH	±2°C/±5% RH	• General case long-term stability • General case intermediate stability
40°C/75% RH	±2°C/±5% RH	• General case accelerated stability
5°C/Humidity not controlled	±3°C	• Refrigerated product long-term stability
–20°C/Humidity not relevant	±5°C	• Frozen product long-term stability

is frequently a part of this evaluation. For example, a firm may wish to change the supplier of the *active pharmaceutical ingredient* (API) in their drug product. This change would require stability testing. Just as in the original application, the results will be evaluated and submitted to regulatory authorities for approval. The magnitude of change and the tolerances established during registration will influence the type of stability program required to support the change.

EXECUTION OF STABILITY STUDIES

As with any good scientific evaluation, stability studies begin with outlining the steps to be taken and the rules for interpreting the results (that is, this plan is the *stability protocol*). Minimally, a protocol should specify the sample storage conditions, testing time points, and packaging configuration and provide reference to the test methods and acceptance criteria against which the test article is being evaluated. Studies performed to support a regulatory submission generally include the criteria for evaluating the data.

Conditions

The ICH Q1A (R2) guidance outlines the generally accepted conditions at which studies are to be performed (Table 25.1). Exceptions to these conditions, with a written justification, can be made depending on the nature of the product and the intended use of the data.

Time Points

The Q1A guidance provides recommended time points for testing. These are typically one, two, three, and six months for accelerated studies, and three, six, nine, 12, 18, 24, and 36 months for long-term studies. The intermediate storage condition is set up to provide additional information should a study fail to meet acceptance criteria during the accelerated study. Like the storage conditions, these may be adjusted as needed to fulfill the purpose of the study.

Packaging Configurations

Each unique presentation of a product must be represented in the commercial support stability program. For example, a drug product that has one strength and is in four package configurations would have four unique presentations of the product. Thus, the four presentations all need be represented on stability.

A stability program can be established by *bracketing*. Bracketing is based on worst-case scenarios (that is, smallest, largest), which are used for the stability study, and the intermediates are assumed to be represented. For example, bracketing can be applied to different container sizes that are made from the same resin, filled with the same product. If a drug product is packaged in the same bottle with three different tablet counts (30-count, 90-count, and 100-count) and in a unit-dose blister, only three presentations (30-count, 100-count, and unit-dose blister) will be on stability. The fourth configuration (90-count), which is bracketed by the 30-count and 100-count, is assumed to be represented. Matrixing can be used. With matrixing, a selected subset of the total number of possible samples with all factor combinations is tested at a specified time point.

Accelerating Studies

ASTM F1980, *Standard Guide for Accelerated Aging of Sterile Barrier Systems for Medical Devices*, provides information for developing accelerated aging protocols to rapidly determine the effects, if any, due to the passage of time on the sterile integrity of the sterile barrier system (SBS), as defined in ANSI/AAMI/ISO 11607–1:2006, and the physical properties of their component packaging materials. The formula to calculate the accelerated aging factor (AAF) is

$$\text{AAF} = Q_{10}^{[(T_{\text{AA}}-T_{\text{RT}})/10]}$$

where

Q_{10} = An aging factor for 10°C increase or decrease in temperature (2 is a common and conservative means of calculating an aging factor)

T_{AA} = Accelerated aging temperature °C

T_{RT} = Ambient temperature °C (normally 23°C)

The formula to calculate the accelerated aging time AAT is

$$\text{AAT} = \text{RT}/\text{AAF}$$

where

RT = Desired real time in days.

Example: An accelerated study is required to evaluate the effect on the packaging materials using a temperature of 50°C for a real time equivalency of 180 days.

$$\text{AAF} = Q_{10}^{[(T_{\text{AA}}-T_{\text{RT}})/10]} = 2.0^{[(50-23/10)]} = 6.50$$

$$\text{AAT} = \text{RT}/\text{AAF} = 180/6.5 = 27.7 \text{ days}$$

The AAT for 50°C for 180 days is 27.7 days.

STABILITY TEST DATA

> Review stability data against specifications and identify trends that can establish, support, or challenge an expiry date. (Evaluate)
>
> **Body of Knowledge III.G.2**

Once data are generated from the stability study, they will need to be evaluated. ICH Q1E *Evaluation of Stability Data* provides guidance for the establishment of retest periods and shelf lives for drug products and drug substances. Stability data for a minimum of three batches of drug product or drug substance should be evaluated. The degree of variability of individual batches affects the confidence that a future production batch will remain within acceptance criteria throughout its retest period or shelf life. Results from physical, chemical, biological, and microbiological tests for critical quality attributes likely to affect the quality and performance of the product should be evaluated. Each attribute should be assessed separately, and an overall assessment should be made for the purpose of determining an expiry date. The retest period or shelf life proposed should not exceed that predicted for any single attribute.

The first step in this process is to compare results against the acceptance criteria. The second step is to look at the trend. Trending is typically performed on sets of data that include the same formulation and packaging configuration.

Based on the stability data extrapolation, companies can infer information about future data and establish dating. This information can aid in establishing acceptance criteria and expiry dating for the new product based on a high degree of confidence. After approval of a new material, the ongoing stability study data will permit the detection of a stability issue, such as a change in impurity levels.

In general, certain quantitative attributes, such as assay and degradation products, can be assumed to be zero-order kinetics during long-term storage; thus, the trending and regression statistics can be applied to these stability-indicating tests. Although the kinetics of other quantitative attributes, such as pH and dissolution, are not known, the same statistical analysis model can be applied.

Stability data for accelerated conditions should be evaluated first, then intermediate conditions if available and then long-term data. For data showing little or no change or variability over time, the proposed expiry date can be up to twice, but no more than 12 months beyond the period covered by the existing data. If the existing long-term data shows change over time or variability among batches, the proposed expiry date should not exceed the shortest time period supported by any batch. If the variability can be attributed to a particular factor such as strength, different expiration dates may be assigned per factor.

For long-term data with variability, statistical analysis may be performed to extrapolate the proposed expiration date up to twice but no longer than 12 months beyond the period covered by long-term data. An appropriate statistical method should be employed to establish with a high degree of confidence a retest period

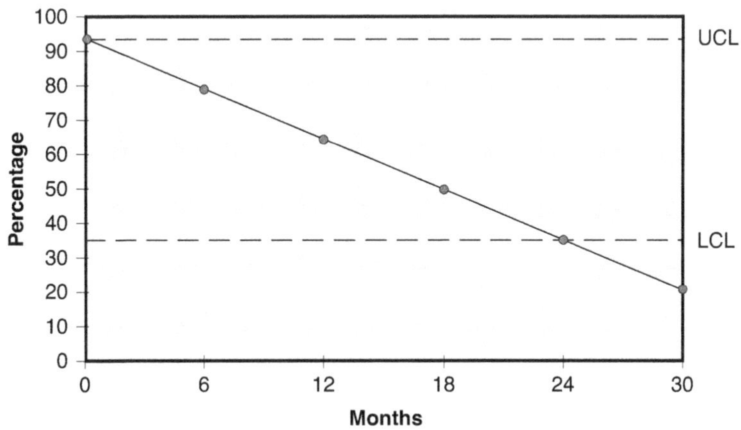

Figure 25.1 Shelf life estimation for assay.

in which a quantifiable attribute will remain with acceptance criteria. For assay or degradation products a linear regression is typically employed to determine the time point at which the 95% confidence limit intersects with the proposed acceptance criterion (Figure 25.1). Batch data is examined individually first. If the estimated retest period for each batch does not exceed the proposed retest period, there is no need to pool the data. If one of the estimated retest periods are shorter than the proposed retest period, then the test data may be combined to estimate a longer retest period if preliminary pooling, covariance tests support this approach.

STABILITY-POINT FAILURE

> Identify the stability-point failure of a product or material and evaluate the implications for regulatory compliance. (Evaluate)
>
> **Body of Knowledge III.G.3**

Stability-point failure occurs when a stored drug product does not conform to specification within the assigned shelf-life period.

In the United States and Japan, specifications for regulatory release testing must be the same as shelf-life specifications. However, more restrictive in-house criteria may be applied for release testing to provide increased assurance that the product will remain within regulatory release criteria throughout its shelf life. The European Union allows different criteria to be used for regulatory release and shelf-life criteria within regulatory guidelines.

Results that do not meet the acceptance criteria or demonstrate adverse trending must be investigated both from a laboratory perspective (was the testing performed correctly?) and from a product perspective (were formulation, manufacturing, packaging, and delivery performed correctly?). Typically, an *out-of-specification* (OOS) procedure will be followed to conduct the investigation. For approved products, and if the stability study is representative of product in commerce, a confirmed OOS result will require notification to the United States Food and Drug Administration (FDA) within three days of the confirmation. Depending on the failure and product evaluation, a company may potentially reduce the expiry dating or recall the product.

Chapter 26

Reserve Samples and Retains

> Describe the various regulatory requirements for retains and reserve samples. (Apply)
>
> **Body of Knowledge III.H**

Reserve and retain samples are representative portions of each lot of active pharmaceutical ingredient (API) and finished product stored under the labeled storage conditions for the purposes of identification or analytical testing should the need arise. Reserve samples are retained for the purpose of future evaluation of quality and not for future stability testing.

MARKETED PRODUCTS

The requirements for reserve samples are specified in 21 CFR 211.170. The US Food and Drug Administration (FDA) requires an appropriately identified reserve sample that is representative of each lot in each shipment of each active ingredient be retained.

The reserve sample consists of at least twice the quantity necessary for all tests required to determine whether the active ingredient meets its established specifications, except for sterility and pyrogen testing. The retention time is as follows:

- For an active ingredient in a drug product, the reserve sample shall be retained for one year after the expiration date of the last lot of the drug product containing the active ingredient.
- For an active ingredient in a radioactive drug product, except for nonradioactive reagent kits, the reserve sample shall be retained for:
  - Three months after the expiration date of the last lot of the drug product containing the active ingredient if the expiration dating period of the drug product is 30 days or less;
  - Six months after the expiration date of the last lot of the drug product containing the active ingredient if the expiration dating period of the drug product is more than 30 days.

- For an active ingredient in an over-the-counter (OTC) drug product that is exempt from bearing an expiration date, the reserve sample shall be retained for three years after distribution of the last lot of the drug product containing the active ingredient.

An appropriately identified representative of each lot or batch of drug product stored under conditions consistent with product labeling in the same container-closure system in which the drug product is marketed or in one that has essentially the same characteristics should also be reserved per 21 CFR 211.170. The reserved drug product sample should consist of at least twice the quantity necessary to perform all required tests. Except for radioactive drug products, the reserved samples should be examined at least once a year for visible evidence of deterioration. Specified retention times are as follows:

- One year after the expiration date of the drug product lot, expect for radioactive or OTC products as described in the following items.
- For a radioactive drug product, except for nonradioactive reagent kits:
  - o Three months after the expiration date of the drug product if the expiration dating period is 30 days or less.
  - o Six months after the expiration date of the drug product if the expiration dating period is more than 30 days.
- For an OTC drug product that is exempt from bearing an expiration date, the reserve sample shall be retained for three years after the lot of the drug product is distributed.

21 CFR 600.13 specifies requirements for the retention of biological product samples. Manufacturers should retain representative material sufficient for examination and testing for safety and potency for at least six months after the expiration date of the product.

European Union (EU) Volume 4, Annex 19, Reference and Retention Samples, provides guidance for the taking and holding of reference and retention samples. EU defines reference samples as a sample of a starting material, packaging material or finished product material stored for the purpose of analysis should the need arise during the shelf life of the product. Retention samples are a fully packaged unit from a batch of finished product stored for identification purposes. For finished products, the retention and reference samples may be regarded as interchangeable. The reference and/or retention samples serve as a record of the batch in the event of a quality complaint, compliance query, or pharmacovigilance report.

The quantity of reference sample material retained should be sufficient to perform two complete sets of analytical control testing. At least one retention sample should be retained for each distinct packaging configuration. Storage conditions should be in concordance with the marketing operation. Reference samples should be stored in a location conveniently available to a laboratory with validated methodology. Retention samples should represent batches distributed within the EEA (European Economic Area) and should be located within the EEA.

Reference and retention samples for finished product should be retained for at least one year after the expiry date. Reference samples of starting materials should be retained for at least two years after release of the product.

CLINICAL TRIAL PRODUCTS

FDA's Division of Scientific Investigations (DSI) and FDA field investigators conduct inspections of clinical and analytical sites that perform bioavailability (BA) and bioequivalence (BE) testing for drug manufacturers seeking approval of a drug product.

Retention samples should be kept at the testing facility where the study was conducted. The study sponsor should provide the testing facility with the reference standard and a supply of the test article sufficient to complete the study and retain the appropriate number of dosage units as reserve samples. The study sponsor should not separate out the samples to be reserved prior to sending the batches to the testing facility. The testing facility will randomly select the reserve samples from the supply sent by the sponsor. This is to ensure that reserve samples are in fact representative of the same batches provided by the study sponsor for the testing.

The testing facility should retain a sufficient quantity of samples to permit FDA to perform five times all of the release tests required in the application or supplemental application. Each reserve sample shall be adequately identified so that the reserve sample can be positively identified as having come from the same sample as used in the specific bioavailability study. Each reserve sample shall be stored under conditions consistent with product labeling and in an area segregated from the area where testing is conducted, and with access limited to authorized personnel.

Each reserve sample shall be retained for a period of at least five years following the date on which the application or supplemental application is approved, or, if such application or supplemental application is not approved, at least five years following the date of completion of the bioavailability study in which the sample from which the reserve sample was obtained was used.

Stability Commitment

If long-term stability data is not available for the shelf life of the drug product at the time of approval, a commitment should be made to continue stability studies post-approval. If the submission did not include stability data from three production batches, material from the first three post-approval production batches should be placed on stability studies for the proposed retest period. Refer to Chapter 25 for more information about stability studies.

Part IV

Infrastructure: Facilities, Utilities, and Equipment

Chapter 27

Facilities

INTRODUCTION

Good manufacturing practices (GMP) were invoked to maintain the quality of pharmaceutical and biopharmaceutical products manufactured for human consumption. The predicate regulatory documents, 21 CFR 210 and 211, outline the minimum GMP for methods to be used in, and the facilities or controls to be used for, the manufacture, processing, packing, or holding of a drug to assure that such drug meets the requirements of the Act as to safety, and has the identity and strength and meets the quality and purity characteristics that it purports or is represented to possess. Failure to comply with the regulations renders a drug adulterated, and the drug, as well as the person(s) and company responsible for the failure to comply, are subject to regulatory action. In GMP, buildings and facilities are the second major topic addressed, indicating their importance in the overall manufacturing of quality products.

BUILDINGS

> Determine and document the requirements for appropriate size and construction of buildings and areas, and the location of control systems. Ensure that construction and location facilitate proper operation and minimize the risk of error and cross contamination including meeting requirements that specify separation of antibiotics, hormones, and toxins. (Apply)
>
> **Body of Knowledge IV.A.1**

DESIGN AND CONSTRUCTION FEATURES (21 CFR 211.42)

Any building or buildings used in the manufacture, processing, packing, or holding of a drug product shall be of suitable size, construction and location to facilitate cleaning, maintenance, and proper operations.

To ensure that products are manufactured safely, building design is the first imperative. Buildings need to be large enough and constructed in a manner to accommodate these processes. They should have adequate space for the orderly placement of equipment and materials to prevent cross-contamination between different components, drug product containers, closures, labeling, in-process materials, or drug products.

Facilities should also be designed to prevent contamination and mix-up through the unidirectional flow of components, drug product containers, closures, labeling, in-process materials, and finished drug products from minimally controlled areas to those in which control is greater. As materials move through the production process, each subsequent area is subject to higher scrutiny and control. For example, the dock where raw materials and components are received will have fewer environmental controls than the area where bulk product is filled into containers. Consideration must also be given to proper storage, removal, and disposal of process waste (liquid and solid) to prevent contamination of other products or adverse environmental impact.

Special consideration must be given to aseptic processing operations and cleanrooms, which require positive pressure differential, transition anterooms, and pass-throughs for people and equipment entering and leaving these areas.

MANUFACTURE AND STORAGE ENVIRONMENT

> Identify requirements for appropriate lighting, ventilation, and drainage to avoid adversely affecting product (either directly or indirectly) during manufacturing and storage. (Apply)
>
> **Body of Knowledge IV.A.2**

Specifically defined areas of adequate size are required for each operation and must be designed to prevent contamination or mix-ups. Specific areas are typically demarcated for:

- Receipt, identification, storage, and holding of components, drug product containers, closures, and labeling pending the appropriate sampling, testing, or examination by the quality control unit before release for manufacturing or packaging
- Holding rejected components, drug product containers, closures, and labeling before disposition
- Storage of released components, drug product containers, closures, and labeling
- Storage of in-process materials
- Manufacturing and processing operations
- Packaging and labeling operations

- Quarantine storage before release of drug products
- Storage of drug products after release
- Control and laboratory operations

Aseptic processing, including floors, walls, and ceilings with smooth, hard surfaces that minimize 90° corners and are easily cleanable; temperature and humidity controls; air supply filtered through high-efficiency particulate air (HEPA) filters under positive pressure, regardless of whether flow is laminar or turbulent; system for monitoring environmental conditions; system for cleaning and disinfecting the room and equipment to produce aseptic conditions; and system for maintaining any equipment used to control the aseptic conditions and assuring they are operating properly prior to working in the area.

LIGHTING (21 CFR 211.44)

The GMP are very specific and succinct regarding lighting. The facilities must have sufficient lighting to ensure that all work can be easily viewed by personnel to allow them to conduct their activities properly. Most companies provide overhead direct lighting, which produces bright, nonglaring light, facilitating visual assessment of each operation.

One area of lighting that overlaps with design and construction is cleaning and maintenance of the lighting fixtures. Since lights are placed above the work area, any particles and/or microorganisms that accumulate on them are potential contaminates that could lead to an adulterated product. Therefore, companies should design lighting systems that facilitate cleaning and maintenance to prevent unwanted contamination, such as lights that are enclosed in metal housings with removable bottom panels. Additionally, cleaning schedules should be generated and implemented. Operators and supervisors should conduct daily visual inspections of the lighting fixtures to ensure that insects or other visible contaminants have not infiltrated the lighting fixtures. An insect finding its way into a light fixture is indicative that the integrity of the fixture has been compromised and requires repair and cleaning.

VENTILATION, AIR FILTRATION, AIR HEATING AND COOLING (21 CFR 211.46)

The GMP requirements stipulate that adequate ventilation shall be provided, and equipment for adequate control over air pressure, microorganisms, dust, humidity, and temperature shall be provided when appropriate for the manufacture, processing, packing, or holding of a drug product. The GMP also require air filtration systems, when appropriate, including prefilters and particulate matter air filters, to be used on air supplies to production areas. If air is recirculated to production areas, measures are to be taken to control recirculation of dust from production. In areas where air contamination occurs during production, exhaust systems or other systems adequate to control contaminants are required. To prevent cross-contamination, air-handling systems for the manufacture, processing, and packing of penicillin are to be completely separate from those for other drug products for human use.

Ventilation, air filtration, and air heating and cooling systems may contribute to contamination, both particulate and microbiological, and these systems should be closely monitored and controlled. As with lighting, a company should generate a schedule for cleaning and maintaining air vents, ducts, and cooling systems, but human observation should be used on a daily basis to ensure proper operation. For example, visual observation of blackened areas on vents is evidence that the frequency of cleaning and maintenance needs to be increased to avoid unwanted contaminants entering the area.

PLUMBING (21 CFR 211.48)

The GMP requirements for water state that potable water shall be supplied under continuous positive pressure in a plumbing system free of defects that could contribute contamination to any drug product. Potable water must meet the standards prescribed in the Environmental Protection Agency's (EPA) National Primary Drinking Water Regulations set forth in 40 CFR 141. Water not meeting such standards is not permitted in the potable water system. Drains must be of adequate size and, where connected directly to a sewer, must be provided with an air break or other mechanical device to prevent back-siphonage.

Supply of water to the facility may significantly contribute to the overall quality of the manufacturing operation. Although a facility cannot control the type of water it receives from a municipality, it should monitor the water on a scheduled basis and take appropriate actions to ensure that the water meets the current EPA regulations. The EPA specifies specific levels of organic, inorganic, and microbial agents that are allowed in municipal water supplies, and typically the municipality reports these levels annually. Pharmaceutical manufacturing facilities should exert increased diligence regarding the requirements, and may adopt a practice of testing the water more frequently than annually to ensure that the incoming supply does not overwhelm its water purification system that removes these agents. This process is critical in assessing the life span of the filters, carbon beds, resins, and ozone or other components used to reduce the amount of agents in the water supply.

Most facilities have installed plumbing that diverts sewage, including liquids used in manufacturing, directly to a closed drain equipped with air breaks downstream from the point of entry. It is good practice to ensure that all drains are capped, even those used for discharging washing solutions after a room is cleaned. In this instance, the drains should remain capped until the room is cleaned, and subsequently recapped when cleaning is concluded.

SEWAGE AND REFUSE (21 CFR 211.50)

The GMP for sewage and refuse state that sewage, trash, and other refuse in and from the building and immediate premises must be disposed of in a safe and sanitary manner. Firms should have systems in place to ensure that sewage and refuse are properly removed and disposed according to applicable regulations. Medical waste should be removed by qualified vendors who provide complete (cradle-to-grave) documentation of removal and ultimate disposal, be it by incineration or other means. Liquid medical waste should be removed by the same vendors, or the firm should have systems and records documenting its inactivation before disposing of it in the municipal waste or sewage system. They must also be in compliance with all EPA regulations before disposal. Nonmedical and nontoxic waste should be removed by general waste haulers, and the firm should have systems and records to document the details of the process.

WASHING AND TOILET FACILITIES (21 CFR 211.52)

The GMP for washing and toilet facilities require that adequate washing facilities be provided, including hot and cold water, soap or detergent, air driers or single-service towels, and clean toilet facilities easily accessible to working areas. The firm is responsible not only for providing washing and toilet facilities but also for cleaning and sanitizing them on a regular basis, as described in written procedures. The firm's training program should include a section devoted to teaching proper washing techniques for all employees after they visit a washroom.

SANITATION (21 CFR 211.56)

The GMP for sanitization are:

- Any building used in the manufacture, processing, packing, or holding of a drug product shall be maintained in a clean and sanitary condition. Any such building shall be free of infestation by rodents, birds, insects, and other vermin (other than laboratory animals). Trash and organic waste matter shall be held and disposed of in a timely and sanitary manner.
- Written procedures are needed to assign responsibility for sanitation, describing in sufficient detail the cleaning schedules, methods, equipment, and materials to be used in cleaning the buildings and facilities; such written procedures shall be followed.
- There shall be written procedures for use of suitable rodenticides, insecticides, fungicides, fumigating agents, and cleaning and sanitizing agents. Such written procedures shall be designed to prevent the contamination of equipment, components, drug product containers, closures, packaging, labeling materials, or drug products, and shall be followed. Rodenticides, insecticides, and fungicides shall not be used unless registered, and used in accordance with the Federal Insecticide, Fungicide, and Rodenticide Act (7 U.S.C. 135).
- Sanitation procedures shall apply to work performed by contractors or temporary employees as well as work performed by full-time employees during the ordinary course of operations.

These regulations provide enough specificity such that a firm can follow them verbatim. Whatever written programs the firm establishes to assure adequate sanitization should be supplemented with proper training of all affected employees as well as frequent monitoring and feedback to ensure compliance.

MAINTENANCE (21 CFR 211.58)

The GMP for maintenance state that any building used in the manufacture, processing, packing, or holding of a drug product must be maintained in a good state of repair. Firms should have written procedures that describe the frequency and type of maintenance they will conduct in the facility. Typically, the more critical areas, such as the aseptic core, will be more frequently inspected and maintained, and the type of maintenance will be more highly scrutinized. In many cases, the batch records will include a section to confirm line clearance and state of maintenance before and after a production run. It may include a section for cleaning and for specific areas that require maintenance. For example, many more areas in an aseptic filling suite will be scrutinized as compared to a bulk processing area. Again, training is an essential component of maintaining a facility, and individuals should be continuously monitored to ensure they are vigilant in keeping the facility suitable for manufacturing operations.

Cleanrooms are an essential function in the processing of pharmaceuticals, biologics, and medical devices that are safe and effective. Cleanrooms provide an environment that minimizes sources of potential contaminants such as dust, microbes, particles, and vapors. Cleanrooms are classified by the maximum number of particles allowed per cubic meter. The controlled dispersion of particulates or contaminates by airflow is essentially how a cleanroom maintains a clean environment. ISO 14644-1 defines a cleanroom as a "room in which the concentration of airborne particles is controlled, and which is constructed and used in a manner to minimize the introduction, generation, and retention of particles inside the room, and in which other relevant parameters, e.g. temperature, humidity, and pressure, are controlled as necessary." Because even a motionless person adds particles to a room, it is important to limit both the number of personnel and the level of activities in the cleanroom.

Two types of cleanrooms exist: unidirectional and non-unidirectional. Each type of cleanroom has its unique advantages and disadvantages. *Unidirectional* cleanrooms are necessary when an ISO class 6 or better cleanroom is required. *Non-unidirectional* cleanrooms are generally less expensive to construct and rely on a high volume of air to disperse contaminates. Air filtration is essential to maintain acceptable particulate/contaminate counts. Two types of filters can be used: *high-efficiency particulate air* (HEPA), which are at least 99.97% efficient, and *ultra-low penetration air* (ULPA), which are at least 99.999% efficient. HEPA filters are used for ISO class 6 to 8 cleanrooms and ULPA filters are used for ISO class 1 to 5 cleanrooms. Cleanrooms are classified by the maximum number of particles allowed per cubic meter (Table 27.1 through Table 27.4). Cleanrooms rely on the capture and control of particulates/contaminants by airflow (Table 27.5).

When it is necessary to control the environment, specifications for parameters such as temperature, humidity, *colony-forming units* (CFUs), and particulates per cubic foot should be established and monitored to remain within specified Alert and Action Limits. No United States Food and Drug Administration (FDA) guidances for these parameters presently exist for environmentally controlled areas such as cleanrooms. Actual parameters will depend on the facility design, local environmental conditions, people, equipment, products and work activities conducted within the cleanroom during operation.

Table 27.1 Cleanroom classification (ISO 14644-1).

ISO class	Maximum permitted number of particles/m³					
	≥0.1 µm	≥0.2 µm	≥0.3 µm	≥0.5 µm	≥1.0 µm	≥5.0 µm
1	10	2				
2	100	24	10	4		
3	1,000	237	102	35	8	
4	10,000	2,370	1,020	352	83	
5	100,000	23,700	10,200	3520	832	29
6	1,000,000	237,000	102,000	35,200	8,320	293
7				352,000	83,200	2,930
8				3,520,000	832,000	29,300
9*				35,200,000	8,320,000	293,000

*Equivalent to normal outside air

Table 27.2 Recommended limits for microbial contamination (ISO 14644-1). Actual limits may be different for each facility based on results established during qualification and continuous monitoring.

ISO class	Microbiological active air action levels (CFU/m³)	Microbiological settle plates action levels (diameter 90 mm; CFU/4 hour)
5	1	1
6	7	3
7	10	5
8	100	50

CFU = colony-forming units

Table 27.3 Maximum permitted number of particles/m³ (*European Union Guidelines to Good Manufacturing Practice*).

Grade	Maximum permitted number of particles/m³			
	At rest		In operation	
	≥0.5 mm	≥5 mm	≥0.5 mm	≥5 mm
A	3,500	0	3,500	0
B	3,500	0	350,000	2,000
C	350,000	2,000	3,500,000	20,000
D	3,500,000	20,000		

Table 27.4 Recommended limits for microbial contamination (*European Union Guide to Good Manufacturing Practices*). Actual limits may be different for each facility based on results established during qualification and continuous monitoring.

Grade	Recommended limits for microbial contamination			
	Air sample CFU/m³	**Settle plates (diameter 90 mm), CFU/4 hour**	**Contact plates (diameter 55 mm), CFU/plate**	**Glove print 5 fingers CFU/glove**
A	< 1	< 1	< 1	< 1
B	10	5	5	5
C	100	50	25	—
D	200	100	50	—

CFU = colony forming unit

Table 27.5 Rule of thumb for recommended air changes per hour based on class of cleanroom (Whyte 2010).

ISO class	Air changes per hour
1	
2	
3	Use unidirectional airflow
4	
5	
6	> 100
7	10–100
8*	2–10

* FDA Guidance for Industry *Sterile Drug Products Produced by Aseptic*

A review of 21 CFR 210, 211, and 820 yields no direct requirements for the use or classification of controlled environments or cleanrooms. These parts of 21 CFR do have references requiring the need for environmental and contamination controls where a lack of control could reasonably be expected to have an adverse effect on product quality. However, FDA has issued a guidance for industry, *Sterile Processing—Current Good Manufacturing Practice,* which recommends a minimum of 20 air changes per hour as acceptable.

*Drug Products Produced by Aseptic Processing—Current Good Manufacturing Practice.**

* Since this document is freely available to all via the FDA website (*www.fda.gov/media/71026/download*), it is recommended that CPGP candidates read the document. It may also be used for reference during the certification exam.

This guidance recommends the use of an ISO class 5 cleanroom (European Union [EU] Grade A) in areas deemed to be critical. A critical area is defined as "one in which the sterilized drug product, containers, and closures are exposed to environmental conditions that must be designed to maintain product sterility." FDA recommends that areas directly supporting aseptic processing activities be at a minimum ISO class 7 while in operation. Consistent with Volume 4 EU *Guidelines to Good Manufacturing Practice Medicinal Products for Human and Veterinary Use, Annex 1 Manufacture of Sterile Medicinal Products*, it is recommended that microbiological contamination (bioburden) counts remain at < 1 at all times.

SPECIFIC ISO SECTIONS RELATED TO CLEANROOMS

ISO 14644-2

Specifications for testing and monitoring to prove continued compliance with ISO 14644-1 provide guidance and direction for testing requirements to ensure the proper operation and compliance for cleanrooms. According to ISO 14644-1, cleanrooms may be classified in one of three states:

- *As-built.* Condition where the installation is complete, with all services connected and functioning, but with no production equipment, materials, or personnel present
- *At-rest.* Condition where the installation is complete, with equipment installed and operating in a manner agreed on by the customer and supplier, but with no personnel present
- *Operational.* Condition where the installation is functioning in the specified manner, with the specified number of personnel present and working in the manner agreed on

The minimum number of sampling points is derived from the following equation:

$$N = \sqrt{A}$$

where

NL = The minimum number of sampling locations (rounded up to a whole number)

A = The area of the cleanroom or clean zone in square meters

Sampling locations may be geometrically distributed throughout the cleanroom during the initial qualification. However, there is little value in monitoring areas where product is not exposed or risk of contamination is low. It is recommended that more sampling points are placed in critical areas that have more activity or where product is directly exposed or stored. A positive control should be included with each monitoring campaign to demonstrate that the test media and incubation conditions are capable of detecting growth. For example, samples taken in an uncontrolled area or initial gowning room should demonstrate positive growth.

Table 27.6 Required and optional testing necessary to ensure proper cleanroom operation and compliance (ISO 14644-2).

Test parameter	Class	Maximum time interval (months)
	Required tests	
Particle count test	≤ ISO 5	6
	> ISO 5	12
Airflow volume or airflow velocity	All	12
Air pressure difference	All	12
	Optional tests	
Installed filter leakage	All	24
Airflow visualization	All	24
Recovery	All	24
Containment leakage	All	24

Table 27.6, derived from ISO 14644-2, lists the required and optional testing necessary to ensure proper cleanroom operation and continued compliance.

ISO 14644-1

This part of ISO 14644 (*Cleanrooms and associated controlled environments—Part 1: Classification of air cleanliness*) covers the classification of air cleanliness in cleanrooms and associated controlled environments exclusively in terms of concentration of airborne particles. Only particle populations having cumulative distributions based on threshold (lower limit) sizes ranging from 0.1 µm to 5 µm are considered for classification purposes. This part does not provide for classification of particle populations that are outside of the specified particle size range, 0.1 µm to 5 µm. Concentrations of ultrafine particles (particles < 0.1 µm) and macroparticles (particles > 5 µm) may be used to quantify these populations in terms of ultrafine descriptors and macroparticle descriptors, respectively. This part of ISO 14644 cannot be used to characterize the physical, chemical, radiological, or viable nature of airborne particles.

ISO 14644-2

This part of ISO 14644 (*Cleanrooms and associated controlled environments—Part 2: Specifications for testing and monitoring to prove continued compliance with ISO 14644-1*) specifies requirements for periodic testing of a cleanroom or clean zone to prove its continued compliance with ISO 14644-1 for the designated classification of airborne particulate cleanliness. These requirements invoke the test described in ISO 14644-1 for classification of a cleanroom or clean zone. Additional tests are also specified to be carried out in accordance with the requirements of this part of ISO 14644. Optional tests, to be applied at the user's discretion, are also identified. This part of ISO 14644 specifies requirements for monitoring of a cleanroom or clean zone (hereafter referred to as an *installation*) to provide evidence of its continued compliance with ISO 14644-1 for the designated classification of airborne particulate cleanliness.

ISO 14644-3

ISO 14644-3 (*Cleanrooms and associated controlled environments—Part 3: Test methods*) specifies test methods for the designated classification of airborne particulate cleanliness and for characterizing the performance of cleanrooms and clean zones. Performance tests are specified for two types of cleanrooms and clean zones: those with unidirectional flow and those with non-unidirectional flow, in three possible occupancy states: as-built, at-rest, and operational. ISO 14644-3 is not applicable to the measurement of products or of processes in cleanrooms or separative devices. See USP<788> Particulate Matter in Injections.

ISO 14644-4

This part of ISO 14644 (*Cleanrooms and associated controlled environments—Part 4: Design, construction and start-up*) specifies requirements for the design and construction of cleanroom installations but does not prescribe specific technological or contractual means to meet these requirements. It is intended for use by purchasers, suppliers, and designers of cleanroom installations and provides a checklist of important parameters of performance. Construction guidance is provided, including requirements for start-up and qualification. Basic elements of design and construction needed to ensure continued satisfactory operation are identified through the consideration of relevant aspects of operation and maintenance.

Application of this part of ISO 14644 is restricted in that user requirements are represented by the purchaser or specifier; specific processes to be accommodated in the cleanroom installation are not specified; fire and safety regulations are not considered specifically, so the appropriate national and local requirements should be respected; and process media and utility services are only considered with respect to their routing between and within the different zones of cleanliness. Regarding initial operation and maintenance, only cleanroom construction—specific requirements are considered.

ISO 14644-5

ISO 14644-5 (*Cleanrooms and associated controlled environments—Part 5: Operations*) specifies basic requirements for cleanroom operations. Aspects of safety that have no direct bearing on contamination control are not considered in this part of ISO 14644, so national and local safety regulations must be observed. This document considers all classes of cleanrooms used to produce all types of products. Therefore, it is broad in application and does not address specific requirements for individual industries. Methods and programs for routine monitoring within cleanrooms are not covered in detail in this part of ISO 14644, but reference should be made to ISO 14644-2 and ISO 14644-3 for monitoring particles, and ISO 14698-1 and ISO 14698-2 for monitoring microorganisms.

ISO 14644-6

ISO 14644-6 (*Cleanrooms and associated controlled environments—Part 6: Vocabulary*) establishes a vocabulary of terms and definitions related to cleanrooms and associated controlled environments.

ISO 14644-7

ISO 14644-7 (*Cleanrooms and associated controlled environments—Part 7: Separative devices [clean air hoods, gloveboxes, isolators and mini-environments]*) specifies the minimum requirements for the design, construction, installation, test, and approval of separative devices in those respects where they differ from cleanrooms as described in ISO 14644-4 and 14644-5. The application of ISO 14644-7 takes into account several limitations: user requirements are as agreed by customer and supplier; application-specific requirements are not addressed; specific processes to be accommodated in the separative device installation are not specified; fire, safety, and other regulatory matters are not considered specifically; and where appropriate, national and local regulations apply.

ISO 14644-8

ISO 14644-8 (*Cleanrooms and associated controlled environments—Part 8: Classification of air cleanliness by chemical concentration [ACC]*) establishes the classification of *air chemical cleanliness* (ACC) in cleanrooms and associated controlled environments in terms of airborne concentrations of specific chemical substances (individual, group, or category) and provides a protocol to include test methods, analysis, and time-weighted factors within the specification for classification. ISO 14644-8 currently considers only concentrations of air chemical contaminants between 100 and 10 to 12 g/m^3 under cleanroom operational conditions. ISO 14644-8is not relevant for application in those industries, processes, or productions where the presence of airborne chemical substances is not considered a risk to the product or process. It is not the intention of ISO 14644-8 to describe the nature of air chemical contaminants. ISO 14644-8 does not give a classification of surface chemical contamination.

ISO 14644-9

ISO 14644-9 (*Cleanrooms and associated controlled environments—Part 9: Classification of surface cleanliness by particle concentration*) establishes the classification of cleanliness levels on solid surfaces by particle concentration in cleanrooms and associated controlled environment applications. Recommendations on testing and measuring methods, as well as information about surface characteristics, are also given. ISO 14644-9 applies to all solid surfaces in cleanrooms and associated controlled environments such as walls, ceilings, floors, working environments, tools, equipment, and products. The *surface particle cleanliness* (SPC) classification is limited to particles between 0.05 µm and 500 µm. Several issues are not considered, including requirements for the cleanliness and suitability of surfaces for specific processes, procedures for the cleaning of surfaces, material characteristics, references to interactive bonding forces or generation processes that are usually time dependent and process dependent, selection and use of statistical methods for classification and testing, and other characteristics of particles, such as electrostatic charge, ionic charges, and microbiological state.

ISO 14698-1

ISO 14698-1 (*Cleanrooms and associated controlled environments—Biocontamination control—Part 1: General principles and methods*) establishes the principles and basic methodology of a biocontamination control (formal system) for assessing and controlling biocontamination when cleanroom technology is applied for that purpose. It specifies the methods required for monitoring risk zones in a consistent way and for applying control measures appropriate to the degree of risk involved. In zones where risk is low it can be used for information.

ISO 14698-2

ISO 14698-2 (*Cleanrooms and associated controlled environments—Biocontamination control—Part 2: Evaluation and interpretation of biocontamination data*) gives guidance on methods for the evaluation of microbiological data and the estimation of results obtained from sampling for viable particles in risk zones for biocontamination control. It should be used, where appropriate, in conjunction with ISO 14698-1.

CLEANROOM OPERATION

Cleanroom Personnel

The cleanroom environment can be contaminated by many possible sources, including personnel, equipment, and materials. Contamination can be generated through outgassing, personal movement, exhaust, make-up air, friction, heat, and static electricity. However, cleanroom personnel generate the most particulate contamination in a clean room. Proper procedures, including equipment and personnel entrance and exit, gowning practices and personnel hygiene, are essential to reduce and/or minimize the human element of contamination.

Gowning refers to the practice of putting on (donning) protective gear, clothing, and uniforms to prevent bacteria and particulates from the ambient environment from entering the clean room. Gowning relies on the proper use of aseptic technique. Aseptic technique requires the use of various barriers and practices to prevent the transfer of microorganisms and other contaminates from personnel and the environment to the products being manufactured, inspected, labeled, or packaged.

Cleanroom garments must not be worn outside the cleanroom environment. Degowning refers to the practice of taking off (doffing) protective gear, clothing, and uniforms when leaving the cleanroom. Proper gowning and degowning is an essential contamination control practice.

Personnel hygiene is also essential to reduce and/or minimize the human element of contamination. Procedures addressing personnel hygiene should delineate employee health, bathing/showering, and clothing requirements.

The following gowning and degowning procedures and hygiene requirements are typical in sterile environments. These procedures are provided as an example. The extent of gowning procedures and hygiene requirements will vary by cleanroom classification, protocol, procedures, and risk of contamination.

Gowning and Hygiene

Example gowning procedure:

1. Walk on the tacky mats to remove any possible contamination on shoe soles.
2. Don the bouffant/hair net, ensuring all hair is contained.
3. Don shoe covers, ensuring to contain all laces and tassels.
4. Move to the clean side of the room.
5. Wash hands using bactericidal soap and dry them thoroughly.
6. Obtain the correct size hood, suit, and boots required for gowning.
7. Don the first sterile gloves following aseptic technique by touching only the inside of gloves. Sanitize the gloves. Sanitize gloves after donning each article, if required.
8. Adjust face mask for a snug facial fit. May be worn over or under hood.
9. Don hood, ensuring a snug fit and proper neck seal.
10. Carefully remove coverall from package by holding the inside neck area. Unfold garment while preventing it from touching the floor, your clothes, or bench. Don using aseptic technique. Exercise care when inserting leg to ensure donning snaps (if used) release and foot clears elastic ankle prior to transferring weight.
11. Put boots (high-top shoe covers) on using aseptic technique. Tuck coverall legs into boots. The first boot donned can now be placed in the "clean" side of gowning area.
12. Don second boot and place the other foot in the "clean" side of the gowning area.
13. Adjust goggles for a snug facial fit.
14. Complete gowning steps by aseptically donning second pair of sterile gloves over the first pair, making sure the cuffs are securely over the sleeves.
15. Perform a final inspection in gowning room mirror prior to entry into the sterile cleanroom.
16. Upon entering the cleanroom area, be sure to walk on the tack mat.

Example routine degowning procedure:

1. Enter gowning area from cleanroom and head directly to the "non-clean" side, usually divided by a bench and/or line.
2. Degown by removing boots, coverall, then face mask, hood, bouffant, shoe covers, and gloves.
3. Place garments into the correct receptacles or dispose of as appropriate.
4. Exit the gowning area.

(Note that this does not apply in the event of an emergency or evacuation where employees exit without degowning.)

Hygiene requirements:

- No personal items, including cell phones, jewelry, keys, watches, jackets, purses, matches, lighters, and cigarettes.
- No eating, drinking, or smoking.
- No gum or candy.
- No cosmetics or heavy use of aerosols, colognes, and perfumes.
- Employees should not enter the cleanroom if they are sick or have a communicable disease.
- Open wounds, sores, and body piercings must be covered.
- Approved skin lotions are sometimes allowed, to reduce skin flaking.
- Employees should bathe or shower daily when scheduled to work in the clean room.
- Employee street clothing should be clean.

Glove Donning Techniques

The closed-gloving technique is the preferred method for donning gloves. This method minimizes contamination of the gown and the gloves. Closed-gloving ensures that the hands only touch the inside of the gown and the gloves, keeping the outside of the gown sterile.

The open-gloving technique is used when changing a glove. This method uses a skin-to-skin and glove-to-glove technique. When donning the glove for dominant hand, the fingers of the nondominant hand touch only the cuff of the glove on the fold (skin-to-skin). To don the glove for the nondominant hand, the gloved fingers of the dominant hand touch only the outside of the glove (glove-to-glove).

The assisted-gloving technique is used when assisting another person with donning gloves. The gloves are held under the cuff and stretched open for the person to insert the hand. The sterile team member points the thumb away from the hand as it is placed into the glove to avoid touching the ungloved hand. When releasing the glove, the assistant pulls the cuff of the glove over the cuff of the gown.

Sources of Contamination

Austin (2000) provides estimated particle emissions for various sources in Table 27.7.

The Austin contamination index shown in Table 27.8 demonstrates estimated particle counts based on personnel activities, and control provided by various levels of gowning. These values can be used to determine relative risk of various activities and sampling strategies. These values are not intended as acceptance criteria. Each facility should establish their own acceptance criteria based on results and product requirements.

Gowning Recommendations

Table 27.9 provides typical cleanroom gowning recommendations based on ISO cleanroom classification. It should be obvious that as the cleanroom requirements become more stringent, the gowning requirements increase. Gowning requirements and procedures will vary by cleanroom classification, protocol, procedures, or other requirements.

Table 27.7 Gowning area particle generators.

Activity	**Number of particles (≥ = 0.3 µm)**
Person emits during garmenting process	3,000,000/min.
Cleanest skin (hands)	10,000,000/ft^2
Employee street clothes	10,000,000 to 30,000,000/ft^2
Floor and bench surfaces	> 10,000,000/ft^2
Garments supplied by cleanroom laundry	1,000,000/ft^2

Source: Philip Austin. 2000. *Encyclopedia of Clean Rooms, Bio-Cleanrooms, and Aseptic Areas*. 3rd ed. Livonia, MI: Acorn Industries.

Table 27.8 Austin contamination index.

Particles ≥ 0.3 µm emitted per minute in garment indicated.					
Personnel activity	**Snap smock**	**Standard coverall**	**Two-piece coverall**	**Tyvek coverall**	**Membrane coverall**
No movement	100,000	10,000	4,000	1,000	10
Light movement	500,000	50,000	20,000	5,000	50
Heavy movement	1,000,000	100,000	40,000	10,000	100
Change position	2,500,000	250,000	100,000	25,000	250
Slow walk	5,000,000	500,000	200,000	50,000	500
Fast walk	10,000,000	1,000,000	400,000	100,000	1,000

Source: Philip Austin. 2000. *Encyclopedia of Clean Rooms, Bio-Cleanrooms, and Aseptic Areas*. 3rd ed. Livonia, MI: Acorn Industries.

Table 27.9 Cleanroom gowning recommendations.

ISO class	4	5	6	7	8
Hair cover	✓	✓	✓	✓	✓
Hood	✓	✓			
Beard cover	✓	✓	✓	✓	✓
Face mask	✓	✓			
Frock				✓	✓
Coverall	✓	✓	✓		
Shoe covers	✓	✓	✓	✓	✓
Boots	✓	✓			
Gloves	✓	✓	✓		

FACILITIES CHANGE CONTROL

> Use various methods to verify that change control practices are in use to maintain the qualified state of the facilities. (Apply)
>
> **Body of Knowledge IV.A.3**

Change control is a system to ensure all changes are documented and evaluated for risk and impact on product prior to approval and implementation. Once the facility systems have been installed and qualified for use, any changes made to the facility or systems should be pre-approved, formally documented, and requalified if necessary to ensure these systems still function properly as designed.

The formal change control standard operating procedure (SOP) should define the review process to include changes to facility drawings (including HVAC ducting and piping diagrams) and equipment layouts. The change control process usually begins with a formal facility change request (FCR) or similar record to document the scope of the proposed change, potential effect of product, materials, or processes, maintenance and calibration activities, and whether the change will require a full or partial requalification/revalidation. It may also include requirements for project planning and management commensurate with the scope and complexity of the change.

A cross-functional team comprising of individuals from facilities, maintenance, engineering, quality assurance, quality control, regulatory, and others as appropriate should be included in the evaluation of the change for potential adverse impact affect product safety and efficacy before implementing the change.

Potential items to consider include new equipment, equipment move within the facility, equipment transfer, major maintenance activity, and change in utilities (air, nitrogen, water, electric, etc.)

Certain changes (such as relocation of a production line or moving the line from a clean room to a lesser controlled area) may also require submission to regulatory agencies.

A best practice is to place a qualified status label on the equipment—for example, IQ performed 8/1/2023, next review due 8/1/2028.

Chapter 28

Utilities

Utilities play an important part in the production of safe and effective biologics, pharmaceuticals, and medical devices. To consistently provide products that are safe and effective, utilities must be properly designed, installed, monitored, and maintained. Utilities include, but are not limited to, electrical power; compressed air; *heating, ventilation, and air conditioning* (HVAC) systems; steam; gases, including medicinal; water, including potable and *water for injection* (WFI); vacuum; and drains. Each of these items provides its own set of unique challenges to properly support production operations and prevent contamination and cross-contamination.

The selection and design of utilities should be made on the premise of properly supporting production operations and preventing contamination and cross-contamination. Considerations should be made for the types of materials used, washing, sterilizing, and depyrogenation. Use of medical-grade materials will help minimize contamination and reduce potential biocompatibility issues.

The United States Food and Drug Administration (FDA) Guidance for Industry, Q7A Good Manufacturing Practice Guidance for Active Pharmaceutical Ingredients, states:

> *All utilities that could affect product quality (e.g., steam, gas, compressed air, heating, ventilation, and air conditioning) should be qualified and appropriately monitored and action should be taken when limits are exceeded. Drawings for these utility systems should be available.*
>
> *Adequate ventilation, air filtration and exhaust systems should be provided, where appropriate. These systems should be designed and constructed to minimize risks of contamination and cross-contamination and should include equipment for control of air pressure, microorganisms (if appropriate), dust, humidity, and temperature, as appropriate to the stage of manufacture. Particular attention should be given to areas where APIs are exposed to the environment. If air is recirculated to production areas, appropriate measures should be taken to control risks of contamination and cross-contamination.*
>
> *Permanently installed pipework should be appropriately identified. This can be accomplished by identifying individual lines, documentation, computer control systems, or alternative means. Pipework should be located to avoid risks of contamination of the intermediate or API. Drains should be of adequate size and should*

be provided with an air-break or a suitable device to prevent back-siphonage, when appropriate.

Once utility systems have been designed and installed, qualification/validation should be done to ensure the system is capable. The company should determine and set operational, cleaning, sanitization, and sterilization parameters for the systems. Appropriate consideration should also be given to maintenance and calibration activities.

Quality-testing methods, along with acceptable thresholds, should be developed and documented in appropriate procedures and work instructions. These procedures should document who will conduct the sampling, how and when samples will be taken, the minimum required sample volume, handling and storage of the samples, chemical, biological, and particulate levels, and how to handle *out of specification* (OOS) results. Frequent monitoring (for example, temperature, pressure, velocity), maintenance, and calibration activities should be conducted according to established procedures to ensure that the system(s) are functioning as intended.

WATER SUPPLY SYSTEMS

> Identify and interpret regulatory requirements for the design of water supply systems including various unit operations (e.g., dechlorination, reverse osmosis, deionization, and distillation), delivery lines, back-flow or back-siphonage prevention, and drainage systems as appropriate for the type of water (potable, purified, and water for injection) needed in various processing steps. (Apply)
>
> **Body of Knowledge IV.B.1**

PLUMBING (21 CFR 211.48)

The GMP requirements for water state that potable water shall be supplied under continuous positive pressure in a plumbing system free of defects that could contribute contamination to any drug product. Potable water must meet the standards prescribed in the Environmental Protection Agency's (EPA) National Primary Drinking Water Regulations set forth in 40 CFR 141. Water not meeting such standards is not permitted in the potable water system. Drains must be of adequate size and, where connected directly to a sewer, must be provided with an air break or other mechanical device to prevent back-siphonage.

Supply of water to the facility may significantly contribute to the overall quality of the manufacturing operation. Although a facility cannot control the type of water it receives from a municipality, it should monitor the water on a scheduled basis and take appropriate actions to ensure the water meets the current EPA standards. The EPA specifies specific levels of organic, inorganic, and microbial contaminants that are allowed in municipal water supplies, and typically the municipality reports these levels annually. Pharmaceutical manufacturing facilities should exert increased diligence regarding the requirements, and may adopt a practice of testing the water more frequently based on product requirements and local environmental conditions (drought, excess rain or flooding). Increased contaminants in the incoming supply can overwhelm the water purification system and shorten the life span of associated filters, carbon beds, resins, and ozone or other components used to reduce the amount of contamination or other agents in the water supply.

Arguably, the most important design consideration for pharmaceutical water systems, is a risk assessment for the level of water quality that is required. Different products require different quality waters. Parenterals require very pure water with no endotoxins that could produce a pyrogenic response in patients. Topical and oral products require less pure water and do not have a requirement for endotoxins. Even with topical and oral products there are factors that dictate different qualities for water. For example, preservatives are marginally effective in antacids, so more stringent microbial limits have to be set. The Pharmaceutical Inspection Convention and Pharmaceutical Inspection Co-operation Scheme (PIC/S) P1 009-3 guidance details a seven-step process for pharmaceutical water systems that includes key design parameters, qualification, inspection, *quality control* (QC) testing, monitoring, maintenance and calibration, and documentation.

Design considerations for water used in the production of pharmaceuticals, including WFI, should follow *good engineering practices* (GEP) using a risk-based approach that is planned and structured throughout the entire life cycle of the system. The system should consider the required water quality and its intended use, the initial water source, and storage considerations. Distillation and reverse osmosis (RO) filtration are the only acceptable methods listed in the USP for producing water for injection (WFI). However, in the bulk pharmaceutical and biotechnology industries and some foreign companies, ultrafiltration (UF) is employed. UF utilizes pressure across a membrane with a pore size in the range of **0.003–0.01 µm** to minimize or remove large molecules like endotoxins. Water shall be supplied under continuous positive pressure in a recirculating system that prevents the possibility of entrainment and contamination. Additional design considerations should include weld quality, passivation, vent filters, suitability of construction materials, slope of pipe works, recirculation velocities and temperatures, check valves to prevent backflow, sanitary joints, capacity relative to demand, valves, draining and flushing, and sampling ports and locations. Distribution systems must also be designed to be sanitized in place (SIP) when necessary to prevent or remove contamination and maintain the system quality.

Another design consideration is the temperature of the system. Traditionally, hot (65–80°C) systems are accepted as self-sanitizing. While the cost of other systems may be less expensive, the cost of maintenance, testing, and addressing potential problems may be greater than the cost of energy saved by a lower temperature system. Whether a system is circulating or one-way is also an important design consideration. Obviously, water in constant motion is less liable to have high levels of contaminants because it is constantly re-purified. A one-way system is basically a "dead-leg" allowing for bacterial growth and endotoxins when not in use, especially supply lines to equipment that is no longer used daily.

The passivation of stainless-steel piping is also important. The passivation process uses acid to coat the surface, creating a thin, transparent oxide film and improving corrosion-resistance. When piping materials are not passivated, iron embedded in the surface during the manufacturing process can corrode and react with other materials, causing stains, discoloration, and product contamination.

Pretreatment of feedwater is recommended by most manufacturers of distillation equipment and is definitely required for RO units. The incoming feedwater quality may fluctuate during the life of the system depending upon seasonal variations and other external factors beyond the control of the pharmaceutical facility. Also, new construction or fires can cause a depletion of water stores in old mains which can cause an influx of heavily contaminated water of a different flora.

Water systems should be designed to operate within these anticipated extremes. However, the only way to know the extremes is to periodically monitor feedwater to capture seasonal and other local variation. If the feedwater is from a municipal water system, reports from the municipality testing can be used in lieu of in-house testing.

COMPRESSED AIR AND GAS SYSTEMS

> Identify and apply regulatory requirements related to compressed air and gas systems including storage, flow regulation, filtration, venting, and purging. (Apply)
>
> **Body of Knowledge IV.B.2**

Design considerations for compressed air and gas systems used in the production of pharmaceuticals should use GEP with a risk-based approach that is planned and structured throughout the entire life cycle of the system. The system should consider the required air or gas quality and its intended use, the initial source, demand capacity and storage considerations. Associated equipment must be suitable for its intended use and be adequately maintained. Piping components should be corrosion resistant and able to withstand the pressures involved.

Compressed air is usually generated on site with filters to remove lubricants, particles, and excess moisture. A significant amount of heat is also generated, which can compromise the system as well as materials stored in the same area. ISO 8573-1 recommends a specification for compressed air be used similar to ISO Class 2. However, caution is required with respect to particles, since this standard also stipulates that no particles larger than 5 μm may be contained in ISO Classes 1–5. Typical stainless steel pipe systems with central filters directly downstream of the central compressor may not meet this requirement. Compressed air networks are operated at flow velocities up to 80 m/s with pressure reducers and valves that can also discharge particles. The limits for particles in air should therefore be determined based on the cleanroom class where used and the specific product requirements. Filters to capture oil may be placed at the compressor and point of use and must be regularly inspected, maintained or replaced to ensure quality. A microbial retentive filter (0.2 µm or smaller) may also be placed at the point of use.

Other gases rely on certification by external suppliers and may be supplied in portable tanks near the point of use or from cryo bulk storage tanks located outside the facility and piped to the point of use. Piping must be compatible with the gas and pressures involved and prevent corrosion and contamination. Unique fittings and labeling should be considered to avoid mix-up if multiple gases are supplied. Gas suppliers should be appropriately qualified and capable of delivering the required certified grade and quality of gas.

UTILITY DESIGN FOR PRODUCTION

> Identify and select utility designs related to production steps (e.g., washing, sterilizing, and depyrogenation) for use with specific materials and processes. (Apply)
>
> **Body of Knowledge IV.B.3**

Utilities for production areas must be designed and selected to accommodate the activities and equipment intended for those areas and comply with all local codes and regulations. Sufficient connections must be provided at appropriate locations for both the intended equipment and current activities as well as flexibility for future modification to accommodate expansion and new process requirements. Distribution systems should have sufficient capacity, pressure, and flow rates for steam, water, compressed air, gas, exhaust venting, and vacuum utilities. Sanitary connections are often required to facilitate cleaning and prevent contamination. Suitable locations must also be provided for electrical connections with appropriate voltage and amperage. Ground Fault Circuit Interruption (GFCI) protection, either at the outlet or at the electrical box, must be provided whenever they are located near a water source. Access to utilities (electrical boxes, drain cleanouts, water shut-offs, etc.) must be located as to be easily accessible for maintenance while not interfering with production equipment or other activities. Large equipment may be hardwired and require dedicated separate shut-offs.

Consideration must also be given to cleaning, sanitation (where appropriate), and maintenance. Utensils and large equipment may require special wet cleaning areas with hot, cold, and purified water connections as well as suitable floor drains. Provisions should also be made for clean in place (CIP) and sterilize in place (SIP) equipment, which usually require flushing with chemicals or steam. Sufficient toilet and rest room facilities should be appropriately sized and conveniently located outside gowning and cleanrooms and should be vented outside under negative pressure as compared to other areas. Gowning rooms may require separate showers, hands-free sinks, and lockers prior to entering the cleanroom. Cleaning equipment and agents should be kept in a dedicated place isolated from active production areas. Provision for capture and disposal of used cleaning agents must also be considered.

UTILITIES DESIGN SPECIFICATIONS

> Review operations of utilities to ensure they meet design specifications. (Apply)
>
> **Body of Knowledge IV.B.4**

The design and layout of utilities and associated equipment in the production areas should be documented in a set of formal and controlled facility drawings showing construction materials, connections, and dimensions, as well as access points and sampling ports where needed. System quality requirements should also be documented, either on the drawings or in a formal design specification that defines the intended use(s) and required quality characteristics. Utilities and associated equipment should be installed per the manufacturer's recommendation unless otherwise necessary. In such case, the reason for the deviation should be explained, including why the intended functions will not be compromised. Further testing of any parameters possibly impacted may also be required.

A set of current drawings and equipment manuals should be maintained for all utilities, equipment, and critical installations. Once the utility systems have been installed, the installation should be qualified and validated to demonstrate the systems operate as intended and all design specifications are met. Validation typically includes a documented *installation qualification* (IQ) to verify the equipment was installed as intended, and an *operational qualification* (OQ) to verify that the "as built" installation conforms to the design drawings and operates as intended. These testing and acceptance criteria are usually specified and approved in a documented IQ/OQ protocol that describes the system and required testing to demonstrate the requirements are met. The protocol should also include a complete listing of equipment covered, instrument calibration requirements, and associated operating or maintenance procedures. A *failure mode effects analysis* (FMEA) or *fault tree analysis* (FTA) may be helpful in identifying critical control criteria to be included in the testing protocol.

This documentation and any test results should be compiled into a validation file that can be consulted during regulatory audits, failure investigations, or future improvement or expansion projects.

UTILITIES CHANGE CONTROL

> Verify that change control practices are used to maintain the qualified state of affected utilities. (Apply)
>
> **Body of Knowledge IV.B.5**

Change control is a system to ensure all changes are documented and evaluated for risk and impact on product quality prior to approval and implementation. Once the utility systems have been installed and qualified for use, any changes

- Out-of-specification (OOS) events
- Frequent out-of-calibration events
- Out-of-tolerance (OOT) events
- Additional load to the system
- Decreased load on the system
- Frequent breakdowns
- New technology

Figure 28.1 Conditions that may require the revalidation of utilities.

made to these systems should be pre-approved, formally documented and requalified if necessary to ensure these systems still function properly as designed. The formal change control standard operating procedure (SOP) should define the review process to include changes to facility drawings (including HVAC ducting and piping diagrams) and equipment layouts. The change control process usually begins with a formal *facility change request* (FCR) or similar record to document the scope of the proposed change, potential effect of product, materials, or processes, maintenance and calibration activities, and whether the change will require a full or partial requalification/revalidation. It may also include requirements for project planning and management commensurate with the scope and complexity of the change. Approvals should include appropriate departments, including operations, maintenance, and quality, before implementing the change. Certain changes (such as changing the method for water purification, use of new technology or increasing or decreasing capacity) may also require submission to regulatory agencies. Figure 28.1 provides examples of some scenarios that may require revalidation.

Chapter 29
Equipment

The basic *good manufacturing practices* (GMP) requirement is to ensure that production equipment is designed, located, and maintained to serve its intended purpose. This requirement helps ensure that active pharmaceutical ingredients (APIs) and drugs of consistent quality are fabricated. Equipment GMP are designed to prevent the contamination of drugs by other drugs, by dust, and by foreign materials such as rust, lubricant, and particles coming from the equipment. Contamination problems can occur due to poor maintenance, the misuse of equipment, exceeding the capacity of the equipment, and the use of worn-out equipment.

Major equipment should be identified with a distinctive number, name, or code that is recorded in batch records. This identification requirement is intended to help document which pieces of equipment were used to make which batches of drug product.

A label indicating the status of maintenance is not required by regulations; however, records demonstrating maintenance should be available. A firm should be able to provide records documenting the date a specific piece of equipment was last calibrated/maintained, the results or action, and when its next maintenance is scheduled. The absence of such documentation or failure to provide these records is considered a GMP deviation.

EQUIPMENT PLANNING

> Review equipment location, design, construction, installation, and maintenance based on the operations to be conducted. (Apply)
>
> **Body of Knowledge IV.C.1**

DESIGN

Equipment should be of appropriate design and adequate size, and the design and construction of equipment should permit cleaning, sanitizing, and inspection of the equipment. Equipment should be designed for closed or contained processes wherever appropriate. Where open equipment is used, or equipment is opened, appropriate precautions should be taken to minimize the risk of contamination. Equipment should be equipped with fittings that can be dismantled and cleaned, allowing validated *clean-in-place* (CIP) equipment to be dismantled for periodic verification. Equipment should be sterilizable/sanitizable, where appropriate, and have filter assemblies designed for easy dismantling. Equipment documentation should include design validation, installation, operation, and performance qualification documents.

Equipment should be suitably located for its intended use. Major equipment (for example, reactors, storage containers) and permanently installed processing lines used during the production of an intermediate or API should be appropriately identified. Fixed pipework should be clearly labeled to indicate the contents and, where applicable, the direction of flow. Equipment should be part of an appropriate, defined maintenance program. Equipment should only be used within its qualified operating range, and dedicated production equipment should be provided where appropriate. Where applicable, chain drives and transmission gears should be enclosed or properly covered. Tanks, hoppers, and other similar fabricating equipment should be equipped with covers.

MATERIALS OF CONSTRUCTION

GMP requirements stipulate that equipment parts that come in contact with raw materials, intermediates, APIs, in-process drugs, or drugs should be designed and constructed so that they are accessible for cleaning or are removable. Equipment parts should be designed and constructed so that surfaces that contact raw materials, intermediates, or APIs do not alter the quality of the intermediates and APIs beyond the official or established specifications. The parts of the production equipment that come into contact with the product must not be reactive, additive, or absorptive to such an extent that it will affect the quality of the product and present any hazard. Surfaces that come in contact with raw materials, in-process drugs, or drugs should be smooth and made of material that is nontoxic, corrosion resistant, nonreactive to the drug being fabricated or packaged, and capable of withstanding repeated cleaning or sanitizing. Equipment made of material that is prone to shed particles or to harbor microorganisms should not contact or contaminate raw materials, in-process drugs, or drugs.

LUBRICANTS

The design of equipment should be such that the possibility of a lubricant or other maintenance material contaminating the drug is minimized. Any substances associated with the operation of equipment, such as lubricants, heating fluids, or coolants, should not contact intermediates or APIs so as to alter their quality beyond the official or established specifications. Any deviations should be evaluated to ensure that the fitness for purpose of the material is not affected detrimentally. Wherever possible, food-grade lubricants and oils should be used.

EQUIPMENT LAYOUT

> Determine the layout of equipment to minimize the risk of errors, to facilitate effective cleaning and maintenance, and to avoid contamination or any other undesired effect on product quality. (Apply)
>
> **Body of Knowledge IV.C.2**

Equipment should be arranged in an orderly manner that permits cleaning of adjacent areas and does not interfere with other processing operations. Equipment should be:

- Located to suit its intended purpose
- Located so that production operations undertaken in a common area are compatible, and so that cross-contamination between such operations is prevented
- Located to optimize the flow of material and to minimize the circulation of personnel (unidirectional flow)
- Installed to prevent any risk of error or of contamination, and to ensure that equipment is not operated where contaminants may fall into the material
- Easily cleaned, sanitized, and inspected
- Located at a sufficient distance from other equipment and walls to permit cleaning of the equipment and adjacent area
- Sealed at the base of immovable equipment along points of contact with the floor

EQUIPMENT CLEANING AND MAINTENANCE

> Review procedures and schedules for equipment cleaning, maintenance, and sanitization (where necessary) to ensure that they meet requirements. (Apply)
>
> **Body of Knowledge IV.C.3**

Equipment and utensils should be cleaned, stored, and, where appropriate, sanitized or sterilized to prevent contamination or carryover of a material that would alter the quality of the intermediate or API beyond the official or other established specifications.

Cleaning GMP require that procedures contain sufficient details to enable operators to clean each type of equipment in a reproducible and effective manner.

Cleaning procedures must strictly follow carefully established and validated methods. These procedures should include:

- Personnel responsible for carrying out cleaning procedures
- A requirement to identify equipment as to contents and cleanliness status
- Intervals for cleaning/sanitizing, including the maximum time that may elapse between the completion of processing and equipment cleaning (dirty hold time), when appropriate
- Instructions for the removal or obliteration of previous batch identification
- When appropriate, instructions for disassembling and reassembling equipment to ensure proper cleaning
- Complete description of agents, materials, and equipment/apparatus to use for cleaning and disinfection, including dilution of cleaning agents used to clean equipment
- Disposal procedures for waste material and debris
- Instructions for storing equipment in clean and dry conditions
- Instructions for the protection of clean equipment from contamination before use
- Inspection of equipment for cleanliness immediately before use

Maintenance

Elements of a maintenance program should include a way to identify equipment with a distinctive name, number, or code. The equipment must be maintained in a good state of repair when in use, and where a potential for the contamination of the drug being fabricated or packaged exists, surfaces must be free from cracks, peeling paint, and other defects. Gaskets must be verified as functional, and the use of temporary devices such as tape is avoided. Equipment parts that come in contact with drugs must be maintained in such a manner that drugs are fabricated or packaged within specifications. Finally, records of maintenance documenting the performance of these activities must be kept.

Schedules and procedures (including assignment of responsibility) should be established for the preventive maintenance of equipment. Defective equipment should, if possible, be removed from production and quality control areas, or at least be clearly labeled as defective. For equipment not requiring maintenance, the firm must be able to support its decision to not include the equipment in the maintenance program.

EQUIPMENT CLEANING VALIDATION OR VERIFICATION

Evaluate the need and methodology for product-contact cleaning validation and verification. Evaluate the difference between single-use, multi-use, and dedicated equipment. (Evaluate)

Body of Knowledge IV.C.4

Regulations require that APIs and drug products be fabricated and packaged in areas that are free from environmental contamination and free from contamination by another drug. Equipment for washing and cleaning should be chosen and used in a manner so as not to be a source of contamination.

The objective of the cleaning validation is to verify the effectiveness of the cleaning procedure for removal of product residues, degradation products, preservatives, excipients, and/or cleaning agents, as well as the control of potential microbial contaminants. In addition, the company needs to ensure that there is no risk associated with cross-contamination of active ingredients.

Equipment should be cleaned at appropriate intervals to prevent buildup and carryover of contaminants (for example, degradants or objectionable levels of microorganisms). A written cleaning (sanitization) process validation program provides some assurance that levels of cleanliness in the plant are maintained. The cleaning validation program should demonstrate that cleaning procedures are robust and effective, establish the maximum time that may elapse between the completion of processing and equipment cleaning (dirty hold time), establish the maximum time that equipment may be held clean before use (clean hold time), and document and justify the amount of carryover from batch to batch.

Cleaning validation is performed on equipment used in continuous or campaign production of successive batches of the same intermediate or API, and non-dedicated equipment cleaned between production of different materials to prevent cross-contamination.

GMP of Cleaning Validation and Analytical Methods

GMP state that a cleaning program must be implemented and be effective in preventing unsanitary conditions. Cleaning procedures for manufacturing equipment are validated based on the cleaning validation guidelines. Residues from the cleaning process itself (for example, detergents, solvents) are removed from equipment during cleaning.

It is necessary to provide evidence demonstrating that routine cleaning and storage does not allow microbial proliferation, and that analytical methods used to detect residues or contaminants are validated. The analytical methods used to detect residuals or contaminants should be specific for the substance or the class of substances to be assayed (for example, product residue, detergent residue, and/or endotoxin) and be validated before the cleaning validation study is done.

In the case of biologic drugs, the use of product-specific assay(s) such as immunoassay(s) to monitor the presence of biological carryover may not be adequate; a negative test may be the result of denaturation of protein epitope(s). Product-specific assay(s) can be used in addition to *total organic carbon* (TOC) for the detection of protein residue.

The analytical method and the percentage recovery of contaminants should be challenged in combination with the sampling method(s) used to show that contaminants can be recovered from the equipment surface, and to show the level of recovery and the consistency of recovery. This step is necessary before any conclusions can be made based on the sample results. A negative test may be the result of poor sampling technique. Acceptance criteria for residues and the choice of cleaning procedures and cleaning agents should be defined and justified, and equipment should be visually examined to verify cleaning.

Cleaning Validation and Sampling, Rinsing, Rinse Samples, and Detergents

Two types of sampling that are generally considered to be acceptable are *direct surface sampling* (swab method) and *indirect sampling* (use of rinse solutions). A combination of the two methods is the most desirable, particularly in circumstances where accessibility of equipment parts can mitigate against direct surface sampling:

- *Direct surface sampling.* Areas hardest to clean but that are reasonably accessible can be evaluated by the direct surface sampling method. This method allows establishment of a level of contamination or residue per given surface area. This method ensures that dried out or insoluble residues can be sampled.

- *Rinse samples.* Rinse samples allow sampling of a large surface area and of inaccessible systems or ones that cannot be routinely disassembled. Limitations include insoluble contaminants or contaminants remaining on the equipment surface.

Indirect testing, such as conductivity and TOC testing, may be of some value for routine monitoring once a cleaning process has been validated. This process could be applicable to reactors or centrifuges and piping between equipment so large it can only be sampled using rinse solution samples.

When detergents are used in the cleaning process, their composition should be known to the user, and their removal should be demonstrated. The manufacturer should ensure that they are notified by the detergent supplier of any changes in the formulation of the detergent. Acceptable limits should be defined for detergent residues after cleaning. The possibility of detergent breakdown should be considered when validating cleaning procedures. Indirect testing such as conductivity and TOC testing may be of some value for assessing the effectiveness of removing the cleaning agent.

Water for injection (WFI) should be used as the last rinse for product-contact equipment used in the fabrication of sterile products. Purified water is considered acceptable as the last rinse for product-contact equipment used in the fabrication of nonsterile products or sterile products for ophthalmic use. Tap water should not be used in the last rinse of any cleaning procedure for product-contact equipment because of the presence of varying levels of organic and inorganic residues, as well as chlorine.

Establishment of Cleaning Validation Limits

No established standard acceptance limits exist for cleaning validation. Companies must establish limits that reflect the capability of the cleaning processes and analytical test methods used. Considerations for cleaning include evaluation of the therapeutic dose carryover, toxicity of the potential contaminant, concentration of the contaminant in rinse and swab samples, removal of cleaning agents, limit of detection of the analytical test method, and identification of critical areas (hardest to clean).

The approach for setting acceptance limits can be product-specific cleaning validation for all products; grouping into product families and choosing a worst-case product; grouping by properties (for example, solubility, potency, toxicity, or formulation ingredients known to be difficult to clean); setting limits on not allowing more than a certain fraction of carryover; and different safety factors for different dosage forms.

Carryover of product residues should meet defined criteria, for example:

- No more than (NMT) 0.1% of the normal therapeutic dose of any product is to appear in the maximum daily dose of the following product.
- NMT 10 ppm of any product is to appear in another product.
- No quantity of residue may be visible on the equipment after cleaning procedures are performed. Spiking studies should determine the concentration at which most active ingredients are visible.
- For certain highly sensitizing or highly potent ingredients (such as penicillins, cephalosporins, or potent steroids and cytotoxics), the limits should be below the limit of detection by the best available analytical methods. In practice, this may mean that dedicated plants are used for these products.

If levels of contamination or residuals are not detected, it does not mean that no residual contaminant is present after cleaning. It only means that the levels of contaminant greater than the sensitivity or detection limit of the analytical method are not present in the sample.

Documentation

Required documentation for cleaning validation includes detailed cleaning procedure(s) documented in *standard operating procedures* (SOPs), and a cleaning validation protocol defining how the cleaning process will be validated. The protocol should include the objective of the validation process; responsibilities for performing and approving the validation study; description of the equipment to be used; the interval between the end of production and the beginning of the cleaning procedure; the number of lots of the same product that could be manufactured during a campaign before a full cleaning is done; detailed cleaning procedures to be used for each product, each manufacturing system, or each piece of equipment; the number of cleaning cycles to be performed consecutively; and any routine monitoring requirement. Other information in the cleaning protocol includes sampling procedures, including the rationale for why a certain sampling method is used; clearly defined sampling locations; data on recovery studies, where appropriate; validated analytical methods, including the limit of detection and the limit of quantitation of those methods; the acceptance criteria, including the rationale for setting the specific limits; other products, processes, and equipment for which the planned validation is valid according to a bracketing concept; change control and revalidation; and approval by management

Manual cleaning procedures are inherently variable. Operators should be trained, monitored, qualified, and periodically assessed on the effectiveness of performing cleaning operations. Manual cleaning methods should be reassessed at more frequent intervals than CIP systems. Cleaning processes should be reassessed at defined intervals, and revalidated as necessary. As a general rule, if cleaning validation has not been performed, equipment should be dedicated to prevent cross-contamination.

EQUIPMENT CHANGE CONTROL

> Verify that change control has maintained the qualified state of equipment. (Apply)
>
> **Body of Knowledge IV.C.5**

A change control system should be in place to ensure that impact to cleaning processes or functions is assessed, documented, and approved. Methods for verifying that a change control system is defined and implemented should include a way to verify that a documented procedure is in place. The procedure should require that changes are properly documented, evaluated, and approved by the quality function. The implementation/effective date of the change should be identified. Significant changes are assessed to determine whether revalidation is necessary. A review of nonroutine maintenance, calibration, or nonconformance records can be used to determine whether changes to equipment have been made using appropriate change controls. Examples of changes that require evaluation and may require revalidation include:

- Changes to the cleaning procedure
- Changes in raw material sources
- Changes to product formulation
- Changes to the production process
- Introduction of new products, new cleaning agents or detergents, or changes to cleaning agents or detergents
- Modifications to equipment

Chapter 30

Qualification and Validation

> Verify that the qualifications and validations of facilities, equipment, and utilities are conducted in accordance with various requirements including factory and site acceptance testing (FAT/SAT), design, installation, operational, and performance qualification (DQ/IQ/OQ/PQ) prior to process validation. (Analyze)
>
> **Body of Knowledge IV.D**

INTRODUCTION

In the pharmaceutical environment, qualification and validation are required for any *good manufacturing practices* (GMP) facility, utility, equipment, or process. The intent of qualification and validation activities is to provide documented evidence that the facility, utilities (for example, water, gases, air), and processes are designed and operate in accordance with GMP requirements. An infinite amount of detail can be provided in each phase of the validation life cycle. Regulations are vague regarding determining specific requirements for what constitutes a qualification or validation document. Industry standards have provided content and activities of the qualification and validation documentation. Figure 30.1 depicts a process validation decision tree used to determine whether and when a process needs validation.

To ensure that the facility, utilities, and equipment have been designed, built, and installed properly and operate as intended, proper qualification and validation activities must occur. The foundation of a successful qualification and validation program begins with design. During the design phase, the specifications, requirements, and critical aspects are developed and documented. *Specifications* are prescribed attributes that are required to provide a quality of work for something to be built, installed, or manufactured. *Requirements* are detailed necessities for the item being built, installed, or manufactured. Requirements can be defined as *user* requirements or as *functional* requirements. *Critical aspects* are specification and design requirements that focus on those attributes that are critical to product quality and patient safety, and these aspects may be outlined within the specification and requirements documents. Another document that is required to

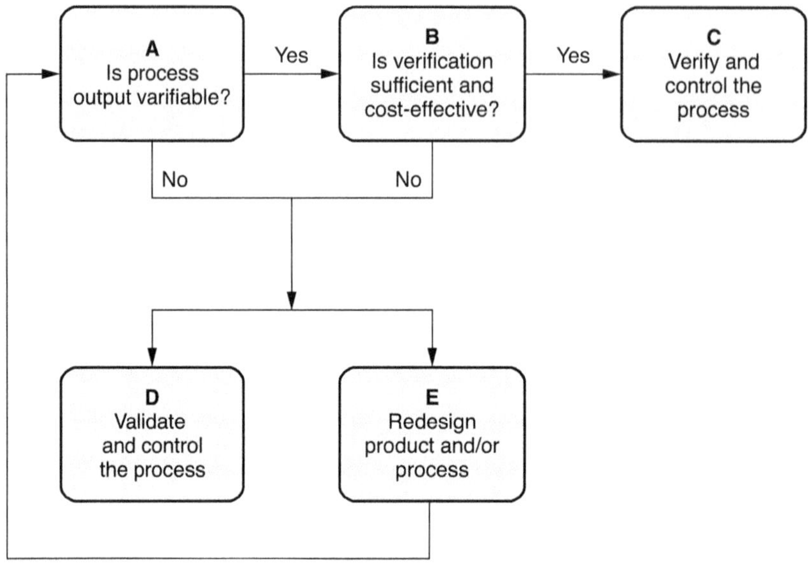

Figure 30.1 Process validation decision tree.
Source: Adapted from GHTF Study Group 3. *Quality Management Systems—Process Validation Guidance*. January 2004.

outline and plan the requirements for the life cycle process is a *validation master plan* (VMP).

VALIDATION MASTER PLAN (VMP) AND MASTER VALIDATION PLAN (MVP)

The VMP/MVP document defines the overall structure, philosophy, and approach of the phases of the validation life cycle. The VMP should define the structure and required documentation for design, qualification, and validation activities, including prerequisites and stage-gate requirements to move on to the next phase in the validation life cycle. The VMP should detail key criteria within each stage and phase, and other qualification and validation requirements and considerations, including lists of products, processes, and systems to be validated, documentation format, required *standard operating procedures* (SOPs), and planning and scheduling. Consideration should also be given to triggers (times, cycles, changes, trends, and so on) for revalidation and/or retrospective validation activities.

The facility, utilities, and equipment should be designed to predetermined requirements and specifications. The critical aspects are used to create and document the design requirements and specifications. The design is documented through *design qualification* (DQ).

DESIGN QUALIFICATION

Design qualification should establish and provide evidence that the equipment is designed in accordance with the requirements of GMP. Design qualification comprises a formal document that requires *quality assurance* (QA) oversight. This document uses critical aspects from approved requirements (user and functional requirements) and other approved specifications as the source documents. Typical items that may be found in the scope of a DQ document include verification that:

- The design will achieve the user requirements.
- The design is GMP and will conform to applicable national standards and guidelines.
- The utility services are appropriate and are qualified or will be qualified.
- All the required support documentation is specified.
- The system will be able to be calibrated.
- The system will be able to be maintained.
- Personnel training requirements will be met.

At the completion of the design phase, the project moves to the next phase—the commissioning and installation qualification phase.

COMMISSIONING AND QUALIFICATION

Once the facility, utility, or equipment has been designed, built, and/or installed, a commissioning program should encompass the additional testing to establish a qualified state, which can be performed through *factory acceptance testing* (FAT), testing that is performed at the vendor's factory before shipping the equipment. This task is important, as equipment that does not meet the requirements of design should not be shipped to the manufacturing facility until it does meet the necessary specifications and requirements. Once the equipment or system is at the manufacturing site and its final destination, as part of a commissioning, a *site acceptance testing* (SAT) exercise may be conducted. As part of commissioning, these activities are prerequisites to site qualification exercises.

The commissioning program should encompass as much testing as possible, to include, but not be limited to, testing the vendor or manufacturer's functional and operational requirements, the design requirements, and other aspects that are critical to quality, business continuity, and safety. This ensures that the systems and equipment are challenged to test their full functionality and operational ability in a noncommercial environment to avoid any potential impact to GMP equipment, areas, and/or product. Qualification activities may be introduced in the commissioning program as long as the program is well documented and QA oversight exists. The commissioning program requires *good documentation practices* (GDP) and *good engineering practices* (GEP) to ensure that the testing activities that are conducted are properly documented and can be used for qualification activities.

Many regulatory guidance documents, such as European Union (EU) Annex 15, United States Food and Drug Administration (FDA) process validation guidance (*Process Validation: General Principles and Practices*), and Pharmaceutical

Inspection Convention and the Pharmaceutical Inspection Co-operation Scheme (PIC/S) P006-3 guidance, require and discuss the elements of *installation qualification* (IQ) and *operational qualification* (OQ). Some additional publications in the last few years include installation and operational qualifications. ASTM E2500 *Standard Guide for Specification, Design, and Verification of Pharmaceutical and Biopharmaceutical Manufacturing Systems and Equipment* introduced concepts of streamlining the traditional IQ and OQ program to eliminate some of the formal documentation requirements and QA oversight that are typically required for formal validation programs. The ASTM E2500 standard proposes the concept of verification, where traditional qualification activities are conducted as verification exercises. The QA oversight is acceptable during the development and approval of requirements, specifications, and critical aspects. The commissioning and verification test requirements and acceptance criteria are defined in a verification plan. QA is not involved at the implementation level of protocol writing and test plan execution as long as the approved verification plan has not been deviated from. Although new standards and publications exist with the intent of providing guidance on performing and documenting commissioning, qualification, and/or verification activities, the relevant intent of installation and operational qualification is that the facility, utility, and equipment/systems are designed properly, installed properly, and operate as intended by design and per GMP requirements.

It is a requirement of the regulatory guidance documents and GMP that the documentation provided by each pharmaceutical company proves a state of control in that the "drug meets the requirements of the Act as to safety, and has the identity and strength and meets the quality and purity characteristics that it purports or is represented to possess." The commissioning and qualification program must have adequate QA governance.

INSTALLATION QUALIFICATION

Elements of the traditional IQ may be incorporated into the commissioning program. The regulatory guidance documents such as EU Annex 15 and FDA process validation guidance require IQ and OQ. Although the installation verification may have occurred during the commissioning stage, the regulatory expectation is that an IQ protocol and report documents that the facility, utilities, and/or equipment were installed properly. The intent of commissioning and/or the IQ is to verify that the facility, utilities, and/or equipment are installed properly. The IQ program includes, but is not limited to:

- Verification of components and parts
- Identification and verification of serial numbers and model numbers
- Identification and documentation of software versions and correct level of installation
- Verification of calibration
- Verification that installation is as specified per appropriate drawings and specifications (including sloping, dimensions, etc.)
- Verification that support utilities conform to national standards and other guidance documents

OPERATIONAL QUALIFICATION

Operational qualification is the documented evidence that the system operates as designed and for its intended use. Commissioning activities can include operational testing, which should include the full range of parameters for which the equipment was designed. The OQ testing may be reduced if full testing was conducted in commissioning, and may include only challenging the ranges or parameters of the intended use of the system. The OQ program includes, but is not limited to:

- Verification of integrated loop testing
- Testing of alarms
- Testing of interlocks and permissive conditions
- Testing of data storage integrity
- Verification of the functionality of the equipment
- Testing of security levels to prevent unauthorized access or changes
- Testing to verify and document power loss recovery
- Verification of procedures (operation, preventive maintenance, and calibration)
- Challenge of the system by stressing the equipment or system to the edge of failure to determine proven acceptable ranges

Once the equipment is qualified, the equipment can be released for *performance qualification* (PQ) and commercial manufacturing. The facility, utility, and equipment will be under a formal change control system and GMP environment. The OQ must be successfully completed before beginning PQ.

PERFORMANCE QUALIFICATION

Performance qualification confirms that the equipment or integrated systems function under normal operating conditions to ensure that the equipment/system functions as intended. During the PQ phase, systems, processes, and utilities are challenged at nominal operating characteristics. This process includes studies that ensure the system functions properly under various conditions, such as evaluating seasonal changes for the water system, evaluating a circuit or piping/equipment configuration for steam in-place validation, or evaluating load patterns for sterilization in autoclaves. The PQ program includes at least three consecutive successful runs or qualifications within the study. The definition of a run or qualification can be designated and predetermined by a description and definition of the run or qualification in the protocol or VMP.

The PQ program is a confirmatory exercise of verification of predefined critical aspects and critical quality attributes. Elements traditionally in a PQ include, but are not limited to:

- Definition of performance criteria and test procedures
- Identification of critical control parameters, critical quality attributes, and key process parameters (for example, predefined specifications and acceptance criteria)

- Determination of the sample size and test intervals (including the rationale to support these requirements)
- Definition of corrective actions if the system does not meet the established criteria

Upon completion and successful verification of PQ, validation of the commercial manufacturing process may begin. The PQ may not be entirely completed but may be released for commercial manufacturing or process validation activities. Stage II PQ for utilities may not be completed (due to required seasonal testing); however, stage I of utility PQ must be completed. The system has demonstrated the required performance characteristics at the end of stage I.

PROCESS VALIDATION

The traditional term *process validation* under the FDA process validation guidance document covers all elements of the validation life cycle (Figure 30.2). The FDA document refers to that activity as a *process performance qualification* (PPQ). For the purposes of this chapter, the term "process validation" will be retained as its traditional activity where the verification of the manufacturing process is completed.

Cleaning validation may be conducted during the PQ phase or the traditional process validation phase in parallel with the manufacturing of the validation batches. The details of the philosophy and approach for process and validation activities must be detailed in the site's VMP. Validated laboratory test methods are required for the execution of process and cleaning validation studies.

After successful completion of the process validation batch campaign, the next stage in the validation life cycle approach is the *operational phase*, or continuous

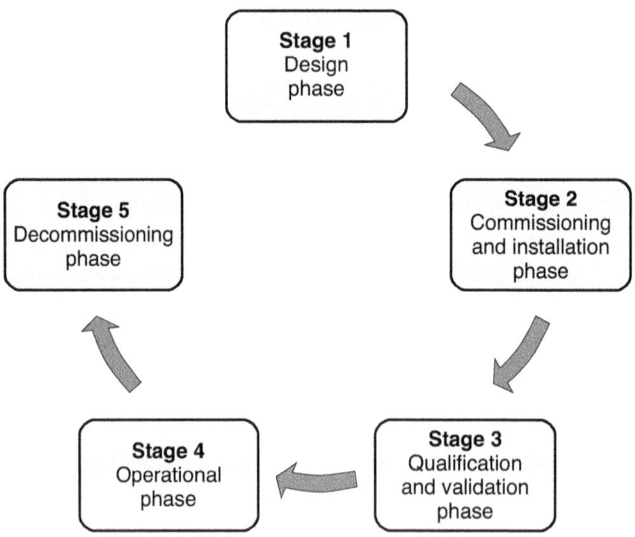

Figure 30.2 Validation life cycle.

Source: Figure courtesy of the ASQ Food, Drug, and Cosmetic Division. Used with permission.

monitoring/sustaining of operations, also termed *continued process verification*. During routine manufacturing, continuous monitoring of the critical quality attributes provides a continuous loop of information. The continuous monitoring aspect of the validation life cycle creates a feedback mechanism that is used to improve the process. Any proposed changes to improve the process must be formally documented in the change control system. Information gained during the continuous monitoring program is feedback into various systems, including updating and improving calibration and preventive maintenance systems.

VALIDATION LIFE CYCLE

Validation is a prospective and confirmatory exercise. The difference between *qualification* and *validation* is that equipment is individually qualified, and the system or process is validated. The intent of validation activities, including PQ and process validations, is to confirm the knowledge gained during design, development, installation, and challenge exercises, and as a confirmatory exercise of a process that is already known and in control. A few types of validation are discussed in some regulatory guidance documents, which include prospective, retrospective, and concurrent. The chapter has discussed and outlined *prospective validation*. Prospective validation is the most preferable and defendable form of validation. This type of validation aligns with the validation life cycle approach and provides the most confidence in the documentation generated that a drug is produced as required by GMP.

Retrospective validation is a look-back approach to validation and an attempt to justify legacy systems based on a retrospective gathering of documentation. Retrospective validation is not normally considered acceptable. There is an expectation of understanding and documenting processes, including the site's legacy processes and equipment. Retrospective validation is difficult to document, as the data gathered must be statistically based and analyzed.

Concurrent validation occurs when manufactured batches are released one lot at a time before the completion of the three (or other designated number) batches for the process validation campaign, and should only be used when drugs are medically necessary and an approved plan that outlines the nature of concurrent release is approved by the relevant regulatory agencies.

Qualification and validation are regulatory-guided programs for GMP pharmaceutical companies. Although these programs are mandated by various regulatory bodies, the intent of a robust qualification and validation program is to ensure that facilities, utilities, equipment, systems, and processes are designed with the end in mind. Requirements are tested and verified to predetermined specifications, and there is verification that the facilities, utilities, or equipment are installed and operate properly. These programs provide documented evidence that what was purchased meets expectations and produces a drug product that meets the requirements of 21 CFR 210.1.

Chapter 31

Maintenance and Metrology Systems

MAINTENANCE PROCEDURES

> Verify that procedures are used for routine and non-routine maintenance of heating, ventilation, air conditioning (HVAC) systems, air and water filters, and other equipment and utilities. (Analyze)
>
> **Body of Knowledge IV.E.1**

Regulations, including 21 CFR Part 211, require the manufacturer to establish and maintain procedures for the maintenance of equipment, utensils, and facilities used in the manufacture, processing, packing, or holding of a drug product.

Procedures for maintenance should be established and follow the manufacturers' recommendations unless otherwise justified. Procedures should describe the frequency, methods, and responsibilities.

When major maintenance activities occur, a documented decision should be made to determine the qualification or validation of the facility, infrastructure, or equipment was impacted by the maintenance event.

Maintenance schedules should be adjusted based on performance. For example, a maintenance interval may be decreased from annually to semiannual based upon gage usage.

Records for preventive and corrective maintenance events should be maintained.

METROLOGY CHANGE CONTROL

> Verify that appropriate calibration and engineering/equipment change control procedures are used and that a metrology program exists for the calibration of instruments that control manufacturing facilities, utilities, and equipment. (Analyze)
>
> **Body of Knowledge IV.E.2**

All instruments that control manufacturing facilities, utilities, and equipment must be maintained to ensure they are functioning as intended. Instrumentation should be uniquely numbered so it can be easily identified during calibration, maintenance, commissioning, and validation. Each instrument should have appropriate unique identification labeling permanently affixed; if the item is too small, the container that holds the device can be appropriately labeled. Associated documentation will link this unique identification to the instrument type, model, and serial number. Documentation includes the date an instrument was put into service, its intended range of use, and the date it was retired.

Based on the instrument category, *installation qualification* (IQ), *operational qualification* (OQ), and *performance qualification* (PQ) may be required when a new instrument, facility, utility, or equipment is introduced or replaced, or if instrumentation is moved from one location to another. A protocol should be written to define the purpose/objective, test procedures to be followed, parameters to be evaluated, and the criteria to demonstrate acceptability.

The IQ is conducted to ensure that the instrument and software, if applicable, were installed in the proper environment and manner as intended by the manufacturer. The OQ is conducted to ensure that the instrument and software, if applicable, operate as designed and generate accurate data. The PQ is conducted to ensure that the instrument and software, if applicable, meet the end user–specific requirements and function as intended in the production environment. A report is generated to summarize the test results for the qualifications and any deviations. Upon completion of the qualification, the instrument can be put into service.

Facilities, utilities, or equipment used in the *good manufacturing practices* (GMP) environment will require routine maintenance and calibrations. The maintenance and calibration schedules can be established based on severity of environment, severity of use, frequency of use, durability of instrument, accuracy of measurement required, precision of measurement required, and history of maintenance and calibration. Each type of instrumentation should have a procedure that specifies its operation, maintenance, and calibration. The procedure must specify what activity should be conducted, what the acceptance criterion is, and the frequency. A log book or software program may be used to track usage, repair, and calibration. All maintenance and calibration information, which includes standards used, must be documented. When an instrument requires repair or calibration, it should be flagged as *out of service* (OOS) in some manner, such as with an "out of service" sticker. If an instrument fails the calibration, an impact assessment must be performed. During the impact assessment, all data generated since the last successful calibration must be evaluated. Upon completion of the repair or calibration, the OOS indicator is removed and the instrument is placed back into service. If the instrument was being calibrated, a new calibration sticker (Figure 31.1), which designates the next calibration due date, is applied.

Calibration standards used for inspection, measuring, and test equipment (IMTE) shall be traceable to national or international standards. If national or international standards are not practical or available, the manufacturer shall use an independent reproducible standard. If no applicable standard exists, the manufacturer shall establish and maintain an in-house standard. The calibration must cover the intended range of use. That is, if a scale is to measure items weighing between 50 grams and 100 grams, the calibration should minimally cover the range. The use of single-point calibrations should be avoided, and if unavoidable, should be justified.

When pre-established requirements for calibration accuracy and/or precision limits are not met, provisions for remedial action should be documented including the determination if there were any adverse effects on materials and products.

MEASUREMENT SYSTEMS ANALYSIS (MSA) OR TEST METHOD VALIDATION (TMV)

Consideration should also be given to ensuring that the inspection, measuring, and test equipment is suitable for its intended purposes and is capable of producing

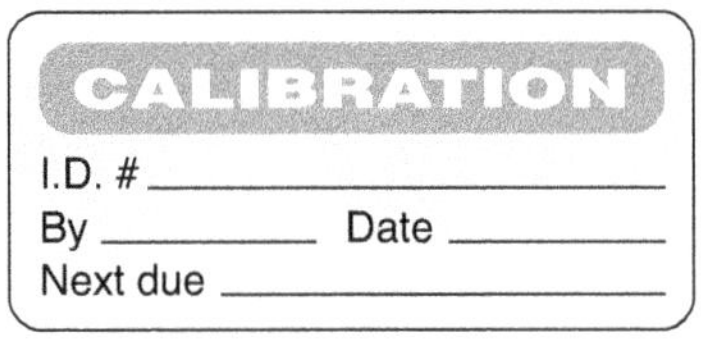

Figure 31.1 Example calibration label.

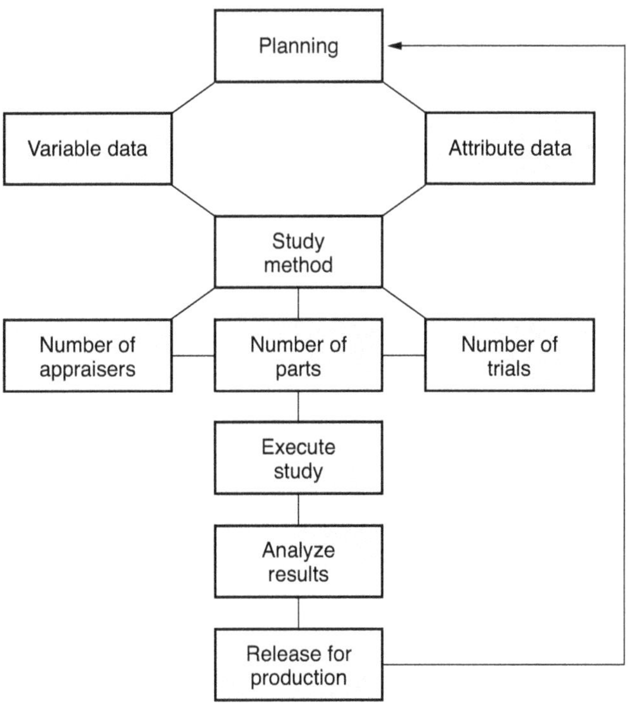

Figure 31.2 MSA/TMV study life cycle considerations.

Source: Mark A. Durivage. 2015. *Practical Attribute and Variable Measurement Systems Analysis (MSA)*. Milwaukee: ASQ Quality Press.

valid results. Suitability is generally assessed through MSA or TMV activities. MSA/TMV studies should be part of a life cycle (Figure 31.2) with decisions based on risk.

Procedures for conducting MSA/TMV studies should also state the environmental conditions under which the measurements are to be made, such as temperature, humidity, and lighting. Additionally, training that demonstrates operator proficiency should be documented.

Chapter 32

General Cleaning, Sanitization, and Sterilization Systems

CLEANING PROCEDURES

> Review cleaning procedures in accordance with cleaning validation whenever validation is required and performed. (Apply)
>
> **Body of Knowledge IV.F.1**

"Washing and cleaning equipment should be chosen and used in order not to be a source of contamination." This statement is a fundamental starting point for any discussion of *good manufacturing practices* (GMP) equipment and parts cleaning. In a pharmaceutical manufacturing environment, rooms and areas are classified based on the risk to the product. Four grades (A, B, C, and D) or classifications (ISO 100 to 100,000) of areas allow various levels of particulates or microbial challenges. These areas extend from cleanest (grade A) to less stringent requirements (grade D).

Sponges and mops used in grades A or B must be of material and design to minimize shedding of particles. Large and small GMP equipment and parts amenable to *clean out of place* (COP) must have facilities or a set of suites equipped with the necessary utilities (for example, water meeting World Health Organization [WHO] drinking water standards, purified water, and possibly water for injection [WFI]). Take-offs and/or hoses from the purified water and WFI loops must have mechanisms (for example, backflow preventers/check valves) or procedures that prevent back-siphonage of cleaning solutions or detergents. Flow of equipment and/or cleaning facility layout must preclude dirty equipment and parts from contaminating or mixing with clean equipment and parts. Annex 1 of the European Union (EU) GMP guidelines section on premises prohibits sinks and drains in grade A or B areas. Therefore, equipment and parts cleaning may not occur in these aseptic core suites. If equipment and parts cleaning suites exist within grade C or D areas, any floor drains should be fitted with traps or water seals to prevent backflow. Sink drains, glassware or equipment washer drains, and autoclave drains must have air breaks between the machine or sink and the drains.

GMP equipment and parts cleaning must be proceduralized to ensure consistency over time and from operator to operator. Pharmaceutical product and *active pharmaceutical ingredients* (APIs) can be contaminated by other pharmaceutical products or APIs, by cleaning agents, by microorganisms, or by other material (for example, airborne particles, dust, lubricants, raw materials, intermediates, auxiliaries). The cleaning process must be qualified and/or validated to demonstrate repeatability and robustness of the cleaning procedure and process. Whenever possible, the cleaning process should be automated to ensure repeatability. In cases where automation is not feasible, very strict requirements must be placed on manual cleaning methods to ensure consistency. These restrictions include identifying and controlling critical control parameters such as cleaning agent concentration, soak and/or scrub time, rinse time, temperature and volume of water, scrubbing parameters, and rinsing parameters.

Dependent on the nature of the product, process soil, and/or active ingredient, cleaning procedures may require cleaning agents and/or sanitizers. Typically, cleaning procedures and cleaning validation studies to confirm the cleaning procedure include only product-contact surface areas. Full understanding of the manufacturing process and equipment train must be attained to determine whether there are other areas in the process or equipment that are not direct product-contact surfaces that may be a source of contamination.

The intent of the cleaning validation program is to provide documented evidence of a robust and repeatable cleaning process that reduces contaminates to predetermined acceptable levels. A robust cleaning validation program is established through the development and approval of a *cleaning validation master plan* (CVMP). The CVMP establishes the philosophy and approach for the site's cleaning program. This document should include the requirements for development, confirmation, and sustainability of the cleaning program.

The CVMP establishes the acceptable residual limits that determine level of acceptability for the cleaning process. These limits should be achievable, verifiable, and based on the most toxic residue and/or the residue that has the highest risk to the product and/or process. These limits can be reflected in equations (for example, 10 ppm, *no observable effect level* [NOEL], therapeutic dose) that calculate maximum allowable carryover of residual. These limits are used during validation of the analytical methods to ensure that they are within the linear range of the method. Cleaning validation methods require full validation under International Conference on Harmonization (ICH) Q2 guidance, including limit of detection, limit of quantitation, specificity, and spike swab and coupon recovery studies. When setting acceptance limits for cleaning agents and sanitizers, the expectation is that no residual exists; ensure that the limits are appropriate. For sanitizers, residue is not acceptable on the surfaces of the equipment, and limits ensure that the cleaning procedure and analytical method can detect that the sanitizer has been removed from the product-contact surfaces.

Four elements are typical to every cleaning validation study in determining the effectiveness of a cleaning procedure and/or cycle. Each of these elements must have predetermined, defined, and approved acceptance criteria. Visual inspection is the first element and protocol test function. Once the equipment is inspected visually for any residual, the microbiological samples are collected to ensure that the surface equipment area is not contaminated by the cleaning validation sampling process. Product and process soil samples follow microbial sampling, and lastly, cleaning agent and sanitizer sampling is performed. These sampling schemes can be adjusted based on the understanding of the cleaning process and potential product impact.

The cleaning procedure is considered robust and repeatable upon completion of three successful runs (the number of runs may be more or less dependent on the knowledge of the process), which indicate that all data met predetermined acceptance criteria, the data were reviewed, and the data were summarized in a final report and approved by the *quality unit* (QU).

To ensure that the cleaning process is sustainable, the CVMP must include requirements that the life cycle is maintained, including requirements on periodic monitoring to assess changes in acceptance criteria, changes in equipment, changes in product or process soil, and changes in the cleaning process. The CVMP outlines requirements for revalidation activities based on changes in the process, equipment, product mix, and acceptance criteria.

SANITIZATION PROCEDURES

> Review sanitization procedures for facilities and equipment and ensure all are in accordance with any required validation studies including details on cleaning schedules, methods, equipment, materials, sanitizers, disinfectants, sporicides, and sterilants. (Apply)
>
> **Body of Knowledge IV.F.2**

The United States Environmental Protection Agency (EPA) licenses chemicals for which its manufacturers claim efficacy in disinfecting inanimate objects. The EPA evaluates disinfectant efficacy chiefly by either of two methods originated by the Association of Official Analytical Chemists (AOAC):

- *Use/dilution method*. An organism is dried to a rod made of glass, stainless steel, polished porcelain, or other nonreactive material. The rod is submerged for 10 minutes or other claim time in a container with the disinfectant being tested (without touching the side of the container). The rod is raised and allowed to drain. A *replication organism detection and counting* (RODAC) plate with agar and the appropriate nutrient is placed on the rod to remove organisms for testing and is then incubated for a set time. Zero growth permits disinfectant to pass the test.

- *Spray method/ hard surface carrier test*. An organism is dried to a measured surface area of a specific material. The disinfectant product is sprayed according to manufacturer's directions, and the surface is left until sufficient contact time elapses, according to directions. The surface is scraped and applied to a RODAC plate of appropriate media, then incubated for a set time. Zero growth permits disinfectant to pass the test.

Regarding the EU requirement for monitoring of disinfectant prepared at *ready-to-use* (RTU) dilution and in its final applicator bottle, unless the site has created and validated a production-like aseptic sterilizing filtration process and validated the sterility through the expiration date, along with validating the container closure of the application container, the full expectation by European Medicines Agency (EMA) is for monitoring of the solution over its shelf life. Alternatively, the supplier of prefilled, presterilized (aseptic filtered or terminally sterilized) vials must have had its validation data audited by the drug manufacturer, including container closure validation. This requirement holds true for liquid, gel-like, and moist pad antiseptics used in the sterile core. The latter, due to their exposure (sheets are pulled like tissues from a container), cannot avoid the requirement for drug manufacturing site monitoring. The alternative, onerous requirement for monitoring entails sampling the disinfectant solution or wipe on a media containing a validated neutralizer of the bactericidal active ingredients.

LITERATURE EFFICACY OF VARIOUS SANITIZERS

Chlorhexidine has a wide spectrum of antibacterial activity against both Gram-positive and Gram-negative vegetative bacteria. A *quaternary ammonium compound* (QAC) has high antimicrobial activity (vegetative cells) if its carbon chain length is C8 to C18 (if aliphatic). These compounds, being positively charged, are also surfactants (that is, have detergent/cleaning properties to get inside crevices). Unfortunately, QAC are not mycobactericidal (for example, active against tuberculosis-causing organisms) or effective against the Gram-negative bacteria *E. coli*, *P. aeruginosa*, and *S. typhimurium*. They are more active at alkaline or neutral pH than at acid pH levels.

Phenolics are effective against both Gram-positive and Gram-negative vegetative bacteria, but only slowly effective against bacterial spores. The virucidal activity of phenolics cannot be generalized (that is, enveloped viruses are more susceptible than other organisms).

Glutaraldehyde (usually 2% v/v supplied alkaline but made acidic before use) possesses sporicidal activity (that is, vegetative cells), and mycobacterial and fungal spore efficacy. It is active against various types of viruses. Formaldehyde, which is a carcinogen, is widely bactericidal, sporicidal, and virucidal; it is effective against protozoa. Iodophors are surface-active agents and surfactants that can solubilize iodine to form compounds containing microbicidal activity over a wide pH range (for example, povidone-iodine solution of 10% w/v). These are sporicidal.

The most common chlorine-releasing compounds are hypochlorites and N-chloro compounds. All of these compounds are irritants and corrosive but have high antimicrobial activity, including sporicidal activity and some mycobacterial activity. These compounds are active against lipid and nonlipid viruses. Sodium hypochlorite (NaOCl) is more active at acid than at alkaline pH. Diluted RTU solutions have short shelf life and must be qualified.

Chlorine dioxide (ClO_2) is an alternative to NaOCl and retains biocidal activity over a wide pH range. Oxine is a sodium chlorite solution that, when acidified, generates chloride dioxide, giving a mixture of chlorite and chlorine dioxide; it is more efficacious against pathogenic bacteria than chloride dioxide alone, and it is virucidal.

Peroxygens include hydrogen peroxide (H_2O_2) and peracetic acid (CH_3CO_3H). Hydrogen peroxide is used at varying strengths (35%, 50%, and 90%) and is bactericidal and a sporicide. H_2O_2 can be used as a sanitizing agent and an antiseptic. Peracetic acid is available commercially as a 15% aqueous solution (35% is potentially explosive). Peracetic acid has a broad spectrum of activity, including bacteria and their spores, molds, yeasts, algae, and viruses.

Ethylene oxide (C_2H_4O) uses its alkylating properties on proteins (used at proper concentration, temperature, and relative humidity) to act as a potent sterilant and sporicide. Ozone (O_3) is another agent that is bactericidal, virucidal, and sporicidal. Sodium hydroxide (NaOH), or *lye*, is active against all microorganisms, including protozoa and prions.

PEST CONTROL

> Review and verify that a pest control program is established and that it uses authorized rodenticides, insecticides, fungicides, fumigating agents, and appropriate traps for pest elimination. (Apply)
>
> **Body of Knowledge IV.F.3**

Any building used in the manufacturing, processing, packaging, or storage of pharmaceutical products must be free of infestation. The pest control program requires appropriate QU oversight. Procedures must describe the use of suitable rodenticides, insecticides, fungicides, or fumigating agents. Rodenticides, insecticides, fungicides, or fumigating agents should be of the appropriate grade, approved by local regulations, and approved by the drug manufacturing site QU. Pest control records must be generated and retained, and a system must exist for the continual capture of crawling and flying pests by an entomologist or pest control expert for identification and analysis. When the latter is contracted by a manufacturing site, a quality/technical agreement must describe all the activities and responsibilities of the contract giver and the contractor. A system should be in place for the reporting/recording of pest sightings by any personnel besides the contractor. This is usually accomplished by the use of a facility map that shows the locations of the bait stations, and recording of any pest activity.

STERILIZATION PROCESSES

> Verify that appropriate sterilization processes are established and validated. (Apply)
>
> **Body of Knowledge IV.F.4**

Sterilization Process overview

There are several methods to provide sterile medical products. All sterilization methods have their own unique advantages and disadvantages. The method used should be based on current scientific data and thoroughly tested to ensure there are no deleterious effects on the products and packaging. It is extremely important to fully validate sterilization processes and continually monitor the process to ensure the required results are consistently maintained.

Gamma Irradiation

Gamma irradiation provides a safe and cost-effective sterilization process of single-use medicinal products and packaging. Cobalt-60 powers gamma irradiators, effectively killing microorganisms throughout the product and its packaging with very little temperature effect and no residues. The radiation dose depends on the type of product, packaging configuration, and the initial bioburden. Dosimetric release allows products to be processed, verified, and immediately released for shipment.

Electron Beam

Electron beam (E-beam) radiation sterilization method is used for processing products of high volume and low-density. E-beam radiation is highly charged stream of electrons generated by accelerators capable of producing continuous or pulsed beams. The medicinal products being sterilized are exposed to E-beam energy, which alters biological and chemical bonds to destroy the DNA and reproductive capabilities of microorganisms.

Dry-Heat Sterilization

Dry-heat sterilization is used for materials that may be damaged by moist heat or that are impenetrable to moist heat. Dry-heat sterilization is nontoxic and does not harm the environment. The dry-heat sterilization process has a slow rate of heat penetration and microbial killing, and may affect some materials due to the high temperatures necessary for the process to be effective.

Filtration

Sterilization by filtration is used to remove bacteria from pharmaceutical fluids that cannot be purified by any other means. A membrane filter with a pore size 0.22 µm (smaller than the bacteria) is used to trap the bacteria. This method is highly dependent upon utilizing proper aseptic techniques.

Steam Sterilization

Steam sterilization (moist heat in the form of saturated steam under pressure) is widely used and the highly dependable. Steam sterilization is nontoxic, rapidly microbicidal, and sporicidal. Steam sterilization may affect some materials due to the high temperatures necessary for the process to be effective.

Ethylene Oxide

Ethylene oxide (EO) is based on a gas diffusion process. Ethylene oxide is capable of sterilizing and rendering products free of viable microorganisms. Ethylene oxide destroys the microbial DNA. The effectiveness of the sterilization process depends on the ability of the ethylene oxide to freely permeate through the product and packaging. All products must be placed in breathable packaging that allows gas to penetrate the sterile barrier and reach all surfaces of the device or product. Products sterilized using this method must be allowed to degas to prevent respiratory irritation and damage.

Typically, after the initial sterilization validation is conducted, routine monitoring is required to ensure the efficacy of the process. Chemical and biological indicators are typically used for steam sterilization processes and irrational indicators (stickers) are used for ionizing radiation processes. Additionally, quarterly dose audits are required for E-beam and gamma irradiation processes.

Chapter 33

Automated or Computerized Systems

Automated or computerized systems are powerful business tools selected to meet business needs for such tasks as research and development, production, and maintenance. These tools can also generate and store evidence for meeting regulatory requirements. As in most industries, automated systems have fostered operational simplicity, efficiency, reliability, repeatability, and accuracy in all business areas from procurement to human resources. Automated systems can increase safety in the monitoring and control of risky processes and regulate processes through real-time feedback.

Regulatory agencies recognize that these systems, under certain conditions, are reliable tools that can provide benefits for all stakeholders. 21 CFR 211.68 and ICH Q7 (Section 5.4) both specifically address the conditions under which automation is considered acceptable. *Good automated manufacturing practices* (GAMP) 5 includes information about automation.

Automated or computerized systems, regardless of whether they are embedded in equipment or products (such as medical devices) or exist as support systems, are company assets that should be subject to all of the scrutiny and control that would be expected with any asset. Pharmaceutical professionals will not necessarily be expected to know the inner workings of a computerized system, but they should know the characteristics of system evolution, maintenance, and regulatory compliant use.

VALIDATION PROCEDURES

> Review procedures for implementation and validation of automated or computerized systems. Verify that critical parameters for their operation and maintenance are controlled and monitored. (Evaluate)
>
> **Body of Knowledge IV.G.1**

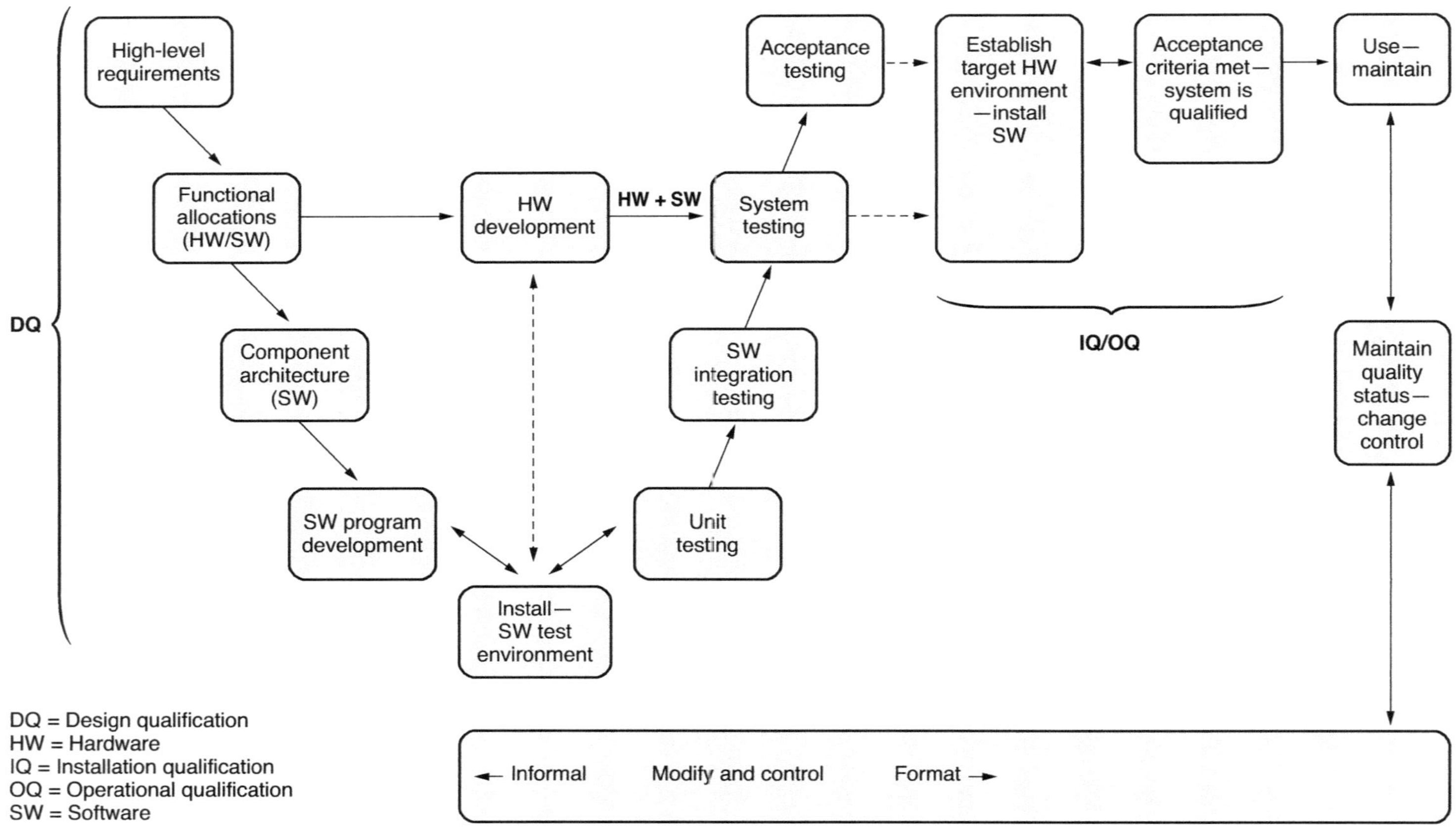

Figure 33.1 Automated and computerized systems are recognized to have a typical life cycle. This life cycle is often represented in a V-diagram format. The far right side of the model eventually leads to system retirement and decommissioning.

Source: Figure courtesy of the ASQ Food, Drug, and Cosmetic Division. Used. with permission.

SYSTEM LIFE CYCLE

All systems evolve through a predictable life cycle (Figure 33.1). Understanding where a system is in its life cycle is important in knowing how it got there and where it is going. The life cycle is a road map with typical characteristics. A number of phases and expectations exist that the pharmaceutical professional should expect when dealing with automated or computerized systems:

- *Identifying the need.* Will an automated or computerized system provide the best solution to the business problem?
- *Proposing solutions.* Solutions can be technical (that is, the computer system is expected to do it all, perhaps without any human intervention), procedural (that is, people will do it all by following a procedure), or, as is most common, a blend of the technical and procedural approaches where procedural steps are supported by the computer system with some human oversight.
- *Deciding how to develop the system with its business use as a major goal.* Should the effort be to build (that is, develop a specific program using in-house or vendor resources), buy (that is, use a program, usually a commercial product that will meet the need), or modify (that is, buy an existing product and customize it to meet the need)?
- *Installing the system.* Have we correctly anticipated the infrastructure and supporting systems needed? How will we replace the old system? How disruptive will it be to ongoing activities?
- *Testing and acceptance.* How much should we test and at what level of detail? Do the test results unequivocally and thoroughly demonstrate that the system can reliably meet the business needs?
- *Training.* Now that the system has been shown to be functionally appropriate, do the people who need to use it know how to use it? Will it be as useful as expected?
- *Change control.* All systems change, and the changes must be controlled. Systems constantly undergo three types of changes: corrective, perfective, and adaptive. Is there a clearly defined and thorough (that is, formal procedure and documentation) methodology for guiding a system from its current state (the as-is) to its desired state (the to-be)?
- *Backup and disaster recovery.* Although hardware and software failures are becoming less frequent, they do happen. All systems bring this inherent risk. Are backup and recovery systems in place so that potential data loss is minimized? Data loss should generally be considered unacceptable, but loss of *good manufacturing practices* (GMP) data should be considered a failure to meet regulatory requirements.
- *Decommissioning.* The system will eventually become outdated, hardware and supporting systems will change, or a better way of meeting the changing business need will be found. The system may

become obsolete, but the data might, for regulatory, legal, or historical reasons, need to be protected and retained. Consideration must be given to addressing the need to replace the system and maintaining (to some working degree) the old system. Not all information and processes can be easily transferred to a new system. Not all new systems can handle existing data formats and processes.

- *Records.* Keeping documented evidence is a regulatory must, and good business practice, but it is not practical to keep everything. Legal considerations, as well as direction from regulatory agencies, should guide what records are kept and for how long.

VALIDATION AND QUALIFICATION

Two important terms associated with but not exclusive to automated systems need to be understood: validation and qualification. In essence, these activities both meet the need for proving that a system is reliable and does what it claims to do, before and during its use, in its intended environment. *Validation* comprises identifying all aspects of testing that need to be completed before a product or process can be deployed for business use. It is common to write a *validation master plan* (VMP) to document the intent, scope, resources, schedules, activities, documentation, test types, and acceptance criteria required to achieve a validated state. Validation plans may address training and methods of change control to be used to maintain a validated state. Software and components will change, but the essential business goals regarding safety, quality, and regulatory compliance must always be met. The VMP provides guidance in achieving and maintaining qualified facilities and validated processes. Figures 33.2 and 33.3 depict examples of a decision tree used to decide whether software requires validation.

Qualification contributes to determining when a system is reliable and stable enough to support validated processes. Typical qualification phases include:

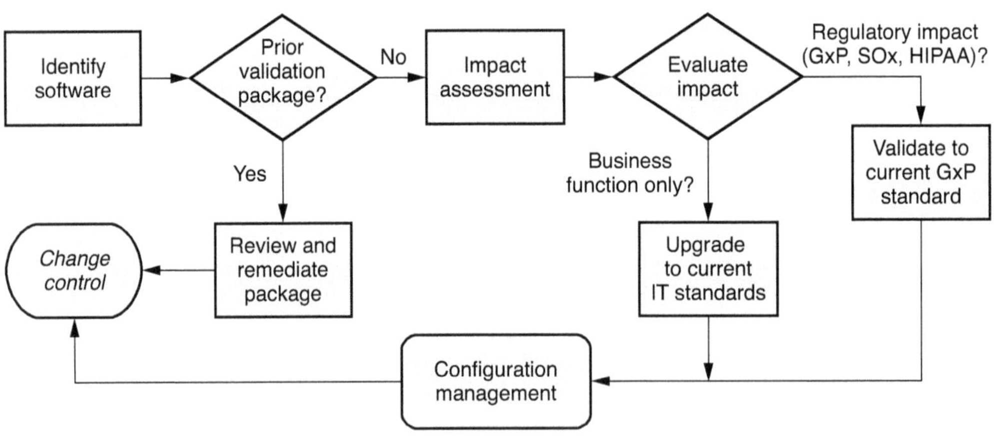

Figure 33.2 Software validation decision tree—legacy software.

Source: Courtesy of John English, HCCP of John T. English LLC—Systems/Regulatory Consulting.

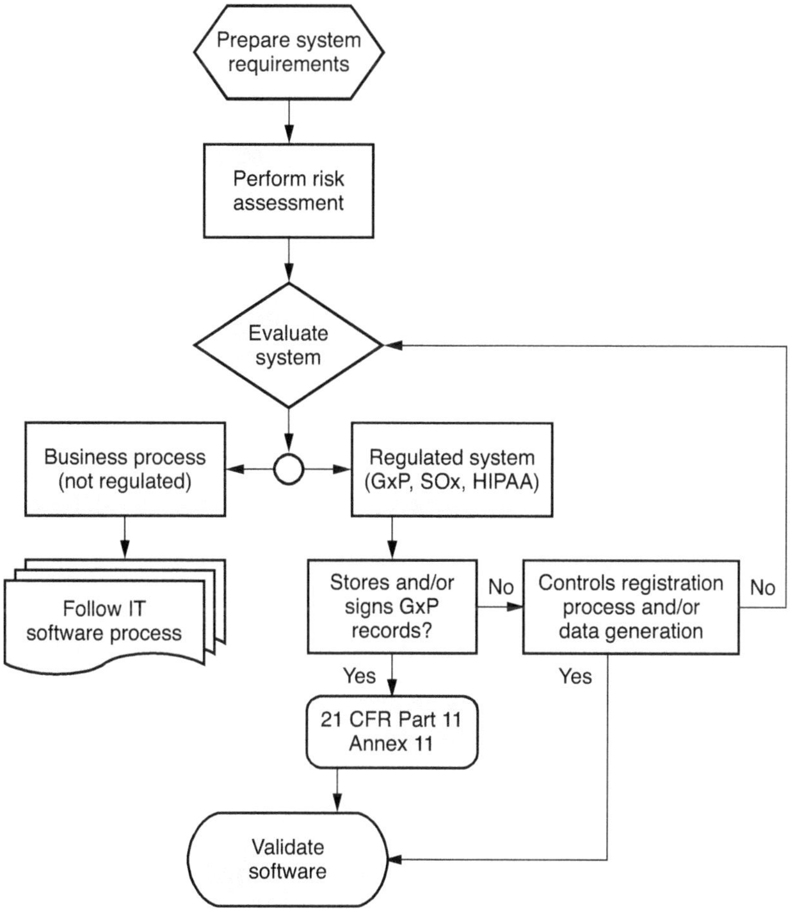

Figure 33.3 Software validation decision tree—new software.
Source: Courtesy of John English, HCCP of John T. English LLC—Systems/Regulatory Consulting.

- *Design qualification* (DQ) is conducted to assure that the system is designed to do what is intended. This phase consists of documenting such topics as user requirements (functional and operational specifications) and includes identifying *critical components,* which are those features that directly impact product and process characteristics such as safety and quality. Critical components can include power supplies, equipment safety features, temperatures, pressures, weights, and alarm features. These features are implemented design features that must remain in control for the process to act predictably. In systems of great complexity where not every possible logic avenue can be practically checked, or every combination of events predicted, designing the system such that critical components or parameters are easily tested and monitored is crucial. If everything cannot be tested, at least test what is critical to the safety and quality of the process or product. The design should facilitate this goal.

- *Installation qualification* (IQ) comprises activities for ensuring that the system is installed safely and that it will function in the environment in which it is intended. IQ, relative to automated systems, includes establishing a complete inventory of components (that is, model numbers, serial numbers, locations, part numbers, revision numbers), making sure that the components are correctly installed (that is, reliable power, correct wiring, proper grounding, working interfaces), making sure that installations are in compliance with vendor recommendations, and doing nominal checks to make sure that information is flowing through the system as expected in the design specifications.
- *Operational qualification* (OQ) proves that the design, its implementation, and supporting components will operate reliably together in a way that will meet the business need. OQ tests the system as it is intended to be used.
- *Component qualification* (CQ) (manufacturing to the correct design criteria) and *performance qualification* (PQ) (consistent performance over time) are other qualification types that are sometimes needed on a specific project. DQ, IQ, and OQ are commonly used for software-intensive systems. IQ, OQ, and PQ are commonly used for equipment. The series DQ, IQ, OQ, PQ might be used for a system that includes both software and equipment. Qualification expectations and acceptance should be documented in a plan.

Addressing these needs will lead a system to be validated (it does or can do what it claims to do) or qualified (it does what, in a business sense, it is needed or expected to do). If a system has been validated and it needs to change, documented evidence must be produced that confirms that the system still performs as effectively as it did when it was originally validated, in spite of the modification.

REGULATORY REFERENCES

Pertinent regulatory references include 21 CFR 211.68, "Automated System Requirements," and ICH Q7, *Good Manufacturing Practice Guide for Active Pharmaceutical Ingredients*. Numerous references can be found concerning written records. 21 CFR 11 does set rules for providing written records in electronic form. In the past, record keeping relied on human observation and interpretation to supply information about which decisions would be made. The constantly improving capabilities, efficiency, and reliability of automated and computerized systems have allowed many industries, including the pharmaceutical industry, to confidently move from paper records (that were subject to human error) to electronic records (that provide greater technical accuracy).

The current regulations are clear and specific concerning what written records the United States Food and Drug Administration (FDA) needs from pharmaceutical manufacturers. In an effort to take advantage of technological improvements, provide improved consistency, and reduce paperwork, industry leaders asked

FDA to set standards by which electronic records and signatures would be acceptable in meeting government regulations for written records. Federal regulations, designated 21 CFR 11, were formulated in response to this request and became law in March 1997. The discussion of electronic records is covered in the following section addressing closed and open systems. Electronic signatures are acceptable in meeting regulatory needs if the signatures have the specific characteristics cited in 21 CFR 11.

OPEN AND CLOSED SYSTEMS

> Distinguish between open and closed computerized systems. (Apply)
>
> **Body of Knowledge IV.G.2**

A few definitions exist for what constitutes open and closed computerized systems, but both systems are centered on who has access to the system and who controls the content. *Open* and *closed* refer to who has jurisdiction over the code (as in open source or closed source systems). Commercial products (*commercial off-the-shelf* [COTS] software) are usually written with proprietary restrictions and are closed systems; the user cannot alter the basic code (or should not, as it will invalidate the warranty and infringe on copyrights). If, however, a computer system is developed in-house (or perhaps through a vendor), the pharmaceutical company will want access to the code so that modifications can be made. When purchasing a custom-developed product from a vendor (also known as a *bespoke system*), it is important to make sure that the agreement ensures that the purchaser will be the recipient of the code. Open systems can become closed systems if the correct restrictions are imposed.

In the pharmaceutical world, open and closed systems are distinctly defined when it comes to using electronic records and electronic signatures to meet regulatory needs. 21 CFR 11, "Electronic Records; Electronic Signatures," provides definitions for closed and open systems. Closed systems are usually local, proprietary workplace type systems that handle unique business needs. These can be systems (that is, automated systems) that operate specific equipment or are timekeeping systems, data gathering systems, or monitoring systems. This information is important to the business and is protected from the outside world. The Internet is an open system; few restrictions are placed on access, and very little checking is done to ensure the integrity (or veracity) of the electronic records. Open systems are public-access systems. Thus, in general, closed and open systems are at opposite ends of the spectrum. Closed systems usually require the user to be granted specific access to the system (usually after some type of background check) and often require a unique log-in and password. Closed systems are strictly controlled. Open systems are for public use. If, in the pharmaceutical industry, a company would like to keep and submit electronic records to meet regulatory needs, a closed system is suggested.

CONFIGURATION MANAGEMENT

> Verify that version control and configuration are maintained and monitored. (Evaluate)
>
> **Body of Knowledge IV.G.2**

All systems change, and pharmaceutical professionals should be familiar with the characteristics of proper change management. Regulatory agencies refer to systems as being either *in control* or *out of control*. Keeping a system in control can be tedious and painstaking, but allowing a system to drift out of control can be fatal to a business. Systems that are prone to unauthorized and uncontrolled modification are a huge risk from both safety and regulatory viewpoints. Simply put, once a system is proven to produce an acceptable product, any change to the system is presumed to change the product. Based on that concept, any modification to a system is assumed to produce a modified (that is, adulterated) product. It is crucial that any change to an automated or computerized system be accurately proposed, and reviewed for its potential impact on the product. Various methods can keep a system, and by extension the product, in control. The way that a system is kept in control is through *change management*. Change management has two elements: change control and configuration management.

Change control is the methodology that defines and reviews changes. Change control evolves during a system's life cycle as the system evolves from the conceptual to productive phases. Basically, change control should not be too restrictive during the development phase, and should be rigorous during the deployment phase. The closer a system gets to the product, the more change control accountability must be demanded. It is acceptable for a developer to try things in the laboratory and to debug an evolving system to make it function, but it is not acceptable to adopt that philosophy in a system that is producing drugs for market. A change control methodology should, at a minimum, document:

- *Who*. Who needs the change? If the change is needed because of a malfunction, who witnessed or found the malfunction? Stakeholders, technical reviewers, approvers, task assignees, customers/patients, *quality assurance* (QA), and so on, are involved with this category.
- *What*. Describe the issues, problems, and enhancements, and what tasks need to be completed to resolve them. What items need to be modified to maintain control? What has been done?
- *When*. When did this impact the process/product, and when does it need to be fixed? If this issue has been ongoing, when did it begin, and how much product was produced during this time? In this situation, follow-up investigations are needed to determine how much product may have been affected.
- *Where*. What physical locations need to change? Is it a local problem, or does this have regional or global impact? If it is a functional change, where else is this function used?

- *How.* How does this impact the system/process/product? To what degree (that is, how severely), and how are the issues resolved?
- *Why.* Why does this affect the process or product? Why did it occur (that is, cause analysis)?

Configuration management is a discipline that can be thought of as software accounting. *Configuration management* collects and provides information about software characteristics over time and provides for protection of entities in terms of backups, vaulting, and archiving. Critical configuration management information includes:

- *Identification.* Configuration item identifiers (for example, part numbers, revisions) must be unique, and their characteristics clearly must be defined. The methods used to maintain the unique identification are referred to as *version control.* The goal is to ensure that a version comprises a unique set of characteristics. If modifications are made to version 3, it is no longer version 3; it needs another unique identifier (based on the chosen version control scheme).
- *Control.* Controlled configuration items must be beyond compromise either physically or electronically, and this goal can be accomplished by physical vaulting (for example, safes, limited access rooms) or by establishing electronic storage areas with limited access. Typically, electronic storage is divided into working and controlled areas. A controlled check-out/check-in scheme is recommended. Configuration items must be retrievable. Off-site storage for physical items is fine, and recommended in potentially dangerous situations, but, if needed to restore functionality, the items must be quickly retrievable.
- *Authorization.* Use must be authorized and clearly scoped. Configuration management programs must have rules to ensure authorized deployment of items, which is usually done by requiring signatures of stakeholders such as managers and QA. The deployments should be strictly defined so that it is known where everything is and when its use began and ended.
- *Accountability.* When and where do we use it, and how long do we need to keep it? Configuration management should be able to answer questions such as "What versions were we running on this piece of equipment three years ago?" In addition, configuration management should maintain an inventory in compliance with company record retention requirements. When items are no longer needed, they should be properly discarded.

TESTING CHANGES

When considering testing needed relative to modifications made to a computerized system, it is helpful to categorize the type of software and what the appropriate level of testing should be. Testing at some level must be conducted to ensure that the system does what it is intended to do (that is, that it is in control). This process is referred to as maintaining the *validated state.* A system that has been

Table 33.1 GAMP classifications for software types, and associated validation expectations.

GAMP class	Category	Validation action
1	Operating systems	Record version
2	Instruments and controllers	Record configuration and calibration
3	Configurable packages	Audit supplier, validate any bespoke code. Apply full life cycle requirements.
4	Systems where the codes, or part of the code, are configurable	Audit supplier and code, validate that any bespoke configurations apply full life cycle requirements
5	Systems using custom or bespoke code that develop predicate rules information	Audit supplier, validate all code, and apply full life cycle requirements

formally tested and accepted for use should maintain that original level of reliable functionality even as it undergoes change. In this regard it is useful to be familiar with the GAMP classifications for software types and their associated validation expectations (Table 33.1).

The characteristics mentioned in regulatory requirements as being important to ensuring that automated systems are in control include access control, data protection, change control, data backup and protection, and audit trail.

Two major aspects of access control are physical and logical access. *Physical access* is meant to limit personnel from restricted areas (that is, doors, locks, signs). Physical restrictions comprise legitimacy checks and should include training considerations (that is, should people who do not know how to properly use the equipment be allowed access). *Logical access* refers to the restrictions placed on computerized systems, usually in the form of identifiers, passwords, biometric restrictors, tokens, and so on. Identification restrictions apply, such as having employees have a unique personnel identifier, and password rules such as changing passwords on a regular basis and not sharing passwords. Biometric methods such as fingerprint and retinal scanners might be used. Logical controls are used to limit access to system functionality and to data, and to restrict system modification.

In general, data protection refers a system's ability to do a number of things (that is, gather, process, and transfer data accurately, maintain data integrity across interfaces, protect data such that it is beyond compromise or unauthorized modification, and retrieve data accurately). Data that can be modified should be subject to audit trails, and the audit trails must travel with the data (the data and their metadata must be linked). Importantly, data that are of regulatory interest and are kept electronically must be able to survive technological changes. It might not be useful to have batch records stored on floppy disks as not all systems are capable of handling that type of physical input. Data protection includes being able to recover data quickly regardless of its media or format. In a simple form, data protection can refer to adhering to proper environmental conditions when storing media.

Backups are created as duplicate files (direct copies of files) kept in case of system failures or other disasters. Using backups to recover files and data is referred to as *restoration*. *Archiving* is a technique of moving data that are no longer actively used to a separate data storage device (or media) for long-term retention. Off-site storage of archived data is common. Data archives comprise older data that might be important and necessary for future reference. Archives may contain data that would be kept for regulatory compliance or needed to support business continuity. Again, these data must be retrievable, regardless of their format.

A system has four major components—hardware, software, documentation, and personnel—and four aspects of system maintenance. Change control (for software and documentation) and training (for personnel) are important aspects of maintenance in general, but hardware has some special considerations. Of these four categories, hardware is the most prone to physical erosion. Hardware must be maintained and calibrated to make certain that it functions as expected. *Calibration* is the act of periodically checking, setting, or otherwise adjusting a device to a specified operational level. Suffice to say that instruments, devices, and other hardware components used in automated systems must be calibrated. Maintenance can be thought of as three types—*reactive* (fix it when it breaks or degrades), *preventive* (change the part, usually on a periodic basis, before it breaks or degrades), or *predictive* (monitor the part and change it when it begins to show out-of-trend behavior).

An audit trail in terms of automated or computerized systems is an electronic accounting of who has accessed a system, when it was accessed, and what changes have been made. The audit trail often exists as a separate file but must be capable of being linked to the file that it describes. The integrity of the audit trail must be maintained as strictly as an actual system file. Keeping an accurate audit trail is an important feature of Part 11 compliance.

Periodic system monitoring may refer to many aspects of the system or the system life cycle. It can broadly refer to access monitoring (that is, who is using the system), resource monitoring (that is, how effectively is electronic traffic handled during use), assessing the state of the system (that is, are instruments properly calibrated), or checking the integrity of the system (that is, have changes had negative impact on the process or the products produced by the process). Relative to change control, periodic reviews are recommended when a system has undergone a number of changes over time. Each change may have maintained the integrity of the system, but unintended interactions between the changes may affect the overall operation of the system. In this case, a revalidation, rerunning the protocols (or portions thereof), would be used to confirm that the system remains in the validated state. Periodic reviews of unchanged operating procedures are common.

SECURITY REQUIREMENTS

> Evaluate on-site, multi-site, and cloud-based computerized systems to ensure they meet regulatory and guidance requirements for key elements such as access control, data protection, electronic signature, change control, data archiving, maintenance, transcription, audit trail, and periodic system monitoring. (Evaluate)
>
> **Body of Knowledge IV.G.4**

Manufacturers must establish controls to ensure that electronic signatures are authentic, trustworthy, and reliable and that these signatures are only used by their genuine owners as required by 21 Part 11 Electronic Records; Electronic Signatures, 21 Part 11 requires controls must be established to prevent and identify fraud and guarantee the identity of persons accessing the system.

21 Part 11 requires manufacturers to establish and maintain clearly written procedures regarding the use of electronic signatures, and train employees trained in the proper use of electronic signature and record systems. Individuals using electronic signature must understand electronic signature are as legally binding as a handwritten signature.

Biometric signatures verify an individual's identity based on measurements of the individual's physical features or repeatable actions and where those features and/or actions are both unique to that individual and measurable.

When biometrics signatures are not used, the manufacturer is required to employ at least two elements of identification. For example, many systems require entry of both a unique username and password. These two elements of identification help to ensure security. Additionally, many systems now incorporate multi-factor validation using a security code emailed or texted to the user to authenticate the user's identity. To maintain a secure system, passwords should not be recycled, periodically reset according to established policies and procedures, and require the use of a combination upper-case and lower-case letters, numbers, and punctuation marks.

The biometric authentication process compares data for the individual's characteristics to that individual's known biometric "master." The reference master is stored in a secure database. The stored data is then compared to the individual's known biometric data to be authenticated.

Biometric signatures must ensure that they cannot be used by anyone other than their genuine owners. Fingerprint scanning devices are an effective method that can be incorporated in biometric signatures because no two fingerprints are alike. In addition to fingerprint scanning devices, facial recognition and retinal scanning technologies are widely used to authenticate system users. Facial recognition and retinal scanning technologies deliver a significantly more reliable authentication experience than two-factor authentication using passwords and usernames.

21 CFR Part 11 requires audit trails. Audit trails are an electronic record of the evidence behind any transaction in an electronic system that ensures all changes are apparent. The audit trail should ensure:

- The original data is maintained.
- The audit trail is protected from deletion or alteration.
- The audit trail is maintained for the life of the record.
- The audit trail is made readily available for audits and inspections.
- The audit trail is generated independently by the computer system.

Audit trails usually include:

- The name of the person, date, and time a record was created, changed, or deleted;

- A means to detect and record any changes to records;
- Measures so that users cannot easily change date or time;

21 Part 11 requires three types of security checks including:

- System operational checks ensure operations or entries occur in the proper sequence, when appropriate.
- System authority checks must be used to control access to electronic record systems and specify levels of authorization for personnel.
- System device checks ensure that, as needed, only data from designated devices such as laboratory instruments and terminals are accepted by the system.

Chapter 34

Business Continuity and Disaster Recovery Planning

SUPPLY CHAIN IMPACT

> Review plans and verify procedures for disaster recovery, record recovery, and business continuity that will guard operations from interruption to the supply chain and ensure data integrity. (Evaluate)
>
> **Body of Knowledge IV.H.1**

AN OVERVIEW OF ISO 22301

ISO 22301 specifies requirements to plan, establish, implement, operate, monitor, review, maintain, and continually improve a documented management system to prepare for, respond to, and recover from disruptive events when they arise. Natural disasters, environmental accidents, technology mishaps, and man-made crises have demonstrated that severe incidents can and will happen, impacting the public and private sectors alike. The challenge goes beyond providing an emergency response plan or using disaster management strategies that were previously used.

ISO 22301:2012, *Security and resilience—Business continuity management systems—Requirements*, is a widely recognized international *business continuity management standard* (BCMS). It was developed by ISO Technical Committee 223. The International Organization for Standardization (ISO) first published this standard in 2012. It cancels and replaces the old BS 25999 business continuity standard, which is obsolete and has been officially withdrawn. The latest version was released in 2019.

The purpose of ISO 22301 is to show individuals how to set up and manage a BCMS. These requirements can be found in seven sections within the standard (Table 34.1). The requirements specified in ISO 22301 are generic and intended to be applicable to all organizations (or parts thereof), regardless of type, size, and nature of the organization. The extent of application of these requirements depends on the organization's operating environment and complexity.

Table 34.1 ISO 22301:2012 requirements for setting up and managing a business continuity management system.

Clause	Topic
1	Scope
2	Normative references
3	Terms and definitions
4	Context of the organization
5	Leadership
6	Planning
7	Support
8	Operation
9	Performance evaluation
10	Improvement

THE PDCA APPROACH

Similarly to ISO 9001 and ISO 13485, ISO 22301 uses what is called the plan–do–check–act (PDCA) cycle (see Figure 34.1) to organize the standard:

- *Plan.* Parts 4, 5, 6, and 7 expect you to plan the establishment of your organization's BCMS;
- *Do.* Part 8 expects you to establish your BCMS;
- *Check.* Part 9 expects you to evaluate your BCMS;
- *Act.* Part 10 expects you to improve your BCMS.

BRIEF OVERVIEW OF KEY CLAUSES OF ISO 22301 BUSINESS CONTINUITY STANDARD

Following Annex L structure of ISO Guide 83, ISO 22301 is organized into 10 main clauses (Table 34.1), and the key activities for each clause are summarized.

Clause 1: Scope

The requirements of ISO 22301 are generic and intended to be applicable to all organizations regardless of type, size, and activities. The application of the requirements depends on the operating environment and complexity.

Clause 2: Normative References

The latest edition and any amendments of ISO 22300, Security, and resilience—Vocabulary apply.

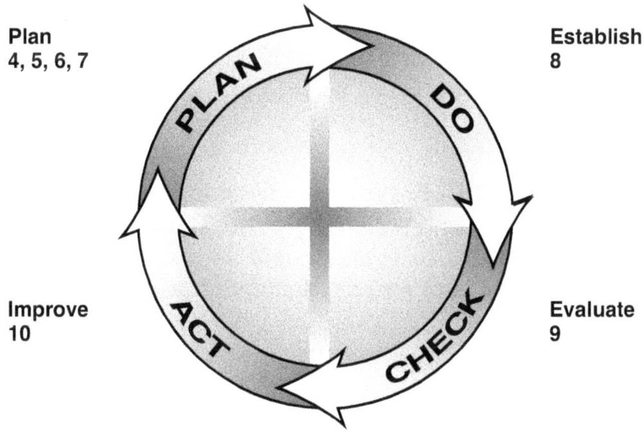

Figure 34.1 Plan–do–check–act cycle applied to ISO 22301:2012.

Clause 3: Terms and Definitions

The terms and definitions provided by ISO 22300 apply. ISO and IEC maintain terminological databases for use in standardization at the following addresses:

- ISO Online browsing platform: available at www.iso.org/obp
- IEC Electropedia: available at www.electropedia.org/

Clause 4: Context of the Organization

Understand your organization, its purpose, and objectives context while understanding the needs and expectations of interested parties in light of legal and regulatory requirements. Organizations should consider how disruptive incidents could impact the organization.

Clause 5: Leadership

Provide leadership and support for your organization and ensure that managers demonstrate their commitment and support and encourage employee involvement. Allocate responsibility and authority for carrying out business continuity roles to the appropriate people within your organization.

Clause 6: Planning

Identify and determine the risks and opportunities that could influence the effectiveness of your organization or disrupt its operation. Define actions and prepare plans to address the risks and opportunities that could influence the effectiveness of your organization or disrupt its operation.

Clause 7: Support

Identify and provide the resources that your organization needs, including procedures and communication tools. Determine the competence requirements of the people under your organization's control who have an impact on its performance, and ensure that people are aware of their responsibilities.

Clause 8: Operation

Plan and develop your BCMS processes by studying potential disruptions and analyzing business risks, and set your priorities. Establish a formal process that your organization can use to evaluate and set business continuity and recovery priorities, objectives, and targets; document, implement, and maintain your priority-setting process.

Clause 9: Performance Evaluation

Determine how you will monitor and measure the performance and effectiveness of your organization. Make sure that your audit program is capable of determining whether your system conforms to requirements.

Clause 10: Improvement

Identify, react to, and evaluate nonconformities when they occur. Implement corrective actions to address causes, and review the effectiveness of your corrective actions. Continuously improve the performance, suitability, adequacy, and effectiveness of your system.

CONTINGENCY PLAN

> Verify the testing and effectiveness of contingency plans as required or proceduralized. (Apply)
>
> **Body of Knowledge IV.H.2**

Contingency plans help organizations identify potential natural and man-made disasters that could disrupt normal operations and develop contingencies if they were to occur. The contingency plan answers the following questions:

- What could happen?
- What is the risk to the company and/or customers?
- What will we do in response?
- What can we do in advance to prepare?

Natural disasters may include avalanches, thunderstorms, lightning, hail, tornados, volcanos, wildfires, winter storms, earthquakes, extreme heat, floods, hurricanes, landslides, and pandemics.

Human-caused disasters may include terror attacks, active shooters, cyberattacks, financial emergencies, nuclear explosions, supplier capacity, labor shortages,

		◀Impact▶		
		Low	Medium	High
◀Likelihood▶	High	Low Risk	Medium Risk	High Risk
	Medium	Low Risk	Medium Risk	Medium Risk
	Low	Low Risk	Low Risk	Low Risk

Figure 34.2 Example contingency risk matrix.

internet outages, bomb threats, facility fire, leadership succession, equipment failures, burglary, chemical spills, industrial accidents, water outages, electrical outages, gas outages, civil unrest, phone outages, IT infrastructure outages, medical emergencies.

To develop a contingency plan, begin by identifying natural and man-made disaster that can reasonably affect the operation of the organization. For example, if the facility is located high in the mountains a tsunami is improbable, but a landslide can reasonably be expected.

Once the potential natural and man-made disasters are identified, the impact or effect to the organization should be assigned. For example, low, medium, or high impact. Additionally, the likelihood of occurrence should be assigned.

The overall risk can then be determined using a simple matrix as shown in Figure 34.2. The level of planning should be commensurate with the level of risk.

The contingency plan should contain information on who to contact in the event of an emergency and the mitigations used to minimize the risk to the organization. It is recommended the contingency plan should be reviewed at a minimum annually during Management Review or after a significant event for effectiveness.

Periodic drills should be conducted to assess the effectiveness of the contingency plan including mock medical emergencies, chemical spills, server restores, and fire drills. After a drill, a postmortem review should occur to discuss what went well and to investigate opportunities for improvement.

Part V

Materials and Supply Chain Management

Chapter 35

Receipt of Materials

INCOMING MATERIALS

> Describe and use processes to receive, label, and store incoming materials (raw materials, bulk chemicals, components, and product labeling) and take appropriate action on deviations (damaged materials, materials from unapproved suppliers, and missing documentation). (Apply)
>
> **Body of Knowledge V.A.1**

Upon the receipt of any material, all documents associated with the shipment or delivery must be reviewed to ensure accuracy and completeness. Starting materials, both *active pharmaceutical ingredients* (APIs) and excipients, should be purchased only from approved suppliers, according to the European Union (EU) *good manufacturing practices* (GMP), Section 5.26, and the World Health Organization (WHO) GMP, Section 14.8.

INCOMING INSPECTION

The EU and WHO GMP have similar requirements. Each container should be inspected for at least integrity of package and seal and for correspondence between the order, the delivery note, and the supplier's labels. If a company receives material made from batches with different supplier lot numbers, each batch must be considered separately for sampling testing and release according to EU GMP Sections 5.27 and 5.28 and WHO GMP Sections 14.9 and 14.10. Containers should be cleaned where necessary and labeled with the prescribed label. Any damage to containers or other problem that could adversely affect the quality of the material should be recorded, investigated, and reported to the *quality unit* (QU) as per WHO GMP Section

14.11. Both GMP (EU GMP Section 5.30 and WHO GMP Sections 14.14 and 17.15) further prescribe that appropriate measures should be taken to ensure the identity of the contents of each container of starting material either by identity testing of each container or validation of the supplier.

US 21 CFR 211.80a–d requires procedures to receive materials, to store them appropriately, and to identify them. US 21 CFR 211.82(a) states, "Upon receipt and before acceptance, each container or grouping of containers of components, drug product containers, and closures shall be examined visually for appropriate labeling as to contents, container damage or broken seals, and contamination."

Written procedures and records are required for the receipt of each delivery of starting material and primary and printed packaging materials. The record of the receipts should include the name of the material on the delivery note and the containers, the in-house name and/or code of material if different, date of receipt, the supplier's name, and, if possible, the manufacturer's name, the manufacturer's batch or reference number, the total quantity and number of containers received, the batch number assigned after receipt, and any relevant comment (for example, state of the containers).

Temperature- or light-sensitive materials should be handled expeditiously. If temperature-sensitive materials are equipped with data loggers, they should be handled per procedure to verify there have not been any temperature excursions.

Deviations should be used to track temperature excursions, use of materials from unapproved vendors, or other atypical events. Some companies also use the deviation system to monitor the receipt of damaged materials, or receipt of materials with incomplete or missing documentation.

Written procedures for the internal labeling, quarantine, and storage of starting materials, packaging materials, and other materials, as appropriate, are needed according to EU GMP Chapter 4, Section 4.19–4.21, WHO GMP 15.32–15.34, 21 CFR 211.80, and 21 CFR 211.82.

IDENTIFICATION AND LABELING OF STARTING MATERIALS

The information that should be recorded on the label for each delivery is prescribed in the EU GMP Section 5.29, WHO GMP Section 14.13, and 21 CFR 211.80d and includes the designated name of the product and internal code reference, the supplier's batch number and control (lot number) provided on the receipt, the status of the contents, and an expiry date or date beyond which retesting is necessary. When fully validated computer systems are used, not all of the above information needs to be legible on the label. Material must be restricted until appropriate testing is performed and disposition obtained from the site's QU.

API GMP FOR RECEIPT OF MATERIALS

In addition to the requirements for starting materials used for drug products, receipt of starting materials used in API manufacturing should comply with specific requirements. Before incoming materials are mixed with existing stocks (for example, solvents or stocks in silos), they should be identified as correct, tested (if appropriate), and released. Procedures should be available to prevent wrongly discharging incoming materials into the existing stock. If bulk deliveries are made in non-dedicated tankers, assurance of no cross-contamination from the tanker is advisable. Means of providing this assurance could include a certificate of cleaning, testing for trace impurities, and/or audit of the supplier. One or more of these processes should be used. Some starting materials may not be tested for compliance because of the hazards involved (for example, phosphorus pentachloride and dimethyl sulfate), and this non-testing is acceptable when a batch *certificate of analysis* (COA) is available from the vendor and when there is a reason based on safety or other valid considerations.

INVENTORY CONTROLS

> Describe and use procedures for documenting inventory transactions, updating changes in material status, allocation, and "stop shipments" for quality holds. (Apply)
>
> **Body of Knowledge V.A.2**

Written procedures are required for receipt, identification, handling, sampling, testing, and approval or rejection of components and drug product containers and closures. Refer to 21 CFR 211 Subpart E—Control of Components and Drug Product Containers and Closures, ICH Q7 section 7 Materials Management.

For paper-based systems, some companies use color-coded labels indicating Quarantine, Released, or Rejected on individual containers and segregate these materials according to status to prevent mix-ups. Inventory transaction logs may be used to track movement of released materials to production, and shipping of product

For fully computerized systems, the labels place on individual containers are used for all inventory transactions. "Random" storage of quarantined and released inventory is permitted, since a qualified computerized system will not allow allocation of rejected or unreleased material for manufacturing. A qualified computerized system will also not allow rejected final product to be released for sale.

For all cases, inventory records demonstrating that written procedures are followed should be available for review or investigations, if needed. These records should be available at all times for paper-based, computerized, and hybrid systems

Chapter 36

Sampling Processes

SAMPLING PLANS

> Review sampling plans for representative sampling, appropriate sample size, and test or inspection criteria. (Apply)
>
> **Body of Knowledge V.B.1**

GENERAL SAMPLING REQUIREMENTS

It is desirable to have written procedures for sampling that include the names of person(s) authorized to take samples, the methods and equipment to be used, the amounts to be taken, and any precautions to be observed to avoid contamination of the material or any deterioration in its quality. Documented sampling procedures and sampling plans approved by the quality unit for all materials should be in place.

The European Union (EU) and World Health Organization (WHO) both have *good manufacturing practices* (GMP) that further prescribe that the sampling procedures should contain the instructions for any required subdivision of the sample, the type of sample container to be used, whether it is for aseptic sampling or normal sampling, labeling, storage conditions, and instructions for cleaning and storage of sampling equipment. Samples should be representative of the material being sampled and should be taken in a manner that does not compromise the integrity of the sampled material or the sample.

Each container of starting materials should be sampled for identification testing unless supplier validation has been done to permit reduced testing. Such reduced testing is not permissible for parenteral product actives or for materials coming from a broker. Acceptable means of validating the supplier processes are listed in EU Annex 8 and WHO Section 17.15. This validation should take account of:

- Nature and status of the manufacturer and the supplier and their understanding of the GMP requirements of the pharmaceutical industry
- *Quality assurance* (QA) system of the manufacturer of the starting material

- Manufacturing conditions under which the starting material is produced and controlled
- Nature of the starting material and the medicinal products in which it will be used

WHO GMP Section 17.16 lists minimum information to be included on the supplier's *certificate of analysis* (COA) as identification, name, and address of the issuing supplier, signature of the competent official and statement of his/her qualifications, name of the material tested, batch number of the material tested, specifications and methods used, test results obtained, and the date of testing.

SAMPLING ENVIRONMENT

> Differentiate and apply the requirements for sampling environment and utensils to the type of the material being sampled. (Apply)
>
> **Body of Knowledge V.B.2**

SAMPLING FACILITIES

The sampling facilities should meet the general requirements described in EU GMP Chapter 3 and WHO GMP Chapter 7 for cleanliness and dust control, and should not result in cross-contamination or sample confusion. The WHO guidance (*WHO Guidelines for Sampling of Pharmaceutical Products and Related Materials*) provides details on facilities and equipment to be used in sampling.

Sampling facilities should be designed to prevent contamination of an opened container, prevent cross-contamination by other materials, product, and the environment, and protect the individual who samples during the sampling procedure. Where possible, sampling should be performed in an area or booth designed for the purpose. To prevent the contamination of a non-hazardous material, sampling should be performed using PPE (personal protective equipment), including laboratory coat, disposable sleeves, and gloves. Hazardous or dust-generating materials may require the use of an extraction or fume hood to minimize exposure to the individual and environment. Some materials should be sampled in special or dedicated environments (for example, aerosol valves, hormones, and penicillins).

The area in which the sample was taken should be recorded in the sampling record.

SAMPLING UTENSILS AND EQUIPMENT

A sampling plan should specify the amount of sample required, the types of sample utensils, types of sample containers to use, and personal protection precautions to follow. Sampling equipment will typically include some measuring device, such as a balance, pipet, or graduated cylinder. Sampling utensils may include spatulas or scoops for sampling solids, dip tubes for sampling liquids, sample "thieves" that are designed to take samples from deep containers of solids. Where reusable sampling utensils are used, cleaning of the utensils needs to be qualified; alternatively, single-use sampling utensils can be used.

The utensils and equipment used may be dependent on the type of material being sampled. For example, bulk gases will require special sample collection tools and containers. USP <1097> Bulk Powder Sampling Procedures discusses in length the considerations to take for sampling bulk powders.

Where sterile materials are individually packaged in one batch, one complete unit may be removed for testing to ensure an opened container is not compromised.

SAMPLE CONTAINERS AND LABELING

Sample containers should not be reactive, additive, or absorptive so as to alter the safety, identity, strength, quality, or purity of the material being sampled, and should provide adequate protection from external factors in storage and use that can cause deterioration. Finally, sample containers should be clean and, where required, sterilized and pyrogen-free.

Samples should be labeled to maintain traceability and identity. WHO Section 17.12 and EU GMP Section 6.13 specify label content for samples, with WHO being more prescriptive. Section 17.12 lists the following requirements for sample label content:

- Name of the sampled material
- Batch or lot number
- Number of the container from which the sample was taken
- Number of the sample
- Signature of the person who has taken the sample
- Date of sampling

TRAINING OF PERSONNEL

Each GMP requires personnel to adequately train for the jobs they perform, including 21 CFR 211.25, EU GMP Chapter 2, and WHO GMP Section 10. EU GMP Annex 8 further specifies that personnel should receive initial and ongoing training in the disciplines relevant to correct sampling. The training should include sampling plans, written sampling procedures, techniques and equipment for sampling, information about the risks of cross-contamination, the precautions to be taken with regard to unstable and/or sterile substances, the importance of considering the visual appearance of materials, containers, and labels, and the importance of recording any unexpected or unusual circumstances.

WHO also states that samplers need to be adequately trained and qualified to perform the sampling operation. Since a sampling technique itself can introduce bias, it is important that personnel carrying out the sampling should be suitably trained in the techniques and procedures used. The training should be documented in the individual's training records.

SAMPLING OF STARTING MATERIALS FOR API MANUFACTURING

At least one test to verify the identity of each batch of material should be conducted, with certain exceptions: hazardous or highly toxic raw materials, other special materials, or materials transferred to another unit within the company's control do not need to be tested if the manufacturer's COA is obtained showing that these raw materials conform to established specifications. Visual examination of containers and labels and recording of batch numbers should help in establishing the identity of these materials. The lack of on-site testing for these materials should be justified and documented.

Samples should be representative of the batch of material from which they are taken. Sampling methods should specify the number of containers to be sampled, which part of the container to sample, and the amount of material to be taken from each container. The number of containers to sample and the sample size should be based on a sampling plan that takes into consideration the criticality of the material, material variability, past quality history of the supplier, and the quantity needed for analysis. Sampling should be conducted at defined locations and following procedures designed to prevent contamination of the material sampled and contamination of other materials. Containers from which samples are withdrawn should be opened carefully and subsequently reclosed. They should be marked to indicate that a sample has been taken.

SAMPLING OF PACKAGING MATERIALS

EU GMP Annex 8 states that the sampling plan for packaging materials should take account of the quantity received, the quality required, the nature of the material (for example, primary packaging materials and/or printed packaging materials), the production methods, and what is known of the QA system of the packaging materials manufacturer based on audits. The number of samples taken should be determined statistically and specified in a sampling plan.

WHO guidance cautions against mixing of printed packaging materials. Only one material should be sampled at a time, and the WHO guidance explicitly states that samples of packaging materials should never be returned to the consignment. It considers sources of nonhomogeneity in packaging material batches and recommends that sampling take these potential sources of nonhomogeneity into account. The sources of nonhomogeneity include materials manufactured on different days or machines, materials manufactured on one machine but at different stations (for example, molding stations), packaging manufactured with different source materials, and change in quality during the process (for example, container wall thickness and text legibility).

US FDA Guidance for Industry "Questions and Answers on Current Good Manufacturing Practice Regulations | Production and Process Controls" question 2 states: "CGMP requirement that containers 'be opened, sampled, and sealed in a manner designed to prevent contamination of their contents' will depend on the purported quality characteristics of the material under sample and the warehouse environment. For containers or closures purporting to be sterile or depyrogenated, sampling should be under conditions equivalent to the purported quality of the material . . . (see 21 CFR 211.94 and 211.113(b)). This is to preserve the fitness for use of the remaining containers or closures as well as to ensure sample integrity, if they are to be examined for microbial contamination. At a minimum, any sampling should be performed in a manner to limit exposure to the environment during and after the time samples are removed (i.e., wiping outside surfaces, limiting time that the original package is open, and properly resealing the original package). Well-written and followed procedures are the critical elements."

FDA also notes that 21 CFR 211.84 permits "a manufacturer to release for use a shipment of containers or closures based on the supplier's certificate of analysis and a visual identification of the containers or closures. Once a supplier's reliability has been established by validation of their test results, a manufacturer could perform the visual examination entirely in the warehouse."

CLEANING

> Ensure the sampling environment is appropriately cleaned and monitored and that sampling utensils are appropriately cleaned or are single-use. (Apply)
>
> **Body of Knowledge V.B.3**

Sampling materials should be checked for cleanliness prior to use. Materials to be sampled should be checked for cleanliness and wiped down prior to entering the sampling area. Sample containers should be wiped clean after sample collection.

Cleaning of sampling materials and sampling containers must be qualified to demonstrate removal of the materials being sampled. For this reason, some firms find it more cost-effective to sample using single-use materials (that is, single-use sample bottles, single-use spatulas, single-use pipettes).

Cleaning of sampling booths should be specified in procedures. Sampling booths should be verified as clean prior to use and should be cleaned between sampling operations. These cleaning procedures should be designed to prevent contamination of the opened container, prevent cross-contamination by other materials, product, and the environment, and protect the individual who samples. Cross-contamination controls may include but are not limited to sampling one container at a time, cleaning between open containers, and changing gloves between operations. A risk assessment of the sampling area may be necessary to ensure that adequate sampling and cleaning procedures are in place for the materials being used.

Sample room cleaning should be defined in procedures to ensure the room remains in a clean and sanitary condition; the requirements in 21 CFR 211.56 and EU GMPs Chapter 3 apply.

Chapter 37

Material Storage, Identification, and Rotation

STORAGE SUITABILITY

> Ensure the storage environment is suitable, controlled, and monitored as required for the type of materials. (Apply)
>
> **Body of Knowledge V.C.1**

Storage locations should meet the general principles outlined for pharmaceutical manufacturing facilities in each of the *good manufacturing practices* (GMP), that is, United States 21 CFR 211, European Union (EU) GMP Chapter 3, and World Health Organization (WHO) GMP Section 12. Additional requirements for storage areas are included in 21 CFR 211 Subpart H, WHO GMP Annex 9, WHO Technical Report Series 908 and Annex 5, and WHO Technical Report Series 937.

Storage areas should be of sufficient capacity to permit the orderly storage of materials and to facilitate cleaning and inspection, and materials should be stored off the floor. Storage areas should be designed to maintain any required label storage conditions, such as controlled temperature and humidity. Records should be maintained of temperature monitoring, and the monitoring equipment should be checked periodically and calibrated. WHO specifies that temperature mapping should show temperature uniformity. Areas should be clean and free from pests. Receiving and dispatch bays should protect materials from the weather and should enable cleaning of the materials before storage.

If quarantine status is maintained by storage in separate areas, these areas should be clearly marked, with access restricted to authorized personnel. Any system replacing physical quarantine should provide equivalent security, particularly for rejected, expired, recalled, or returned materials or products. The status of all materials should be identified.

Highly active materials, narcotics, and hazardous materials should be stored in dedicated areas and subject to additional safety and security measures. Materials should be handled and stored to prevent mix-ups and cross-contamination. They should be stored under conditions that assure that their quality is maintained, and stock should be appropriately rotated.

Printed packaging materials should be stored under secure conditions with limited access. Cut labels and other loose printed materials should be stored and transported in separate closed containers to avoid mix-ups. Packaging materials should be issued for use by designated personnel in accordance with defined procedures.

Rejected materials and products should be identified and controlled to prevent their use before a final disposition by the *quality unit* (QU). Broken or damaged items should be segregated and withdrawn from usable stock.

LABELING OF STORED MATERIALS

> Confirm that the identification label for stored materials contains the required information. (Apply)
>
> **Body of Knowledge V.C.2**

Each container should be identified, and the status of the material should be controlled either by physical labeling and segregation or by electronic systems that provide equivalent security. Starting materials should be appropriately labelled with at least the following information prior to storage:

- Product name and internal code, where applicable
- A batch number given at receipt
- Where appropriate, the status of the contents (for example, in quarantine, on test, released, rejected)
- Where appropriate, an expiry date or a date beyond which retesting is necessary.

When fully computerized storage systems are used, the above information need not necessarily be in a legible form on the label.

Incoming materials, intermediate product, and finished product should be physically or administratively quarantined until they have been released for use or distribution. The types of controls used should be defined in written procedures.

Modern methods of tracking products through the supply chain include barcodes and radio-frequency identification (RFID) technology. Barcode technology uses optical machine-readable coded bars. Barcodes are represented by varying the widths and spacing of parallel lines. Quick Response Code (QR) is the trademark for a type of matrix barcode (or two-dimensional barcode) that has gained popularity in recent years. RFID uses an electromagnetic field to identify and track products. RFID uses a special tag that contains electronically stored information. The RFID tags can be passive or active. Passive tags collect energy from a special reader using radio waves. Active tags use a battery. RFID tags may be discretely embedded in the tracked object.

Any electronic or computerized system used labeling and tracking of stored materials must be qualified for its. Qualification is performed to demonstrated that the system is capable of working as designed accurately, consistently, and reproducibly. A written description of the labeling system should be available and maintained.

STOCK ROTATION

> Define and use stock rotation requirements such as first-in/first-out (FIFO) and first-expired/first out (FEFO). (Apply)
>
> **Body of Knowledge V.C.3**

WHO GMP specifies that materials and product should permit batch segregation and stock rotation by a first-expire, first-out rule (FEFO) (WHO Annex 2, WHO good manufacturing practices for pharmaceutical products: main principles). US FDA 21 CFR 211.86 specifies that drug product components, drug product containers, and closures be rotated so that the oldest approved stock is used first, unless otherwise appropriate. Periodic stock reconciliation should be performed by comparing actual and recorded stocks, and any significant discrepancies should be investigated. Partly used containers should be securely closed and resealed to prevent contamination. Materials from opened containers should be used before those in unopened containers. Damaged containers should not be used unless the quality of the material has been shown to be unaffected. The QU should be notified of damage, and any actions taken should be documented.

First-expired/first-out (FEFO) and first-in/first-out (FIFO) are common terms used in logistics and inventory management. FEFO is used describes a way of dealing with perishable products, or products with a specified expiry date. The FIFO method uses goods which acquired or manufactured first are sold, used, or disposed of first.

RETEST DATES VS. EXPIRATION DATES

> Describe the difference between retest dates and expiration dates. (Understand)
>
> **Body of Knowledge V.C.4**

Stocks should be checked periodically for obsolete and outdated materials. Procedures should be in place to prevent the issuance and use of outdated materials. Approved procedures should be in place for control of returned goods. All returns should be placed in quarantine pending disposition by the QU. Products returned from the market and that have left the control of the manufacturer should be destroyed; they may be considered for resale, relabeling, or recovery in a subsequent batch only after they have been critically assessed by the QU in accordance with a written procedure.

Upon receipt, some materials may not have an expiration date from a vendor, but a firm will assign an expiration date to ensure that material is re-evaluated to determine its suitability for use.

US FDA 21 CFR 211.87 specifies that "components, drug product containers, and closures shall be retested or reexamined, as appropriate, for identity, strength, quality, and purity and approved or rejected by the quality control unit in accordance with § 211.84 as necessary, e.g., after storage for long periods or after exposure to air, heat or other conditions that might adversely affect the component, drug product container, or closure." Any retest or reexamination needs to be justified and documented.

Expiration dates assigned by a vendor denote a shelf-life for the item.

MIX-UP RISK

> Identify potential sources of mix-up and identify methods to minimize their risk (material segregation, labeling, special storage for rejects, control of material returns, lot-control methods, and special processes for materials with similar names). (Analyze)
>
> **Body of Knowledge V.C.5**

Mix-up risk is minimized where fully electronic computerized systems are used to manage inventory with bar codes or RFID. These systems allow for "random" or virtual control of materials within a warehouse where only one batch is stored in one location.

Mix-ups happen when partial implementation of computerized systems are used, or when full paper systems are used. US 21 CFR 211.94 specifies: "There shall be written procedures describing in sufficient detail the receipt, identification, storage, handling, sampling, testing, and approval or rejection of components and drug product containers and closures; such written procedures shall be followed." Incoming receipt, inventory controls, distribution practices, identification, personnel training, and qualification procedures are among the items required to prevent mix-up. These procedures should minimally describe:

- Material segregation
- Material identification (distinctive codes/traceability)
- Material status
- Labeling
- Recordkeeping
- Double-checks of critical steps
- Quarantine segregation
- Stock rotation
- Returned materials storage
- Reject material storage
- Training and qualification

EU GMPs require a separate, restricted areas for rejected materials.

Rejected materials should be identified and controlled under a quarantine system designed to prevent their unauthorized use in manufacturing.

Chapter 38
Shipping and Distribution

A finished pharmaceutical's labeling and/or pharmacopeia monograph provide specific directions for the temperature and humidity conditions of storage and shipment to the consumer, which is ultimately based on stability data. The United States Pharmacopeia (USP) General Notices section on preservation, packaging, storage, and labeling provides working definitions for the temperature ranges/storage devices designated as *freezer, cold, cool, controlled cold temperature, room temperature, controlled room temperature, warm, excessive heat,* and *protection from freezing.*

TEMPERATURE-SENSITIVE REQUIREMENTS

> Identify special requirements for temperature-sensitive products including tertiary packaging, design, and monitoring devices. (Analyze)
>
> **Body of Knowledge V.D.1**

Warehousing/storage temperatures should be controlled and monitored using calibrated monitoring devices and records of temperature, and alarms, where applicable, should be maintained. Refrigerators and freezers used to store drugs should:

- Be well maintained;
- Be equipped with alarms;
- Be free from frost buildup;
- When combined, be a two-door unit with separate freezer compartment and door;
- Allow for adequate air distribution and orderly storage within the chamber; storage practices and loading configurations should not lead to the obstruction of air distribution;
- Have sensors for continuous monitoring, and alarms located at the points representing the temperature extremes.

Table 38.1 Temperature conversion formulas.

$t_C = \frac{5}{9}(t_F - 32)$	Convert Fahrenheit to Celsius
$t_F = \frac{9}{5}t_C + 32$	Convert Celsius to Fahrenheit
$t_K = t_C + 273.15$	Convert Celsius to Kelvin
$t_C = t_K - 273.15$	Convert Kelvin to Celsius

If an alarm is associated with the refrigerator or freezer, the functionality of the alarm should be checked periodically at the upper and lower set points. Warehouses must undergo temperature profiling using a suitable number of temperature-recording instruments placed strategically (for example, based on sun-facing walls, ceiling height). Temperature profiles should occur over at least three consecutive 24-hour periods and during days of extreme outdoor winter cold and summer heat. Products should not be stored in areas shown by temperature mapping to present a risk.

The calculation of a *mean kinetic temperature* (*MKT*) is necessary for compliance with 21 CFR 203.32, "Drug Sample Storage and Handling Requirements" (part of 21 CFR 203, "Prescription Drug Marketing"). *MKT* is a single derived temperature that, if maintained over a defined period, would afford the same thermal challenge to a pharmaceutical product as would have been experienced over a range of both higher and lower temperatures for an equivalent defined period. In other words, *MKT* is a calculated, fixed temperature that simulates the effects of temperature variations over a period of time. It expresses the cumulative thermal stress experienced by a product at varying temperatures during storage and distribution. The calculation of *MKT* is given in USP <1150> and usually requires 52 weeks of continual temperature monitoring or weekly high-temperature and weekly low-temperature recordings. Table 38.1 provides formulas for changing degrees Fahrenheit and Celsius to degrees Kelvin and degrees Kelvin to Fahrenheit and Celsius.

The formula is

$$MKT = \frac{H / R}{-\ln \dfrac{e^{-H/RT_1} + e^{-H/RT_2} + \ldots + e^{-H/RT_n}}{n}}$$

where

ΔH = The heat of activation, 83.144 kJ mole^{-1} (typically from 60 to 100 kJ/mole for solids and liquids)

R = The universal gas constant 8.3144×10^{-3} J mole^{-1} K^{-1}

T = The temperature in degrees K

n = The total number of *equal* time periods over which data are collected

ln = The natural log

e = The natural log base

Note: Ambient temperature is generally accepted as 25–30 °C (77–86 °F).

The $\Delta H/R$ term can be simplified as follows:

$$\frac{\Delta H}{R} = \frac{83.144 \text{ kJ mole}^{-1}}{8.3144 \times 10^{-3} \text{ J mole}^{-1} \text{ K}^{-1}} = 10,000\ ^{\circ}\text{K}$$

The simplified MKT formula becomes

$$MKT = \frac{10,000}{-\ln \frac{e^{-10,000/T_1} + e^{-10,000/T_2} + \ldots + e^{-10,000/T_n}}{n}}$$

Example: The label conditions of a medicinal product require storage at ambient temperature 25–30 °C. Temperature readings were recorded on three consecutive days at 18:00 hours (more data points are generally required). On the second day, there was a malfunction in the HVAC system resulting in higher than normal temperatures. The temperature readings are as follows:

Day 1: 26 °C or 263.15 °K

Day 2: 35 °C or 308.15 °K

Day 3: 27 °C or 300.15 °K

$$MKT = \frac{10,000}{-\ln \frac{e^{-10,000/T_1} + e^{-10,000/T_2} + \ldots + e^{-10,000/T_n}}{n}}$$

$$MKT = \frac{10,000}{-\ln \frac{e^{-10,000/263.15} + e^{-10,000/308.15} + e^{-10,000/200.15}}{3}} = 298.09\ ^{\circ}\text{K}$$

To convert back to degrees Celsius:

$$298\ ^{\circ}\text{K} — 273.15 = 29.94\ ^{\circ}\text{C}$$

The *MKT* is 29.94 °C. Therefore, the thermal stress to which the product was exposed is acceptable for the stated label storage conditions of 25–30 °C.

SPECIAL REQUIREMENTS

Determine specific product requirements and apply them to routine shipping processes. (Apply)

Body of Knowledge V.D.2

Various temperature measurement technologies exist that are distinguished by their fragility and their measurement precision. USP <1118> describes several temperature measuring devices: alcohol or mercury thermometer, *infrared* (IR) device, *resistance temperature detector* (RTD), thermistor, thermocouple, data loggers, and time-temperature integrators (typically single use). Temperature and humidity monitoring devices should be calibrated at predetermined intervals. Single-use monitoring devices should be qualified.

Selection of a shipping container and/or box should be based on:

- Storage and transportation requirements of the drugs
- Space required for the amount of drugs to be transported
- Anticipated external temperature extremes
- Estimated maximum length of time required for transportation of the drugs, including any in-transit storage

REPORT REQUIREMENTS

> Analyze shipping reports and transportation requirements in accordance with good distribution practices. (Analyze)
>
> **Body of Knowledge V.D.3**

SUPPLY CHAIN SECURITY

> Identify and apply the various means to secure the supply chain including tamper-evident seals, shipping manifests, verification of documentation, barcoding, radio frequency identification (RFID), and serialization. (Apply)
>
> **Body of Knowledge V.D.4**

Measures should be in place to prevent unauthorized persons from entering and/or tampering with vehicles and/or equipment and to prevent the theft or misappropriation thereof. Monitoring data recorded during vehicular transport should be reviewed upon receipt of the pharmaceutical products to assess maintenance of any special storage conditions. Any deviations from storage conditions that are considered to be acceptable (that is, small, brief excursions) should be determined in consultation with the marketing authorization holder and/or manufacturer.

At a port of entry into a country, consignments of pharmaceutical products should be stored under suitable conditions for as short a time as possible. All reasonable steps should be taken by importers to ensure that products are not mishandled or exposed to adverse storage conditions at wharves or airports. Where necessary, persons with pharmaceutical training should be involved with the customs procedures, or should be readily contactable.

Any counterfeit or suspected drugs found in the pharmaceutical supply chain should be segregated immediately from other pharmaceutical product. The marketing authorization holder and relevant competent authorities should be informed immediately.

21 CFR 211.132 requires *over-the-counter* (OTC) retail drug products sold in the United States (except a dermatological, dentifrice, insulin, or lozenge product) to have a tamper-evident (primary and/or secondary) package. Such a package has one or more indicators or barriers to entry that, if breached or missing, can reasonably be expected to provide visible evidence to consumers that tampering has occurred. Each retail package of a drug product having a tamper-evident feature(s) must bear a statement that identifies all tamper-evident feature(s) and any capsule sealing technologies, is prominently placed on the package, and is so placed that it will be unaffected if the tamper-evident feature of the package is breached or missing.

Chapter 39

Traceability and Sourcing

TRACEABILITY REQUIREMENTS

> Define and differentiate the requirements for traceability of incoming materials, intermediates, and finished drugs. (Apply)
>
> **Body of Knowledge V.E.1**

Traceability refers to the ability to determine (trace) what materials have been brought into the facility, which materials, equipment, and personnel were used to manufacture a specific batch of finished product, and where the finished goods were distributed for commercial sale. To achieve complete traceability, companies must have systems in place to record the receipt and use of materials, by unique identifier (either a batch or control number), that go into the manufacture of the drug product, and systems for identifying where the finished product batches were distributed. Good documentation is key to meeting traceability requirements and ensuring, in the case of a recall, that all impacted product is identified and can be removed from the field, if necessary. The United States Food and Drug Administration (FDA) *good manufacturing practices* (GMP) regulations and EU GMPs illustrate the concept of traceability throughout a finished dosage form facility.

Raw Materials and Components

Before a raw material or component arrives at the receiving dock of the pharmaceutical company, it is imperative that the company has done its homework on the material and the vendor that supplies the material. Establishing the quality and integrity of the raw materials and components that are used in manufacturing and packaging the drug product begins with an understanding of how these materials are manufactured and packaged by the vendor. It is important to qualify vendors to ensure the receipt of quality raw materials and components. It is also important to verify, through the qualification process, that the vendor has mechanisms in place to trace the manufacture of its products, and that the product label is a unique identifier that can be traced forward into the manufacture of the finished dosage form.

Proper receipt of raw materials and components is crucial from a traceability perspective. When materials and/or components arrive on the dock, a verification of the vendor's bill of lading should be conducted. Upon receipt, GMP regulations require that the finished dosage form manufacturer assign a unique code (batch or control number) to each lot of incoming materials for each shipment received. The vendor's unique identifier (batch or lot number) should be recorded to allow traceability back to the vendor in the event of a problem. The regulations direct the pharmaceutical company to verify the vendor's packaging, labeling, and tamper-evident seal (if applicable) to ensure that the material received is legitimate and not counterfeit. Finally, it is important to understand that multiple shipments of the same batch of material require unique batch or control numbers to be assigned to each receipt. It is not acceptable to assign the same batch or control number to multiple receipts of the same vendor-supplied material or component batch.

Product Manufacturing and Packaging

GMP regulations require that each manufactured batch is assigned a unique identifier, typically a batch or lot number, that is specific to the product being manufactured. As raw materials are weighed and dispensed for use in manufacturing, the batch or control number of the raw materials must be recorded. As these materials are charged into the manufacturing equipment, their identity (by batch or control number) and quantity must be recorded on the manufacturing batch documentation. It is important to note that the addition of raw materials to a batch is typically viewed as a critical step and, as a result, requires two individuals to verify that the correct materials were added and to record the correct batch information on the associated documentation.

As manufacturing progresses, appropriate documentation of the manufacturing process (for example, the equipment used, personnel involved) is needed. The use of processing aids (for example, filters or tubing) should be documented. If individual batches are blended together to produce a single, larger batch for further processing, there should be traceability to the individual batches that were used to manufacture the larger batch. Batches should only be combined in this manner according to preestablished procedures and validated processes. Batches that failed to meet required specifications should not be combined with batches that have met specifications in hopes of producing a larger batch that meets the predetermined specifications for the product.

Water is a raw material that is commonly used in a number of pharmaceutical dosage forms and manufacturing processes. For documentation of process water use from continuous flow loops, a specific batch number or unique identifier cannot be easily applied to the water that is used in the manufacturing process. As a result, traceability with respect to water is typically point-of-use and time-based. Any issues that arise with the quality of the water have the potential of impacting all of the batches that used the water since the date of the last acceptable testing results.

Finally, like the requirements during manufacturing, each batch or control number of the packaging and labeling components must be recorded on the batch documentation to meet traceability requirements.

Finished Dosage Form

GMP regulations require the identification of each packaged lot of product with a batch or lot number that permits the determination of the history of the manufacturing and packaging operations, as well as the control and testing of the batch.

Finished Dosage Form Distribution Records

GMP regulations require the maintenance of distribution records containing, at a minimum:

- Name and strength of product
- Name and address of the supplier
- Name and address of the consignee
- Name and address of the customer
- Batch or lot number of the product shipped
- Quantity shipped
- Date shipped

These distribution records enable the tracking of each finished drug product. Distribution records should include shipments to commercial customers, transfers to internal departments (for example, samples for *quality control* [QC] testing), as well as samples furnished to physicians or other healthcare providers. A system should be in place to account for 100% of the packaged finished dosage form and allow for complete notification of customers in the event of a product recall.

Records should be available for review by health authorities for inspection purposes for the period stated in national legislation where the product is distributed.

BIOLOGICAL AGENT REQUIREMENTS

> Identify and apply the requirements related to biological agents such as bovine spongiform encephalopathy (BSE) and transmissible spongiform encephalopathy (TSE). (Apply)
>
> **Body of Knowledge V.E.2**

Transmission spongiform encephalopathies (TSE) came to the forefront in the mi-1980s with the identification of the bovine form (*bovine spongiform encephalopathy* [BSE]) in British cattle. This discovery of mad cow disease, and the risk of transmission to humans, highlighted the need for greater controls over the sourcing of animal-derived materials for use in the pharmaceutical industry.

The ideal situation when sourcing raw materials for the manufacture of drug products would be to avoid the use of bovine-derived materials, or materials from other animal species in which TSE naturally occurs. Therefore, when a pharmaceutical company has a choice between materials of animal or nonanimal origin, the materials of nonanimal origin are clearly preferred. Unfortunately, sourcing materials of nonanimal origin is not always possible, and, as a result, it is critical for a pharmaceutical company to take steps to minimize the risk of transmitting animal TSE through their drug products.

The European Medicines Agency (EMA) identifies mechanisms to minimize the risk of transmitting TSE:

- Sourcing materials produced from animals residing in geographic areas of low risk for the disease
- Cleaning and control procedures that minimize the risk of cross-contamination between production batches
- Removal or inactivation of TSE agents during production processing

Of these activities, the well-controlled sourcing of animal-derived materials is the most critical as this prevents the introduction of TSE into the drug manufacturing facility. To ensure that the materials used in the manufacture of the finished dosage form do not contain these types of biological agents, defined steps are recommended (see Figure 39.1).

GMP regulations place responsibility for ensuring the quality of the materials used to manufacture drug products squarely on the shoulders of the pharmaceutical company. The globalization of the pharmaceutical supply chain, and the increasing use of materials manufactured in emerging countries (from a pharmaceutical perspective) such as India and China, has made this responsibility increasingly difficult to achieve.

Over the last few years, much debate and discussion has taken place regarding the establishment of a drug pedigree to ensure the integrity and authenticity of each and every drug product that enters the pharmaceutical supply chain. Much of this discussion has arisen as a result of the increase in counterfeit pharmaceutical products entering the supply chain, which puts consumers and patients at significant risk. It is imperative that every organization involved in the pharmaceutical

- Understand the source of all the materials that are used in the manufacture of the drug product and identify those that are animal derived.
- For animal-derived products, understand the country of origin of the starting materials and ensure that the country is rated as low risk.
- Understand the controls that the supplier of the animal-derived material has in place to minimize the introduction of these agents into their facility.
- Obtain a BSE/TSE statement from each supplier of animal-derived materials that demonstrates their commitment to ensure that their materials are free from these types of agents.
- Closely monitor the suppliers of animal-derived materials as the risk of TSE cannot be entirely eliminated.

Figure 39.1 Steps to ensure that materials used in manufacture of finished dosage forms do not contain harmful biologic agents.

supply chain does their part to minimize or eliminate the introduction of counterfeit materials or products into it.

PEDIGREE AND SOURCING REQUIREMENTS

> Identify and apply requirements for maintaining pedigree and sourcing details for active pharmaceutical ingredients (APIs), biological starting materials, excipients, intermediates, and finished products. Document the supply chain from raw materials through wholesale or retail to end user. (Apply)
>
> **Body of Knowledge V.E.3**

A typical supply chain for a pharmaceutical company is shown in Figure 39.2. Inputs (for example, raw materials, packaging components) into the finished dosage form manufacturing process typically are purchased from outside suppliers. Once the finished dosage form is manufactured, packaged, tested, and released, it is shipped to a wholesaler who may store it for a period of time. The wholesaler may sell the product to a chain drug store (which requires the product to pass through the chain drug store warehouse before reaching the pharmacy) or directly to a pharmacy customer. The finished product typically passes through a number of hands before it reaches the end customer or patient.

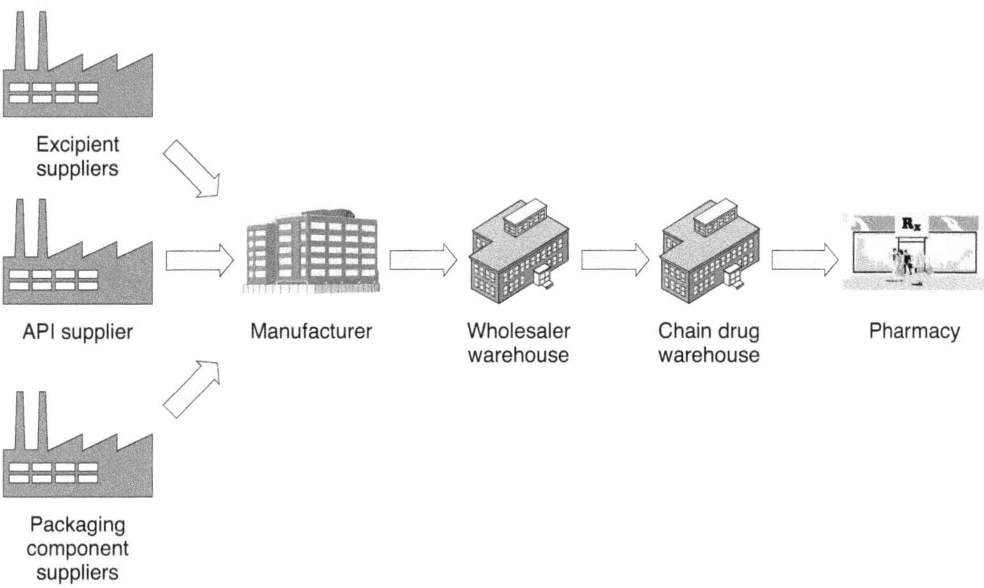

Figure 39.2 Typical pharmaceutical supply chain.
Source: Figure courtesy of the ASQ Food, Drug, and Cosmetic Division. Used with permission.

Chapter 40

Salvaged/Returned Goods and Destruction

Good *manufacturing practices* (GMP) regulations were designed to ensure that pharmaceutical products that are manufactured and released for distribution are safe, pure, and effective, and that they have a level of quality commensurate with their use in humans or animals. This level of quality needs to be maintained throughout the distribution network to ensure that products are not negatively impacted during transit and subsequent storage. When pharmaceutical products are returned to the company, or salvaged after a natural disaster, it is imperative that appropriate measures are taken to verify the quality of the product before reconditioning or releasing the product back into the supply chain. It is critical that returned or salvaged product that is ultimately rejected is disposed of appropriately.

DISPOSITION

> Review salvaged and returned goods and evaluate them for disposition. (Evaluate)
>
> **Body of Knowledge V.F.1**

Drug products that are returned to the pharmaceutical company, for whatever reason, need to be identified during the receiving process and segregated from other salable product until an appropriate disposition is made. They must be handled according to written procedures that take into account any specific storage requirements and the time elapsed since the product was originally dispatched. Returns should also be conducted in accordance with national law and contractual arrangements between the parties. Documentation of the receipt of returned goods, including the product name, strength, lot number, reason for the return, and quantity returned, must be recorded. If the reason for the drug product return implicates other batches, an investigation should be conducted to determine the extent of the issue.

After receipt, returned goods need to be evaluated to determine an appropriate disposition (for example, return to stock, reprocessing, or rejection). This evaluation should consider the length of time outside the control of the pharmaceutical company, the likely conditions under which the product was distributed or held before return, and the condition of the drug product, its container, carton, or labeling. In the case of an error in shipment, or if the drug products have been in the distribution channel for a brief time, this evaluation may be limited to a physical examination of the returned product. In this case, if the physical examination demonstrates that the product was not damaged during the short period of time that it was outside the control of the company, it can be returned to stock as salable product.

In situations where the product has been in the distribution channels for some time, or there is a question about how the product was stored, additional testing may be necessary to ensure that the product meets appropriate standards of safety, identity, strength, quality, and purity. In general, products that have been outside the control of the pharmaceutical manufacturer for some time should not be returned to salable stock unless:

- Products are in their original unopened and undamaged primary and secondary packaging containers and are in good condition;
- Products have not expired and have not been recalled;
- Records demonstrate that medicinal products have been transported, stored and handled in compliance with their specific storage requirements;
- Remaining shelf life is suitable for continued distribution.

European regulations require that returned goods be examined and assessed by a person authorized to do so. The company needs to weigh the benefit of returning the product to stock against the cost of additional verification testing to ensure that the product has not been negatively impacted. In addition, the pharmaceutical company should look closely at the labeling of the returned drug product before returning it to stock to ensure that the product labeling includes the latest safety information. Reprocessing of returned drug product is allowed provided that the subsequent drug product meets the appropriate standards, specifications, and characteristics of the original drug product. Reprocessing is a viable option where an update to the labeling is warranted to ensure that the latest safety information is available for patient use.

In all cases, the holding, evaluation, inspection, and testing, if necessary, to disposition a returned product should be documented. Release of returned product back into the supply chain should be done only after review of all of the relevant information by the *quality unit* (QU), *qualified person* (QP), or EU GDP Responsible Person (RP). The activities surrounding returned goods, and the functional areas involved in the process, need to be outlined in a formal procedure, and records of the return process need to be maintained.

Finally, returned drug products may serve as an indication of quality problems related to the product itself or how the product was handled during the distribution process. As a result, it is imperative that the root cause for the return be identified and documented so that the return data can be reviewed as part of the annual product review process.

SALVAGED DRUG PRODUCTS

When it is clear that pharmaceutical products have been subjected to improper storage conditions (for example, extremes in temperature or humidity) or exposed to atypical conditions (for example, smoke or chemical fumes) due to natural disasters, fires, accidents, or equipment failures, they should not be salvaged and returned to the marketplace.

Whenever it is unclear or there is some level of doubt regarding whether the drug products have been subjected to these types of conditions, salvaging operations may be conducted, but only when it is evident from analytical testing that the drug products meet all applicable standards of identity, strength, quality, and purity, and it is evident from an inspection of the facility where the salvaged drug products were held that the drug products were not subjected to improper conditions as a result of the natural disaster or accident.

In the case of salvaged drug products, a preponderance of evidence that demonstrates the acceptability of the products is required. Without a substantial body of evidence, salvaged drug products should be rejected and disposed of appropriately. The outcome of the evaluation of salvaged drugs should be documented. *Reverse logistics* is the term used to describe the process of returning products back through the distribution network for reimbursement, product assessment, and appropriate disposal. Recognizing that the effective management of the reverse logistics process is essential for the company and the environment, some pharmaceutical firms have turned to third-party reverse logistics companies to handle all aspects of the returned goods process. In general, the product that flows through these processors will ultimately be rejected, and therefore no evaluation of the product is deemed necessary.

Like any third-party service provider that conducts a GMP operation on behalf of the pharmaceutical company, these providers must be assessed to ensure that they are capable of doing the work, are providing adequate data back to the pharmaceutical company for quality evaluation, and are disposing of product in compliance with the appropriate laws and regulations.

DESTRUCTION FACILITIES AND PROCESSES

> Determine the destruction requirements for materials including suitable facilities and processes. (Apply)
>
> **Body of Knowledge V.F.2**

Public awareness and concern are growing over the disposal of pharmaceutical products and the potential for adverse impact on the environment. An article published in 2002 by the US Geological Survey ("Pharmaceuticals, Hormones, and Other Organic Wastewater Contaminants in US Streams, 1999–2000: A National Reconnaissance") outlined the presence of pharmaceuticals and other organic wastewater contaminants in streams across the country. This report highlighted the need for improving the handling and disposal of pharmaceutical waste. Proper disposal of potentially dangerous pharmaceutical products is now of critical importance not just in the United States but around the world.

All drug products that are rejected by the pharmaceutical company need to be disposed of in accordance with applicable laws and in a manner that protects consumers as well as the environment. Many drug products are considered hazardous waste and must be disposed of by companies that are licensed to handle this type of material. In the United States, the Environmental Protection Agency (EPA) is responsible for defining hazardous chemicals and the procedures to be followed to properly dispose of these materials. Drug formulations are considered hazardous waste if they fall into one of two primary categories: listed hazardous waste or characteristic hazardous waste. It is critical to understand the EPA (or other relevant environmental regulatory agency) requirements as they pertain to a company's specific products, that the products are assessed to determine the appropriate category for disposal, and, ultimately, that they are disposed of in accordance with the laws and regulations that govern the operations in the area. The complexity of such laws has increased the use of third-party vendors by pharmaceutical companies, but it is important to understand that these vendors need to be adequately assessed and qualified by the pharmaceutical company to ensure that they are abiding by the laws for handling and disposal of returned drug products.

Controlled substance drug products are another special category that requires very tight control over their inventory and destruction. The destruction of controlled substance products requires specific documentation (Form 41) for the Drug Enforcement Agency (DEA) in addition to the documentation that is required to be maintained for GMP purposes. The disposal of controlled substances often requires witnessing the actual disposal to ensure that the product does not inadvertently reenter the supply chain.

The processing of pharmaceutical product returns and salvaged drug products can be a daunting task and requires knowledge of laws and regulations outside the GMP regulations. Proper disposition and disposal of pharmaceutical products ensures that consumers receive drugs that have the required quality, and that they, and the environment, are not inadvertently harmed by inadequate disposal of rejected or expired drug products.

Part VI

Sterile and Nonsterile Manufacturing Systems

Chapter 41

Master Batch and Completed Batch Records

Clearly written, detailed batch records are critical to ensuring product quality. The *master batch record* (MBR) accurately describes the production processes required to make the product the same way every time, in order to meet the defined product specifications. MBRs are written and approved by the department that manufactures the product and are approved by the *quality unit* (QU). MBRs should be written to precisely describe the required information for trained operators to execute consistently.

REQUIRED ELEMENTS

> Review batch records for required elements including proper issuance, sections on yields, critical manufacturing step verification, processing instructions, and hold times. (Apply)
>
> **Body of Knowledge VI.A.1**

MASTER BATCH RECORD STRUCTURE

Batch records are the primary documentation that capture real-time performance of production records. Batch records are a key part of what is used to confirm that all expected and required actions have been completed within parameters to produce product that meets specifications and complied with quality standards.

The *master batch record* (MBR) is the version-controlled documentation necessary to trace the complete cycle of manufacture of a batch of product, from the dispensing of materials through all processing, testing, and subsequent packaging to the dispatch for sale or supply of the finished product. This documentation includes quality control, quality assurance, and environmental data relevant to the intended manufacturing.

The *master production record* (MPR) may be segmented on intended manufacturing and testing stages, each part controlled separately.

The *production batch record* (PBR) is issued for the manufacturing of one (or more) batches from the MPR and is compiled during the manufacturing operation.

Table 41.1 Master Production / Production Batch records comparison.

Paper System	Electronic System	Description
Master production record	Master recipe	Contains product name or designation, recipe designation or version, formulas, equipment requirements or classes, sequence of activities, procedures, normalized bill of materials (quantity per unit volume to produce)
	Work instructions	Additional detailed instructions—may include electronic SOPs or SOP references
	Critical process parameters	Required process parameters that are to be checked or monitored or are to be downloaded to other systems such as automation
Production batch record	Control recipe	A master recipe dispatched or otherwise made available in manufacturing-related areas for execution. Includes master recipe information with the addition of schedule, specific quantity to make, actual target bill of materials quantities, and other data for the batch and production instance
	Electronic production record	A store of data and information created by systems or entered by personnel during execution of control recipes
		May be located in one or more systems or databases
		Data may or may not be stored in human readable format
	Production report	Data and information in human readable format, presented either in electronic or paper format for activities, such as review, disposition, investigation, audit, and analysis

The MBR and PBR may be controlled in the document management system, within a manufacturing execution system (MES)/electronic batch record (EBR) platform or some hybrid of the two. Parts may also be found within the LIMS, data historian and other electronic systems. A critical part of building the MBR is ensuring the correct connections between the MBR and data existing in specific electronic platforms. Table 41.1 provides a comparison between MBRs and PBRs.

ISSUANCE

In most instances, it is a regulatory authority expectation that master manufacturing records used during manufacture of product are controlled by the quality unit, including the distribution to production.

When there is a validated computerized system, these records can be maintained and controlled by production with quality management oversight, which ensures their proper control, including the traceability and reconciliation.

Each copy of a PBR must be assigned a unique batch identity number upon issuance.

A MBR intended to be printed (including paper-on-glass) contains a section for documenting issuance of *production batch records* (PBRs) generated from the MBR. This section provides signature spaces for the person who generated the PBR and for a second person who verifies the PBR. Their signatures indicate that the PBR is of the correct, current version and that it is an accurate, legible copy of the MBR.

YIELD

Include the expected (theoretical) yield at key processing steps in the MBR. Actual yields are tracked in the PBR to ensure that product is not unexpectedly lost in processing. Production yields should be predictable, and within acceptable tolerances, if the manufacturing process is in control. Calculation of percentage of theoretical yield is performed by one person and verified by a second person. If the percentage of theoretical yield is not within the defined range, an investigation is conducted to determine the reason for the unexpected result.

CRITICAL STEP VERIFICATION

Significant processing steps that are critical to product quality should be verified by a second person to ensure that they are performed in accordance with the batch record, creating product that meets predetermined specifications. For example, addition of materials to a batch of product is performed by one person and verified by a second person.

Standard Product Considerations

Packaging and labeling requirements must be a part of the master manufacturing records of active pharmaceutical ingredients or active pharmaceutical ingredient intermediates sold as finished products.

For the manufacturing of sterile and aseptic products, critical environmental and hygiene monitoring as well as sterilization requirements must be identified in the MBR.

For the manufacturing of cold-chain products, time-out-of-refrigeration (TOR) requirements during manufacturing steps must be part of the MBR and must be reported.

PROCESSING INSTRUCTIONS

The instructions in the MBR for making the product should include identifying information describing the product, a list of all of the materials that are used to make the product, including quantity, and the location and equipment used to make the product. The identifying information can include the name of the product, the processing step (for example, fermentation of product X), the concentration or strength of the product, and an identification number.

The MBR includes processing steps in the order in which they must occur. Identify any critical processing parameters that could impact product quality and the allowable ranges around these parameters that must be maintained, such as pressure, temperature, elapsed time, and speed in revolutions per minute. Example: Set centrifuge to run for 15 ± 1 minutes at 7500 ± 200 rpm.

The MBR includes in-process samples that are taken to check product quality against defined acceptance criteria. Conductivity, or pH, is frequently checked during processing to determine whether the levels are appropriate before performing the next step. Example: pH acceptance criterion is 6.8 ± 0.2.

The processing instructions include packaging and labeling requirements. The primary packaging (product container) and secondary packaging (cartons, shipping boxes, and shrink-wrap) are included in the list of materials required for the batch. Labeling text is documented to ensure that the product is accurately identified. Requirements for the label material are included, which describe a type of label that will adhere securely to the product while remaining legible throughout storage and distribution of the product to the customer.

Storage conditions of the intermediate or finished product, with any storage time limits, are included in the batch record to ensure that the product is stored under conditions (temperature, humidity, protection from light) that maintain the quality of the product.

HOLD TIMES

Any time limits for processing steps, hold times between steps, and maximum length of time to perform the total process that could impact product quality are included in the batch record. Sampling and testing requirements to measure product attributes to determine whether the product meets specifications are included in the MBR. When a PBR is issued from the MBR, a unique identification number is assigned to the PBR, often called a *batch number*. The issuer signs and dates the PBR to indicate it is an accurate, legible reproduction of the approved MBR, and a second person reviews the PBR to confirm its accuracy.

RECORD PROCESSING REQUIREMENTS

> Confirm that batch records meet requirements for execution, review, and disposition decisions. (Apply)
>
> **Body of Knowledge VI.A.2**

COMPLETED BATCH RECORDS

As the PBR is used to make a batch of product, several items need to be documented at the time the significant processing steps are performed:

- Date (and time, if important) of execution, with the date and initials of the persons who performed and checked the step
- Equipment used

- Weight, lot number, and actual quantity of materials used
- Actual measured values of the critical processing parameters
- Samples collected and results for in-process and release tests
- Actual yield where theoretical yields are identified
- Description of packaging, labeling, and labels, as well as a sample of the label
- Documentation of any deviation and its related investigation if the deviation has potential impact to product

REVIEW AND DISPOSITION OF PROCESSING BATCH RECORDS

The responsibility for review and disposition of the product by examination of the PBR should be defined in a written procedure. Production personnel are responsible for reviewing PBRs and ensuring they are completed and signed and that any related deviations are documented and investigated.

QU personnel are responsible for reviewing and approving the completed batch records and releasing or rejecting product. Part of the QU's review is to ensure that any deviations and related investigations that could impact product, and any *out-of-specification* (OOS) test results, are satisfactorily addressed. The QU ensures that the product complies with established specifications before any product is released from the organization's control.

Chapter 42

Production Operations

GENERAL

Production operations must comply with *good manufacturing practices* (GMP) to manufacture quality product that meets regulatory requirements. GMP for manufacturing operations include requirements for the prevention of cross-contamination during production, process validation, and environmental conditions, based on the types of manufacturing performed. These GMP should be written and approved, with procedures and documentation for production operations. The GMP for production operations include:

- Access to production premises should be restricted to designated personnel;
- Before any processing operation is started, steps are taken and documented to ensure that the work area and equipment are clean and free from any raw materials, products, product residues, labels, or documents not required for the current operation;
- At all times during processing, all materials, bulk containers, major items of equipment, and the rooms used are labeled or otherwise identified with an indication of the product or material being processed, its strength, and the batch number or status. The processing status of major units of equipment should be indicated either on the individual units of equipment or by appropriate documentation, computer control systems, or alternative means;
- Contemporaneous and complete batch production documentation is necessary;
- Process steps must be verified by a second person. Process steps can be performed using a validated computer system;
- Other critical activities should be witnessed or subjected to an equivalent control;
- Any deviation from instructions or procedures is avoided. Any deviation should be documented and explained. Any critical deviation should be investigated. Qualified personnel write a report that describes the deviation, the investigation, the rationale for disposition,

and any follow-up activities required. The report is approved by the *quality control* (QC) unit;

- Rejected materials and products are clearly marked as such and are either stored separately in restricted areas or controlled by a system that deters their utilization in manufacturing or processing activities for which they are not suitable. These materials are either returned to their vendors or, where appropriate, reprocessed or destroyed. Actions taken are recorded.

Provided that changeover procedures are validated and implemented, non-medicinal products may be fabricated or packaged/labeled in areas or with equipment that is also used for the production of pharmaceutical products.

PROCESS VALIDATION

Process validation includes the validation and security of computerized or automated data handling systems. Validation studies are conducted in accordance with predefined protocols. A written report summarizing recorded results and conclusions is prepared, evaluated, approved, and maintained.

The manufacturing processes for human drugs, veterinary drugs, or biologics must comply with the current good manufacturing practice (cGMP) regulations to ensure that the products are safe and effective. While the fundamental principles of cGMP remain consistent across various categories, there are nuanced differences and specific guidelines for human drugs, veterinary drugs, and biologics to reflect the uniqueness of the specific product category. The manufacturing processes vary vastly based on the *dosage form* (the physical form in which a drug is produced and dispensed) and *route of administration* (the location at which the drug is administered) to deliver the drug in the right package.

Dosage forms could be liquid (elixirs, emulsions, gels, lotions, mixtures, solutions, or syrups), solid (tablets, capsules, lozenges), or semisolid (ointments, creams, paste, gel). Route of administration could be oral, parenteral, transdermal, sublingual, transdermal, intranasal, intraocular, rectal, or vaginal. The manufacture of different types of dosage forms will require unique types of equipment and automation based on the process complexity, batch sizes, and the level of automation.

Human drugs that contain one or more active ingredients and numerous excipients are chemically synthesized, with known structures, considered small molecules, and are relatively stable. Biologics, on the other hand—including vaccines, blood, and blood components, allergenics, somatic cells, gene therapy, tissues, and recombinant therapeutic proteins—are large molecule substances, relatively unstable, not easily identified or characterized, and require controlled processes and complex technology. Biological products tend to be heat sensitive and susceptible to microbial contamination, making it necessary to use aseptic principles from initial manufacturing steps, which is also in contrast to most conventional human drugs.

Manufacturing of veterinary drugs needs to consider the varied animal species, wide range in body sizes, and potential differences in drug administration route. Human and animal drugs can be manufactured on the same manufacturing lines and must be manufactured to the same current good manufacturing practice (cGMP) standards.

APPLICATION FACTORS

> Describe and differentiate the requirements for manufacturing processes according to their application: human drugs, veterinary drugs, or biologics. (Apply)
>
> **Body of Knowledge VI.B.1**

Human Drugs

Production requirements for human drugs are regulated by 21 CFR 211, while production requirements for biologics are regulated by 21 CFR 600, 601, and 610. Animal drugs are regulated by 21 CFR. Sterile drugs are susceptible to particulate, pyrogenic, and microbiological contamination. Due to the health hazard associated with the use of contaminated sterile products, special precautions are required in the production of these products. The skill, training, and competency of all personnel involved are critical. *Quality assurance* (QA) is important, and the production must follow carefully established and validated methods of preparation and sterilization.

To maintain product sterility, it is essential that the environment in which aseptic operations (for example, equipment setup, filling) are conducted be maintained in a controlled manner and at an appropriate quality. Activities conducted in critical areas include manipulations (for example, aseptic connections, sterile ingredient additions) of sterile materials before and during filling and closing operations.

High-efficiency particulate air (HEPA)–filtered air should be supplied in critical areas at a velocity sufficient to sweep particles away from the filling/closing area and to maintain unidirectional airflow during operations. The velocity parameters established for each processing line should be justified and appropriate to maintain unidirectional airflow and air quality under dynamic conditions within the critical area.

Supporting clean areas can have various classifications and functions. Many support areas function as zones in which nonsterile components, formulated products, in-process materials, equipment, and container/closures are prepared, held, or transferred. These environments are soundly designed when they minimize the level of particle contaminants in the final product and control the microbiological content (bioburden) of articles and components that subsequently are sterilized.

The nature of the activities conducted in a supporting clean area determines its classification. The United States Food and Drug Administration (FDA) recommends that the area immediately adjacent to the aseptic processing line meet, at a minimum, class 10,000 (ISO 7) standards under dynamic conditions. Manufacturers can classify this area as class 1000 (ISO 6) or maintain the entire aseptic filling room at class 100 (ISO 5). An area classified at a class 100,000 (ISO 8) air cleanliness level is appropriate for less-critical activities (for example, equipment cleaning). European Union (EU) GMP recommend that the supporting areas cascade, so that a supporting area from grade A is grade B, which cascades down to grade C, and again down to grade D.

VETERINARY DRUGS

The manufacturing process for veterinary drugs must be manufactured to the same current good manufacturing practice (cGMP) standards as human drugs, and be controlled to preserve the drug's identity, strength, quality, and purity as well as to show that the drug product can be consistently produced from batch to batch. Additional requirements for Animal Drugs are included in Subchapter E, 21 CFR Parts 500–599.

Veterinary drugs must be labeled in accordance with 21 CFR Part 201.105. The manufacturing process must ensure that the labeling contains all necessary information to use the drug safely and effectively, including the risks associated with the drug.

BIOLOGICS

The cGMP requirements for biological products is considered as complementary to the general requirements for drug products. However, the specific requirements for each of the quality system elements could include specific requirements. Starting materials play a critical role in the manufacturing process as their source and origin would impact the quality of the finished product. The risk of contamination of starting materials with adventitious agents such as those causing transmissible spongiform encephalopathy (TSE) must be assessed. Controls must also be in place where human or animal cells are used that include the number of generations between the seed lot or cell bank and the finished product with adoption of master cell bank (MCB) and working cell bank (WCB).

Due to the potential inherent variability in the biological processes, it is essential to have controlled manufacturing processes. The use of closed systems to control asepsis and contamination must be considered where practicable. Logical and unidirectional flow of material, personnel, and processes, as well as the use of clean-in-place and sterilization-in-place systems, should be considered wherever possible. Air-handling systems should be designed, constructed, and maintained in a manner to control cross-contamination. Processes must be developed for appropriate biosafety considerations and decontamination (to remediate any accidental spillages of live organisms).

Process validation and cleaning validation must be established using a quality risk management approach to identify, control, and mitigate risks across the manufacturing process.

PROCESS OPERATIONS

> Understand and differentiate between continuous and batch manufacturing. (Apply)
>
> **Body of Knowledge VI.B.2**

Continuous manufacturing and batch manufacturing are two distinct approaches in the production processes.

In continuous manufacturing, the production process runs continuously without interruption. Raw materials are added, and the product is continually processed without the need for distinct batches. This approach offers advantages such as increased efficiency and reduced production changeover times. This approach is suitable for products with stable, well-defined, and well-controlled processes.

In contrast, batch manufacturing involves manufacture of products in discrete batches. Each batch goes through the entire pre-established production process before the next one begins. This process allows easier quality control as each batch can be tested and evaluated independently. This approach provides flexibility in handling different formulations but takes long processing times given the start-stop processes, including line clearances, and long hold times between steps.

Both types of manufacturing are subject to the same current good manufacturing practices (cGMP) and quality control standards; however, the process monitoring in continuous manufacturing tends to be more automated and more frequent than in batch manufacturing.

The development, implementation, operation, and life cycle management of continuous manufacturing is detailed in the International Conference on Harmonization (ICH) Q13, Continuous Manufacturing of Drug Substances and Drug Products.

Nonsterile Manufacturing and Manufacturing of Liquids, Creams, and Ointments

FDA guidance is found in sections of 21 CFR regarding topical drug products, and similar information can be found in Eudralex Volume 4, Parts I and II and Annexes.

UTILITY REQUIREMENTS

> Identify the facility and utility requirements that are appropriate for different production environments and product types including sterile vs. nonsterile manufacturing, solid and semisolid dosage forms, liquids, creams, ointments, and combination products. (Analyze)
>
> **Body of Knowledge VI.B.3**

All utilities with the potential to impact product quality (for example, water, compressed air, heating, ventilation, and air conditioning) must undergo qualification. This includes installation qualification (IQ) as a documented verification that the utilities and equipment, as installed or modified, comply with the approved design the manufacturer's recommendations and/or user requirements; operational qualification (OQ) as a documented verification that the utilities and equipment, as installed or modified, perform as intended throughout the anticipated operating ranges; and performance qualification (PQ) as a documented verification that utilities and equipment as connected together can perform consistently effectively and reproducibly based on the approved process method and specifications. Procedures must be in place for appropriate monitoring, and action should be taken when limits are exceeded. Recertification, requalification, and revalidation must be performed as appropriate. Documentation, including drawings, should be retained.

Several utilities used in the manufacturing environments include, for example, water systems, heating, ventilation, and air conditioning (HVAC), compressed air, and electric supply.

Pharmaceutical water is the most widely used ingredient in drug manufacturing, and also in cleaning applications. USP <1231> *Water for Pharmaceutical Purposes*, classifies pharmaceutical water into four categories, namely:

- Bulk Monographed Water and Steam (includes purified water, water for injection, water for hemodialysis, and pure steam)
- Sterile Monographed Waters (includes sterile purified water, sterile water for injection, bacteriostatic water for injection, sterile water for irrigation, and sterile water for inhalation)
- Drinking Water
- Other Non-Monographed Waters (includes ammonia-free water, carbon-dioxide-free water, distilled water, freshly distilled water, deionized water, deionized distilled water, filtered water, high purity water, deaerated water, oxygen-free water, and water for bacterial endotoxins testing)

Heating, ventilation, and air-conditioning (HVAC) play an important role in maintaining an appropriate manufacturing environment compliant to the principles of current good manufacturing practices (cGMP). A well-designed HVAC system prevents contamination and cross-contamination and provides comfortable working conditions for the operatives. Some of the other design factors to be considered include temperature, relative humidity, room pressure, air changes per hour, pressure differentials, and pressure cascades to correspond with the manufacture and storage of the pharmaceutical products.

There are two types of HVAC systems—once-through (conditioned air is supplied to the cleanroom and discarded through return filters) and recirculated (the conditioned air from the cleanroom goes back into the HVAC system and is reprocessed and reused). HVAC consists of several components, namely dampers, cooling coil, blower, heating coil, coarse filters, HEPA filters, risers, blowers, and exhaust. The ISO 14644 series of standards covers the classification of air cleanliness in cleanrooms and associated controlled environments. The ISO classes of air cleanliness per ISO 14644-1, *Cleanrooms and associated controlled environments—Part 1: Classification of air cleanliness by particle concentration* is provided Table 42.1.

Filters of different porosity are used in the HVAC, coarser ones to arrest big particles and to prevent clogging on the downstream HEPA filters. The conditioned air is finally delivered to the manufacturing spaces through the HEPA filters, which remove at least 99.97% of aerosols 0.3 micrometers (μm) in diameter. The specification used in the European Union, the multi-part EN 1822, *High Efficiency Air Filters (HEPA and ULPA)*, from which ISO 29463, *High efficiency filters and filter media for removing particles from air* (all parts), is derived, defines several

Table 42.1 ISO Classes of Air Cleanliness by particle concentration.

ISO Class number (N)	Maximum allowable concentrations (particles/m^3) for particles equal to and greater than the considered sizes, shown below[a]					
	0,1 µm	0,2 µm	0,3 µm	0,5 µm	1 µm	5 µm
1	*10*[b]	d	d	d	d	e
2	100	*24*[b]	*10*[b]	d	d	e
3	1000	237	102	*35*[b]	d	e
4	10,000	2370	1020	352	*83*[b]	e
5	100,000	23,700	10,200	3520	832	d, e, f
6	100,000	237,000	102,000	35,200	8320	293
7	c	c	c	352,000	83,200	2930
8	c	c	c	352,000	832,000	29,300
9g	c	c	c	35,200,000	8,320,000	29,3000

[a] All concentrations in the table are cumulative; e.g. for ISO Class 5, the 10 200 particles shown at 0,3 µm include all particles equal to and greater than this size.
[b] These concentrations will lead to large air sample volumes for classification. Sequential sampling procedure may be applied; see Annex D.
[c] Concentration limits are not applicable in this region of the table due to very high particle concentration.
[d] Sampling and statistical limitations for particles in low concentrations make classification inappropriate.
[e] Sample collection limitations for both particles in low concentrations and sizes greater than 1 µm make classification at this particles size inappropriate, due to potential particle losses in the sampling system.
[f] In order to specify this particle size in association with ISO Class 5, the macroparticle descriptor M may be adapted and used in conjunction with at least one other particle size. (See C.7.)
[g] This class is only applicable for the in-operation state.

classes of filters by their retention at the given most penetrating particle size (MPPS): Efficient Particulate Air filters (EPA), HEPA, and Ultra Low Particulate Air filters (ULPA). The HEPA filters in pharmaceutical applications should have an EN 1822 classification of at least H13 or equivalent.

Compressed air is extensively used in the pharmaceutical industry and could be broadly classified as *direct contact* for applications like tablet coating, drying, or hardening process; and *indirect contact* for powering production machinery and conveyors). Direct contact increases the risks of contamination to the product, while indirect contact has a minimal product risk. Appropriate risk assessment must be performed. According to the FDA's *Guidance for Industry Sterile Drug Products Produced by Aseptic Processing—Current Good Manufacturing Practice (2004)*, "a compressed gas should be of appropriate purity (for example, free from oil) and it's microbiological and particle quality after filtration should be equal to or better than that of the air in the environment into which the gas is introduced." The multi-part standards on ISO 8573, *Compressed Air*, include purity classes of compressed air with respect to particles, water, and oil, and identify gaseous and microbiological contaminants.

Electrical power, along with the support of an *uninterruptible power supply* (UPS), is a fundamental utility in pharmaceutical manufacturing for continuous and reliable operation of various processes.

SANITIZATION AND PROTECTION

> Identify various production operations that require gowning, sanitization, hygiene, and other product-protective steps. (Apply)
>
> **Body of Knowledge VI.B.4**

Clean areas should be cleaned thoroughly according to documented procedures. Detailed hygiene programs should be established. They should include procedures relating to health, hygiene practices, and clothing of personnel. Hygiene procedures should be understood and followed in a very strict way by all persons whose duties take them into the production and control areas. The hygiene program should be promoted by management and discussed during training sessions. Unsanitary practices in the production and storage areas should be prohibited, such as eating, drinking, chewing gum, or smoking, as well as storing food, drink, smoking materials, or personal medication in those areas. Requirements concerning personal hygiene, with an emphasis on hand hygiene, are outlined and are followed by employees. Personnel should be instructed to use the hand-washing

Table 42.2 Cleanroom gowning requirements as specified by EU GMP, ICH, and Health Canada.

Grade A and B (Class 100, ISO 5)	Grade C	Grade D
Headgear totally encloses the person's hair, as well as any beard or mustache; the headgear is tucked into the neck of the suit; a face mask is worn to prevent the shedding of droplets.	The person's hair, as well as any beard or mustache, is covered.	The person's hair, as well as any beard or mustache, is covered.
Trouser bottoms are tucked inside the footwear, and garment sleeves are tucked into the gloves.	A one- or two-piece trouser suit, gathered at the wrists and with a high neck, and appropriate shoes or overshoes are worn.	Protective clothing and appropriate shoes or overshoes are worn.
The protective clothing sheds virtually no fibers or particulate matter and retains particles shed by the body.	The protective clothing sheds virtually no fibers or particulate matter.	
Sterilized, non-powdered rubber or plastic gloves and sterilized or disinfected footwear are worn.		
Outdoor clothing should not be brought into changing rooms.	Outdoor clothing should not be brought into changing rooms.	

facilities. Direct skin contact should be avoided between the operator's hands and raw materials, primary packaging materials, intermediate or bulk drug, or exposed product, as well as with any part of the equipment that comes into contact with the products. Houseplants are not permitted in fabrication, packaging/labeling, and storage areas or in any other areas where they might adversely influence product quality. Requirements concerning cosmetics and jewelry worn by employees are outlined and are observed by employees. Soiled protective garments, if reusable, are stored in separate containers until properly laundered and, if necessary, disinfected or sterilized. Steps should be taken to ensure as far as is practicable that no person affected by an infectious disease or having open lesions on the exposed surface of the body is engaged in the manufacture of medicinal products. Personal hygiene procedures, including the use of protective clothing, apply to all persons entering production areas. The hygiene program clearly defines clothing requirements and hygiene procedures for personnel and visitors. It must be the manufacturer's responsibility to provide instructions ensuring that health conditions that can be of relevance to the quality of products are brought to the manufacturer's attention. In general, any unhygienic practice within the manufacturing areas or in any other area where the product might be adversely affected should be forbidden.

In addition, some jurisdictions require medical examinations, particularly at time of recruitment. After the first medical examination, examinations should be carried out when necessary for the work performed and the employee's personal health.

Gowning procedures and practices should be documented based on the cleanroom classification. Table 42.2 provides cleanroom gowning requirements as specified by EU GMP, ICH, and Health Canada.

Chapter 43

In-Process Controls

IN-PROCESS TESTING

> Identify appropriate tests for each step in the manufacturing process and review results. (Analyze)
>
> **Body of Knowledge VI.C.1**

General Requirements

In-process testing requirements for the US and EU pharmaceutical markets are crucial to ensure the safety, efficacy, and quality of pharmaceutical products during the manufacturing process. These requirements aim to detect any deviations or abnormalities that may occur during various stages of production and help to identify and rectify any potential issues early on, thereby minimizing the risk of producing substandard or unsafe products.

For the US pharmaceutical market, in-process testing is governed by the regulations set forth by the Food and Drug Administration (FDA) in 21 CFR §211.110, "Sampling and Testing of In-process Materials and Drug Products" and through the adoption of ICH Q7 guideline. The FDA requires pharmaceutical manufacturers to conduct in-process testing to monitor critical quality attributes and ensure compliance with current good manufacturing practices (cGMP). The specific requirements for in-process testing may vary depending on the type of product being manufactured, but generally include the following:

1. Sampling and testing: Manufacturers must establish appropriate sampling plans to ensure representative samples are collected during the manufacturing process. These samples are then subjected to various tests to assess critical quality attributes, such as potency, purity, identity, and uniformity.

2. Specifications: In-process testing should be performed against predetermined specifications or acceptance criteria. These specifications define the acceptable limits for various quality attributes. If the test

results fall within these limits, the process can proceed. Otherwise, corrective actions need to be taken.

3. Analytical methods: Validated analytical methods should be used for in-process testing. These methods should be sensitive, specific, accurate, and precise to ensure reliable and reproducible results. Manufacturers must adhere to the FDA's guidelines for method validation.
4. Process control: In-process testing serves as a tool for process control. It allows manufacturers to identify any deviations from the established manufacturing process and take corrective actions promptly. Process control includes monitoring critical process parameters, such as temperature, pressure, and pH, to ensure consistency and reproducibility.
5. Documentation: Accurate and detailed documentation of in-process testing activities is essential. Manufacturers must maintain comprehensive records of test results, including raw data, calculations, and observations. These records serve as evidence of compliance with regulatory requirements and can be reviewed during inspections by regulatory authorities.

Examples of common in-process testing and controls include:

- Tablet or capsule weight variation
- Disintegration time
- Adequacy of mixing to assure uniformity and homogeneity
- Dissolution time and rate
- Clarity, completeness, or pH of solutions
- Vial weight checks
- Filter integrity checks (aseptically produced products)
- Identity checks
- Moisture content
- Residual solvents

Similarly, in the EU pharmaceutical market, in-process testing requirements are governed by the European Medicines Agency (EMA) and are outlined in the EU good manufacturing practice (GMP) guidelines included in Eudralex Volume 4. The requirements are similar to those in the United States, with a focus on ensuring product quality and patient safety.

Additionally, in-process materials shall be tested for identity, strength, quality, and purity as appropriate, and approved or rejected by the quality control unit, during the production process (for example, at commencement or completion of significant phases or after storage for long periods). Rejected in-process materials must be identified and controlled under a quarantine system designed to prevent their use in manufacturing or processing operations for which they are unsuitable.

CRITICAL PROCESS PARAMETERS (CPPS)

> Monitor critical process parameters (CPPs). (Analyze)
>
> **Body of Knowledge VI.C.2**

Critical process parameters (CPPs) are the key variables or factors that have a significant impact on the quality, safety, and effectiveness of a pharmaceutical product during its manufacturing process. These parameters must be closely monitored and controlled to ensure consistent product quality. The specific critical process parameters may vary depending on the nature of the pharmaceutical product and the manufacturing process. However, some common examples of critical process parameters in the pharmaceutical industry include:

- Temperature: The temperature at various stages of the manufacturing process can affect the chemical reactions, solubility, stability, and microbial growth of the product.
- Pressure: Controlled pressure levels are crucial in processes like compression, filtration, and sterilization to ensure proper product formation and integrity.
- pH: The pH level plays a vital role in maintaining the stability and efficacy of pharmaceutical formulations. It can influence the solubility, bioavailability, and chemical reactions of active ingredients.
- Mixing and homogenization: Proper mixing and homogenization of ingredients are critical to ensure uniform distribution and consistency of the product, preventing issues like content non-uniformity or phase separation.
- Time: The duration of various process steps such as heating, cooling, drying, or reaction times can impact the product's quality, purity, and stability.
- Flow rate: In processes involving fluid handling, controlling the flow rate is necessary to ensure proper mixing, dissolution, and uniform distribution of components.
- Environmental conditions: Parameters like humidity, air quality, and cleanliness of the manufacturing area can affect product sterility, contamination risks, and overall quality.

These are just a few examples, and depending on the specific manufacturing process, there may be additional critical process parameters to consider. It is essential for pharmaceutical manufacturers to identify, monitor, and control these critical process parameters to ensure consistent product quality and compliance with regulatory standards.

Monitoring critical process parameters is crucial to ensure the quality, safety, and efficacy of pharmaceutical products. Both the United States and EU have specific requirements for monitoring these parameters. Below are the general requirements in each region.

United States

In the United States, the Food and Drug Administration (FDA) sets guidelines for monitoring critical process parameters in the pharmaceutical industry. The FDA requires manufacturers to establish and maintain control over critical process parameters to ensure consistent product quality. The specific requirements include:

1. Identification of critical parameters: Manufacturers must identify the critical process parameters that significantly affect product quality. These parameters may include temperature, pressure, humidity, pH, mixing speed, and others.
2. Setting acceptance criteria: Acceptance criteria should be established for each critical process parameter. These criteria define the acceptable range or limits within which the parameter should be controlled.
3. Monitoring techniques: Manufacturers must use appropriate monitoring techniques to measure and track critical process parameters. This may involve the use of sensors, probes, data loggers, or automated systems to continuously or periodically monitor the parameters during production.
4. Data collection and analysis: Accurate and reliable data should be collected and analyzed to ensure that critical process parameters are within the established acceptance criteria. This data helps identify any deviations, trends, or potential issues that may affect product quality.
5. Documentation: Comprehensive documentation of critical process parameter monitoring activities is required. This includes recording the measured values, time of measurement, and any actions taken to address deviations or out-of-specification results.

European Union

In the EU, critical process parameter monitoring requirements are outlined in the Good Manufacturing Practice (GMP) guidelines issued by the European Medicines Agency (EMA). The requirements are similar to those in the US and include:

1. Identification and control: Manufacturers should identify critical process parameters and establish appropriate control measures to ensure product quality and consistency.
2. Monitoring techniques: Suitable techniques should be employed to monitor critical process parameters. This may involve the use of calibrated instruments, automated systems, or visual inspection, depending on the specific parameter being monitored.
3. Sampling and testing: Manufacturers should establish sampling plans to collect representative samples for testing critical process parameters.

These tests help confirm that the parameters are within the specified limits.

4. Data analysis and trending: Collected data should be analyzed and trends should be evaluated to identify any potential issues or deviations from the established control limits. This enables proactive measures to be taken to maintain process control.

5. Documentation: Thorough documentation of critical process parameter monitoring activities, including protocols, records, and reports, is necessary to demonstrate compliance with regulatory requirements.

International Harmonized Guidance

The International Council for Harmonisation of Technical Requirements for Pharmaceuticals for Human Use (ICH) provides guidelines and recommendations for the pharmaceutical industry worldwide. The ICH guidelines emphasize the importance of identifying and controlling *critical process parameters* (CPPs) to ensure product quality and consistency.

According to ICH guidelines, CPPs are defined as parameters that need to be controlled within predefined limits to ensure that the product meets its desired quality attributes. These parameters have a direct impact on the critical quality attributes (CQAs) of the product.

The ICH Q8 (R2) guideline, titled "Pharmaceutical Development," highlights the significance of understanding and controlling CPPs during the development and manufacturing of pharmaceutical products. It stresses the need for a systematic approach to identify CPPs based on scientific knowledge and risk assessment.

Additionally, the ICH Q9 guideline, titled "Quality Risk Management," provides a framework for identifying, assessing, and controlling risks associated with CPPs. It emphasizes the importance of incorporating risk management principles into the identification and control of CPPs.

Overall, the ICH guidelines emphasize the importance of identifying and controlling CPPs throughout the product lifecycle to ensure consistent product quality, safety, and efficacy. These guidelines serve as a valuable resource for pharmaceutical manufacturers in implementing robust quality systems and complying with global regulatory requirements.

PROCESS CAPABILITY

> Understand the importance of conducting process capability studies, calculate C_p and C_{pk}, and monitor process capability. (Apply)
>
> **Body of Knowledge VI.C.3**

While control charts tell us about the historical performance and stability of a process, these charts do not tell us anything about the ability of the process to meet customer requirements. In order to determine if a process is capable of conforming to specifications, we must compare the process performance, including both location and variability, with the specifications.

There are three potential process capability indices for variables data: C_r, C_p, and C_{pk}. Additionally, there are three process capability indices for variables data: P_r, P_p, and P_{pk}. The main difference between the "C" and "P" indices is that C_r, C_p, and Cpk are used with sample data to determine if the process is capable of meeting customer specifications (short term), whereas P_r, P_p, and P_{pk} are used with population data to determine if the process is capable of meeting customer specifications (long term). These capability indices (short and long term) assume normality of the underlying distribution. Figure 43.1 depicts a capable process versus a not capable process. Figure 43.2 depicts a process with control issues.

The following risk-based interpretation of C_p and C_{pk} values are commonly used and accepted:

- $C_p < 1$ unacceptable (fallout rate greater than 2699 ppm)
- $1 \le C_p \le 1.33$ marginal (fallout rate of 63–2699 ppm, process improvement activities should be considered)
- $C_p \ge 1.33$ acceptable for low-risk process outputs (fallout rate of less than 63 ppm)
- $C_p \ge 1.50$ acceptable for medium-risk process outputs (fallout rate of less than 7 ppm)
- $C_p \ge 1.67$ acceptable for high-risk process outputs (fallout rate of less than 1 ppm)

These indices compare the process statistical control limits to the product specifications to determine the process capability. Figure 43.3 demonstrates the relationship between statistical control limits and product specifications.

Figure 43.4 is a graphical representation of process capability ratios compared to the statistical control limits and part specification limits. There are three ways to increase process capability: decrease the process variation, center the process, or increase the product specification limits.

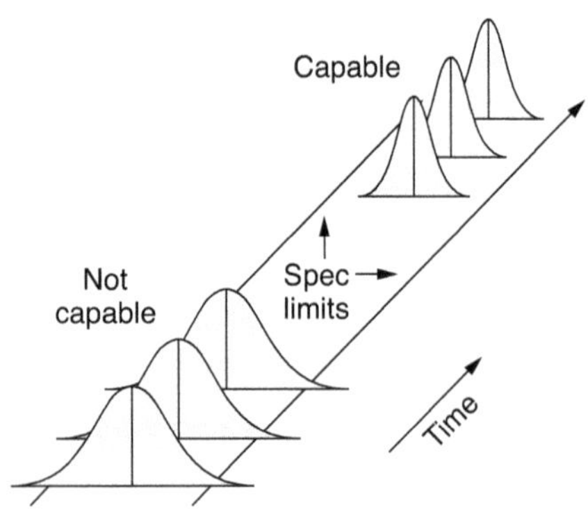

Figure 43.1 Capable process versus not capable process.

Source: Adapted from *Innovative Control Charting* (Wise and Fair 1997). Used with permission.

Short-term capability	Desired ratio
$C_r = \dfrac{6s}{USL-LSL}$	<1
$C_p = \dfrac{USL-LSL}{6s}$	>1
$C_{pk} = \text{Minimum of } \dfrac{\overline{X}-LSL}{3s}, \dfrac{USL-\overline{X}}{3s}$	>1

Long-term capability	Desired ratio
$P_r = \dfrac{6\sigma}{USL-LSL}$	<1
$P_p = \dfrac{USL-LSL}{6\sigma}$	>1
$P_{pk} = \text{Minimum of } \dfrac{\mu-LSL}{3\sigma}, \dfrac{USL-\mu}{3\sigma}$	>1

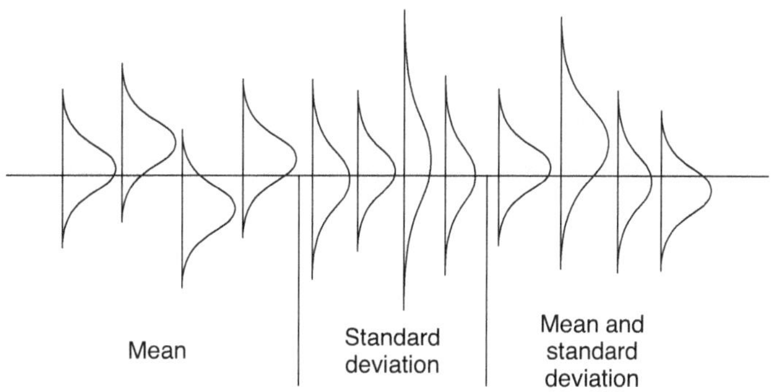

Figure 43.2 Process control issues.

Source: Practical Engineering, Process, and Reliability Statistics, 2nd Ed (Durivage 2022) used with permission.

SPECIFICATION LIMITS

> Assess specification limits in relation to registration or compendial requirements. (Evaluate)
>
> **Body of Knowledge VI.C.4**

Routinely assessing specification limits in the pharmaceutical industry is a regulatory requirement to ensure product quality, safety, and efficacy. Regulatory

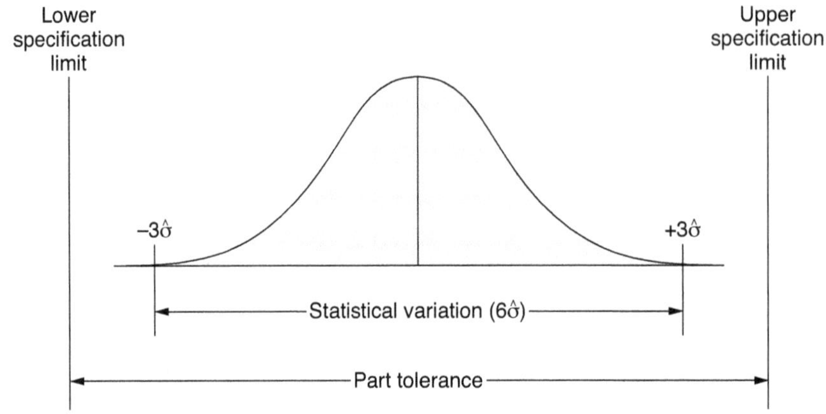

Figure 43.3 Relationship between statistical control limits and product specifications.
Source: Practical Engineering, Process, and Reliability Statistics, 2nd Ed (Durivage 2022) used with permission.

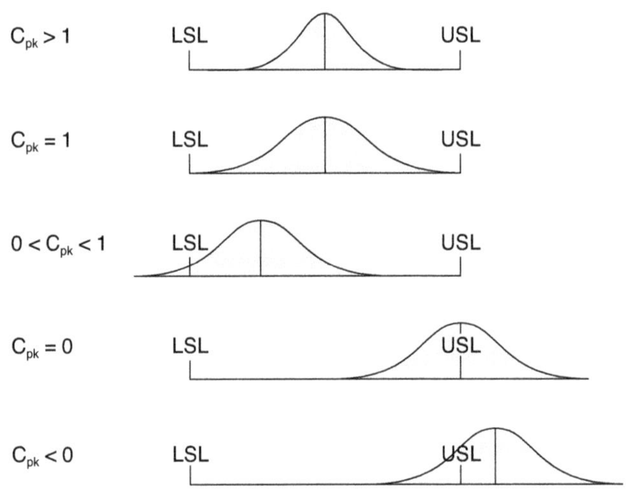

Figure 43.4 Graphical relationship of process capability ratios.
Source: Practical Engineering, Process, and Reliability Statistics, 2nd Ed (Durivage 2022) used with permission.

bodies such as the U.S. Food and Drug Administration (FDA) and the European Medicines Agency (EMA) require pharmaceutical manufacturers to establish and maintain specifications for their products. Some specific regulatory requirements related to assessing specification limits include:

- Current good manufacturing practices (cGMPs): Both the FDA in the United States and the EMA in the European Union have established cGMP regulations that require pharmaceutical manufacturers to have documented specifications for their products, including critical quality attributes and performance parameters.

- Stability testing: Regulatory guidelines mandate that pharmaceutical manufacturers conduct stability studies to assess the shelf life and stability of their products. These studies involve monitoring critical product attributes over time and comparing them against predetermined specification limits.
- Process validation: Pharmaceutical manufacturers are required to validate their manufacturing processes to ensure consistent product quality. This includes establishing appropriate specification limits for critical process parameters and regularly monitoring and assessing them during routine production.
- Regulatory submissions: When seeking regulatory approval for a new pharmaceutical product or making changes to an existing product, manufacturers must provide comprehensive data on specification limits and demonstrate that the product meets the established criteria.

Additionally, routinely assessing specification limits is required as part of annual product reviews where manufacturers ensure departures were appropriately investigated and documented and that the process is not unintentionally drifting or changing over time. Regularly assessing specification limits helps to identify any deviations from the desired quality parameters, enabling timely corrective actions to maintain product quality and safety. Benefits of regularly assessing specifications and their limits include:

- Regulatory compliance: The pharmaceutical industry is heavily regulated, with stringent guidelines and standards set by regulatory authorities such as the FDA (in the United States) and the EMA (in the EU). These standards often include establishing specification limits for critical product attributes, such as potency, impurities, dissolution, and content uniformity. Manufacturers must demonstrate that their products meet these limits to gain regulatory approval for marketing and distribution. Routine assessment of specification limits ensures compliance with these regulations and guidelines, minimizing the risk of non-compliance and potential regulatory actions.
- Compendial standards: Compendial bodies, such as the United States Pharmacopeia (USP) and the European Pharmacopoeia (Ph. Eur.), establish standards for pharmaceutical products that serve as a reference for quality testing. These compendial standards often include specific specification limits that products must meet to be considered compliant. Manufacturers are expected to test their products against these limits and provide evidence of compliance when submitting their products for registration or compendial verification.
- Process optimization: Regularly evaluating specification limits allows for the optimization of manufacturing processes. By analyzing data and identifying areas where specifications can be improved or modified, companies can enhance process efficiency, reduce variability, and minimize waste or rejections.

- Batch-to-batch consistency: Pharmaceutical products are often produced in batches, and maintaining consistency between batches is crucial. Routine assessment of specification limits helps to ensure that each batch meets the required quality standards and aligns with registration and compendial requirements, minimizing variability and ensuring consistent product performance.
- Product stability and shelf life: Pharmaceutical products have specific shelf life and stability requirements to ensure their effectiveness and safety throughout their intended lifespan. Regularly assessing specification limits helps to monitor and control parameters that can affect product stability, allowing for appropriate adjustments to maximize shelf life and maintain product integrity.
- Risk mitigation: By routinely assessing specification limits, potential risks and deviations can be identified early on. This allows for proactive measures to be taken to mitigate risks, prevent product failures, and ensure patient safety.
- Documentation and reporting: When submitting products for registration or compendial verification, manufacturers must provide detailed documentation and reports that include information on specification limits and the results of testing against those limits. These documents serve as evidence of compliance and support the evaluation of product quality by regulatory authorities or compendial bodies.

Overall, routinely assessing specification limits in the pharmaceutical industry is essential for maintaining product quality, ensuring regulatory compliance, optimizing processes, and mitigating risks. By establishing and assessing specification limits, manufacturers can demonstrate the quality, safety, and efficacy of their products, facilitating the registration and approval process. It helps to safeguard patient health and trust while upholding the highest standards of quality and safety in pharmaceutical manufacturing.

FDA Guidance—Powder Blends and Finished Dosage Units

Guidance was drafted to assist manufacturers in meeting requirements of 21 CFR 211.110 to demonstrate the adequacy of mixing and ensure uniformity of in-process dosage units. This guidance describes the procedures, sampling, and testing that should be performed during development of the product, including performing an assessment of and validating powder mix uniformity, correlating powder mix uniformity with stratified in-process dosage unit data, and correlating stratified in-process samples with final product.

Section VI, "Verification of Manufacturing Criteria," describes methods for establishing in-process acceptance criteria. The resulting data are used to establish stratified sample locations. At least 10 locations during capsule filling or tablet compression should be identified to represent a batch.

Time Limits

Both 21 CFR 211.111, "Time Limitations on Production," and EU GMP Part II, (ICH Q7) 8.2 "Time Limits," establish time limits to assure the quality of the product, where appropriate. Time limits should be met to ensure the quality of intermediates and product. Deviation from time limits may be acceptable if it does not compromise the quality of the product. Acceptability of deviation must be justified and documented; however, time limits may be inappropriate when processing to a target value, such as pH, or drying to specified values. Manufacturers should have the ability to establish production controls using in-process parameters based on desired end points measured using real-time testing or monitoring equipment.

Microbiological Contamination

Both 21 CFR 211.113, "Control of Microbiological Contamination," and EU GMP Part II, (ICH Q7) 8.51 and 8.52, state that a company needs written procedures designed to prevent contamination by objectionable microorganisms. Operations should be conducted in a manner that prevents contamination by other materials. Handling of *active pharmaceutical ingredients* (APIs) after purification should include precautions to avoid contamination.

Contamination Control

Eudralex Volume 4, Part I, Production 5.18, and EU GMP Part II, (ICH Q7) 8.5, address contamination control. Residual material carryover into successive batches of the same intermediate or API may be acceptable if there is adequate control. Carryover should not result in carryover of degradants or microbial contamination that can adversely affect the API impurity profile.

21 CFR 211.115 REPROCESSING

Written procedures are needed for reprocessing batches that do not conform to standards or specifications. The procedures should include steps to ensure that the reprocessed material conforms to specifications. These procedures need to include review and approval by the QU.

COMBINATION PRODUCTS

FDA does not have GMP regulations for combination products. Each constituent part (drug, device, biologic) is subject to specific GMP regulation if it is marketed separately. For combination products sold as one unit (for example, drug-filled syringes), both sets of regulations are applicable to the whole product after joining. FDA recommends speaking with the Agency during product development to discuss how best to achieve compliance to regulations for combination products.

BIOTECHNOLOGY PRODUCTS

Bioreactor fermentation parameters should be specified and monitored. Review of parameter profiles should verify that run parameters are consistent or have an established pattern from batch to batch. Parameters may include growth rate, pH, waste by-product level, viscosity, addition of chemicals, density, mixing, aeration, foaming, and culture purity. Written procedures are needed to assure proper aseptic techniques during cell inoculation and absence of adventitious agents, including criteria for rejecting contaminated runs. The procedures should describe what investigations and corrective actions will be performed in the event that growth parameters exceed established limits.

Once the fermentation process is completed, the desired product is harvested and, if necessary, refolded to restore configurational integrity, and then purified. For recovery of intracellular proteins, cells must be disrupted (lysed) after fermentation. After disruption, cellular debris can be removed by centrifugation or filtration. Centrifugation can be open or closed. The adequacy of the environment must be evaluated for open centrifugation. Ultrafiltration is commonly used to remove the desired product from the cell debris. The porosity of the membrane filter is calibrated to a specific molecular weight, allowing molecules below that weight to pass through while retaining molecules above that weight. For recovery of extracellular protein, the primary separation of product from producing organisms is accomplished by centrifugation or membrane filtration. Initial separation methods can be used after centrifugation to concentrate the products.

Further purification steps primarily involve chromatographic methods to remove impurities and bring the product closer to final specifications. Chromatographic separation techniques are multistage separation methods in which the components of a sample are distributed between two phases, one of which is stationary, while the other is mobile. The stationary phase may be a solid or a liquid supported on a solid or a gel. The stationary phase may be packed in a column, spread as a layer, or distributed as a film. The mobile phase may be gaseous or liquid or supercritical fluid. The separation may be based on adsorption, mass distribution (partition), or ion exchange, or may be based on differences in the physicochemical properties of the molecules, such as size, mass, or volume.

The purification process is primarily achieved by one or more column chromatography techniques, such as affinity chromatography, *ion-exchange chromatography* (IEC), gel filtration, *hydrophobic interaction chromatography* (HIC), and reverse-phase *high-performance (pressure) liquid chromatography* (HPLC).

Various types of filtration methods, such as diafiltration, ultrafiltration, and microfiltration, may be used in the purification of vaccine products. Some of the filters used may be single-use and some may be multiuse. The filters are usually placed within a filter-housing apparatus. The criteria used for the evaluation of the column purification should be applied to the filter housings and the multiuse filters.

Viral Clearance

For products derived from cells or source material of human or animal origin, viral inactivation/removal should be performed in accordance with the process in the approved license application. In some manufacturing operations, there will be a specific viral inactivation/removal step; in other operations, viral inactivation/ removal will be accomplished by a step or steps in the manufacturing process that are not specifically considered to be viral inactivation/removal steps. In some instances, more than one viral clearance step is used for a given product. Pre- and post-viral inactivation/removal steps (with the exception of products such as albumin, which are virally inactivated in final containers) should be completely separated. Separate areas with a dedicated *air-handling unit* (AHU) or single-pass air should be used for those steps that occur after viral clearance procedures.

Heat treatment is one method of clearing infectious agents from biologics. Heat treatment is sometimes referred to as *pasteurization*, and heating equipment such as large water baths may be referred to as *pasteurizers*. Technically, however, pasteurization is heating at 63 °C for 30 minutes, which is not sufficient to render plasma derivatives virally inactive.

The parameters specified in the batch record should be achieved such that the validated process for viral inactivation/removal is accomplished. Changes made to the process that do not require submission of a supplement to the Center for Biologics Evaluation and Research (CBER) should be validated.

Inactivation

If the active ingredient is a killed or inactivated version of a live bacteria or virus, the methods for inactivation will have been established and reviewed during product approval. Either heat or chemical treatment may be used for inactivation. The manufacturer should have validated the process and followed the validated procedures during production. All inactivation parameters should be monitored and the appropriate testing performed with acceptable results.

Manufacturing and Aseptic Processing

Aseptic processing in-process controls include environmental monitoring methods, such as surface monitoring, that can be performed with touch plates, swabs, and contact plates, active and passive air monitoring, container weight variation, fill weight, leak testing, established microbial specifications for in-process testing for the lots made, and aseptic connections and transfers. Procedures should be in place for interruption of the fill, should it occur, and for limiting access to controlled and classified areas. Filters should be evaluated before use to assure they meet specifications, and integrity testing should be performed on filters post-fill; results should be in keeping with the manufacturer's and validated specifications. Written procedures for gowning should be in place and followed.

Some bulk products are held after sterile filtration and before filling. The holding period and storage conditions should have been validated. For lengthy filling operations, time limits should be set and validated to assure that the duration of the fill does not affect the potency of the product and its susceptibility to microbial contamination.

Lyophilization

Lyophilization, or freeze-drying, is a process in which water is removed from a product after it is frozen and placed under a vacuum, allowing the ice to change directly from solid to vapor without passing through a liquid phase. The process consists of three separate, unique, and interdependent processes: freezing, primary drying (sublimation), and secondary drying (desorption). The lyophilization process generally includes dissolving the drug and excipients in a suitable solvent, usually *water for injection* (WFI), sterilizing the bulk solution by passing it through a 0.22-micron bacteria-retentive filter, filling into individual sterile containers and partially stoppering the containers under aseptic conditions, and transporting the partially stoppered containers to the lyophilizer and loading them into the chamber under aseptic conditions. The solution is frozen by placing the partially stoppered containers on cooled shelves in a freeze-drying chamber or prefreezing in another chamber. A vacuum is applied to the chamber, and the shelves are heated to evaporate the water from the frozen state. The final step is complete stoppering of the vials, usually by hydraulic or screw-rod stoppering mechanisms installed in the lyophilizers.

It is desirable after freezing and during primary drying to hold the drying temperature (in the product) at least 4 °C to 5 °C below the eutectic point of the product. The lyophilizer should have the necessary instrumentation to control and record the key process parameters, including shelf, product, and condenser temperature, and chamber and condenser pressure. Other process controls that should be specified in manufacturing procedures are time, temperature, and pressure limits. The monitoring of product temperature is particularly important for those cycles for which there are atypical operating procedures, such as power failures or equipment breakdown.

Chapter 44

Dispensing and Weighing Controls

Dispensing and weighing controls should be such that the identity, purity, and quality of the raw materials are maintained throughout the process. In addition, the controls should be such that there is complete traceability of the materials. The requirements for weighing controls are covered in 21 CFR 211.101.

STAGING AREAS

> Review product dispensing and after-dispensing staging areas to determine if they meet requirements. (Apply)
>
> **Body of Knowledge VI.D.1**

A clearly marked staging area for materials that need to be weighed is necessary. This area should be large and free-flowing enough to maintain adequate separation between materials and for operators to easily take one item at a time to the weigh room. Components for each batch of product need to be segregated from each other. Material being returned to inventory after removal of a required component should be staged in a separate area. These areas should be clean and with a good flow to prevent mixing of chemicals. The material names, item code, and lot number should be clearly visible at all times in order for operators to easily pick the items they need.

DISPENSING MATERIALS

> Identify the requirements for using weighing equipment and handling utensils for dispensing raw materials or intermediates including proper cleaning, labeling, and environmental controls, based on the type of material and manufacturing process being used. (Analyze)
>
> **Body of Knowledge VI.D.2**

WEIGHING EQUIPMENT

Clearly written procedures for the use and maintenance of weighing equipment should be available for training the operators. The procedures dictate the scales to be used for each range of weights. Weighing and measuring devices should be of suitable accuracy for the intended use, and they are to be calibrated to ensure accurate results within appropriate ranges. Weighing scale calibration is performed every three to six months as per the procedures. In addition, a daily weight check in the range in which the scale is used is performed with a National Institute of Standards and Technology (NIST)–calibrated weight set, and the information is logged in. The weight set should be calibrated annually. These procedures ensure that the scales are accurate when they are used for weighing of materials for commercial batches of pharmaceutical products.

DISPENSING UTENSILS

Utensils used for dispensing materials should be nonreactive to the material. If the utensils are reusable, they should be clean, sanitized, and dry before each use. The cleaning process needs to be validated to ensure that no cross-contamination occurs. Clean utensils should be stored properly in clean, labeled areas/bins and separated in the weigh room from used utensils that need cleaning.

If weighed material needs to be transferred to a secondary container, the secondary container should be clean, sanitized, dry, and nonreactive. The secondary container must be labeled immediately with the component name or item code, control or lot number of component, weight or measure transferred to the container, and batch number of product for which this component was weighed.

WEIGH ROOM

The weigh room should be separated from the rest of the production area to allow for easy material handling and dispensing without the material flying and settling on or otherwise contaminating other processes in the production area. Proper environmental controls (ISO 14644) need to be maintained as per the requirements of the raw materials that are to be weighed and the stringency of the product being manufactured. Even in the least stringent pharmaceutical product manufacture, at a minimum, the area should be ISO 14644 class 8. Only one chemical is taken into and handled in the weigh room at a time. Materials needed for each batch of product are weighed, reconciled, and removed from the weigh room before starting on a second batch. Material weighed for each product batch is separated from other batches. The pallets used to store the weighed material are labeled with the batch number, product name, pallet number, and total number of pallets. The pallets are staged in a labeled designated area for further batching.

WEIGHING PROCESS

Weighing, measuring, or subdividing operations for components must be adequately supervised. Each container of component dispensed must be examined by a second person to ensure that the component was released by the *quality control*

(QC) group, the weight or measure is correct as per the production batch record, and the containers are properly identified as per 21 CFR 211.101.

Electronic processes can be used to increase controls over the weighing process, but these processes need to be validated. The most common electronic process is attaching a printer to the weigh scales to print the tare, gross, and net weights dispensed. The printers can be programmed with readers that can read bar codes of the items being weighed. These printers can be made to communicate with the material resource planning software to reconcile the material weighed from inventory. These processes do decrease the sources of errors if the programs are all validated completely to ensure that they are free of ambiguity.

AUDIT

Periodic audit of production batch records should be conducted to ensure that the dispensing process retains all traceability of materials used in any batch of product. Ensure that when multiple lots of one component are used in a production batch, there is information about the quantities of each lot used in the batch record.

Chapter 45

Requirements for Critical Unit Processes

PROCESS PARAMETERS

> Use required critical process parameters (CCPs) for unit processes such as sterilization or sterilizing filtration, aseptic filling, depyrogenation, lyophilization, other drying processes, tablet granulation and compression, terminal sterilization, and cream or ointment emulsification. (Apply)
>
> **Body of Knowledge VI.E.1**

Critical process parameters for pharmaceutical production are determined by characterizing the range within which operation must occur to ensure the quality of the final product and begin by defining the critical quality attributes (CQAs) of the drug product. CQAs are physical, chemical, biological, or microbiological properties or characteristics that should be within an appropriate limit, range, or distribution to ensure the desired product quality is met. CQA definition starts with defining a *quality target product profile* (QTPP), which specifies the attributes and characteristics (for example, dosage form, stability requirements) that will ensure quality, safety, and efficacy of the finished dosage form for a specific product. Then, each attribute or characteristic is evaluated for severity of harm resulting from failure to meet that quality attribute before applying any risk controls or mitigations. Once the CQAs are defined, process parameters are evaluated for the impact their variability has on a CQA and produce an output separating operational parameters from CPPs, which should be monitored or controlled to ensure the process produces the desired quality finished dosage form.

HAZARD ANALYSIS AND CRITICAL CONTROL POINTS (HACCP)

Critical process parameters (CPPs) can be identified through the use of *hazard analysis and critical control points* (HACCP) analysis. HACCP is a process control system designed to identify and to prevent microbial, chemical, and physical hazards in production. It includes steps designed to prevent problems before they occur and to correct deviations as soon as they are detected. Such preventive control systems with documentation and verification are widely recognized by scientific authorities and international organizations as the most effective approach available for producing products safely. HACCP consists of seven principles: hazard analysis, critical control point identification, establishment of critical limits, monitoring procedures, corrective actions, record keeping, and verification procedures.

Hazard Analysis

Plants determine the food safety hazards and identify the preventive measures the plant can apply to control these hazards.

Critical Control Point Identification

A *critical control point* (CCP) is a point, step, or procedure in a food process at which control can be applied and, as a result, a food safety hazard can be prevented, eliminated, or reduced to an acceptable level. A food safety hazard is any biologic, chemical, or physical property that may cause a food to be unsafe for human consumption.

Establishment of Critical Limits

It is important to establish critical limits for each CPP. A *critical limit* is the maximum or minimum value to which a physical, biological, or chemical hazard must be controlled at a critical control point to prevent, eliminate, or reduce it to an acceptable level. *Design of experiments* (DOE) is a statistical tool that can be used to determine the process ranges and limits. Critical limits and CPPs feed into the development of a control strategy defining the monitoring and responses for each CPP which forms the basis of process validation activities.

Corrective Actions

Corrective actions are to be taken when monitoring indicates a deviation from an established critical limit. A plant's HACCP plan needs to identify the corrective actions to be taken if a critical limit is not met. Corrective actions are intended to ensure that no product injurious to health, or otherwise adulterated as a result of the deviation, enters commerce.

Record Keeping

HACCP regulation requires that all plants maintain certain documents, including its hazard analysis and written HACCP plan, and records documenting the monitoring of critical control points, critical limits, verification activities, and the handling of processing deviations.

Verification Procedures

Procedures for verifying that the HACCP system is working as intended are required. *Validation* ensures that the plans do what they were designed to do;

that is, they are successful in ensuring the production of safe product. *Verification* ensures that the HACCP plan is adequate, that is, working as intended. Verification procedures may include such activities as review of HACCP plans, CCP records, and critical limits, and microbial sampling and analysis.

VALIDATION STUDIES

> Explain and evaluate validation studies, specifically the methodologies and acceptance criteria required before implementing critical unit processes. Explain and evaluate validation studies such as requirements for aseptic processes including process simulations ("media fills") and temperature controls. (Evaluate)
>
> **Body of Knowledge VI.E.2**

PROCESS VALIDATON

In 2011, the US FDA published a completely revised guidance for process validation studies defining a new three stage framework for demonstrating manufacturing processes for drug products are capable of consistently producing drugs products meeting their specifications which is rapidly becoming the standard of process validation worldwide for pharmaceutical products. It is built on fundamental tenants that quality, safety, and efficacy must be designed and built into the product therefore testing in-process and final product samples alone is insufficient to assure product quality and each stage of manufacturing must be appropriately controlled.

As described in this guidance, this framework is comprised of three stages:

- Stage 1: Process Design. The commercial manufacturing process is based on knowledge gained through development and scale-up activities. Identification and control of CPPs provide assurance that CQAs will be met.
- Stage 2: Process Qualification. The process design is evaluated to determine if the process is capable of reproducible commercial manufacturing both within a lot and lot to lot variations. This activity used to be referred to as *process validation.*
- Stage 3: Continued Process Verification. Ongoing assurance is gained during routine production that the process remains in a state of control and provides an ongoing means to evaluate emerging trends for their impact in real time as materials, equipment, etc. potentially drift or change over time.

PROCESS PERFORMANCE QUALIFICATION PREREQUISITE PROGRAMS

The process qualification program should be built upon a solid foundation of prerequisite programs. This process traditionally has been accomplished through the application of *good manufacturing practices* (GMP). These conditions and practices are now considered to be prerequisite to the development and implementation of effective HACCP plans. Prerequisite programs provide the basic environmental and operating conditions that are necessary for the production of drug products. Common prerequisite programs may include, but are not limited to:

- *Facilities.* The establishment should be located, constructed, and maintained according to sanitary design principles. Linear product flow and traffic control minimize cross-contamination.
- *Supplier control.* Each facility should ensure that its suppliers have in place effective GMP and safety programs. These may be the subject of continuing supplier quality agreements and supplier system verification.
- *Specifications.* Written specifications for all components, products, and packaging materials are needed.
- *Production equipment.* All equipment should be constructed and installed according to sanitary design principles. Preventive maintenance and calibration schedules should be established and documented.
- *Cleaning and sanitation.* All procedures for cleaning and sanitation of the equipment and the facility should be written and followed. A master sanitation schedule should be in place.
- *Personal hygiene.* All employees and other persons who enter the manufacturing plant should follow the requirements for personal hygiene.
- *Training.* All employees should receive documented training in personal hygiene, GMP, cleaning and sanitation procedures, personal safety, and their role in the HACCP program.
- *Chemical control.* Documented procedures must be in place to ensure the segregation and proper use of chemicals in the plant. These include cleaning chemicals, fumigants, and pesticides or baits used in or around the plant.
- *Receiving, storage, and shipping.* All raw materials and products should be stored under sanitary conditions and the proper environmental conditions, such as temperature and humidity, to assure their safety.
- *Traceability and recall.* All raw materials and products should be lot-coded and a recall system should be in place so that rapid and complete traces and recalls can be done when a product retrieval is necessary.
- *Pest control.* Effective pest control programs should be in place.

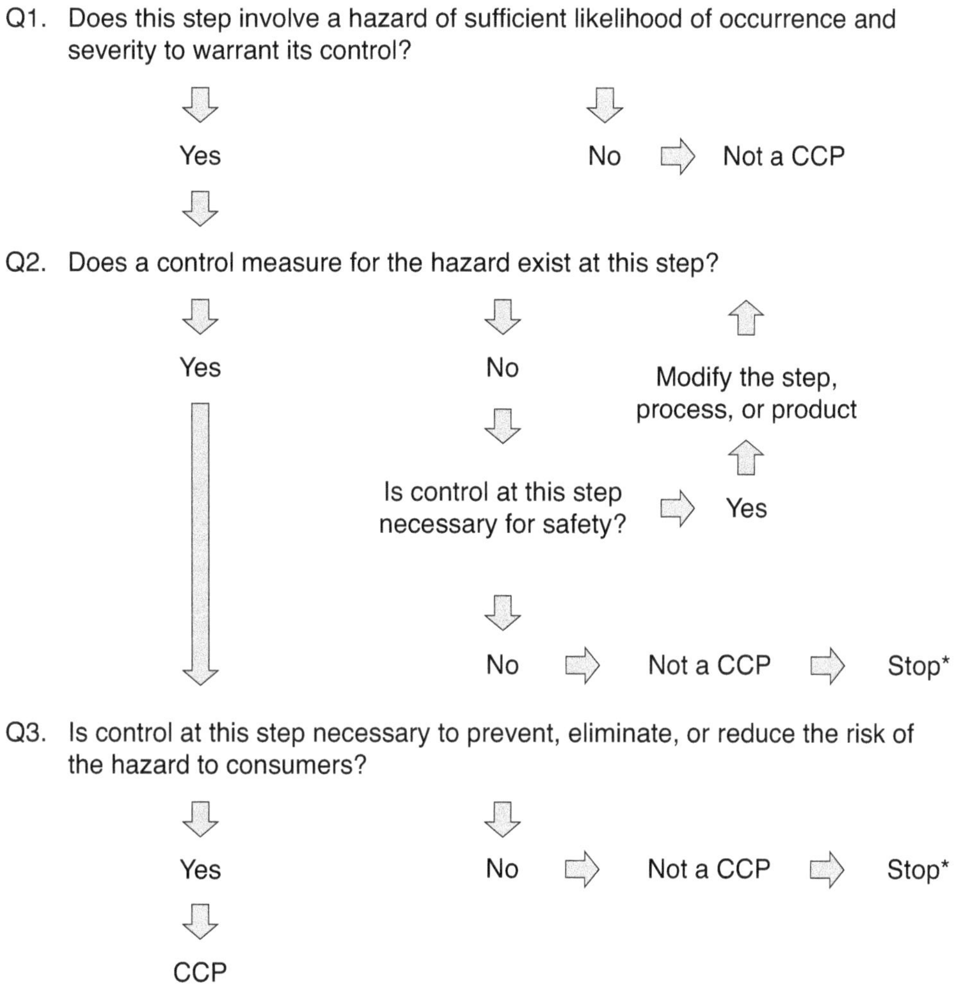

Figure 45.1 Critical control point decision tree example number 1.

Other examples of prerequisite programs might include *quality assurance* (QA) procedures; *standard operating procedures* (SOPs) for sanitation, processes, and product formulations; glass control; procedures for receiving, storage, and shipping; labeling; and ingredient handling practices. Figures 45.1 and 45.2 can be used to determine whether a process parameter is a CCP.

ASEPTIC PROCESSES

For aseptic manufacturing, there are additional validation requirements that are designed to confirm the ability of the aseptic procedures to prevent and control microbial contamination during drug product filling and lyophilization activities. This exercise is known as aseptic process simulations (a.k.a., media fills) and involves using a sterile microbiological growth medium as a surrogate to formulated product and running standing filling and lyophilization processes. At the

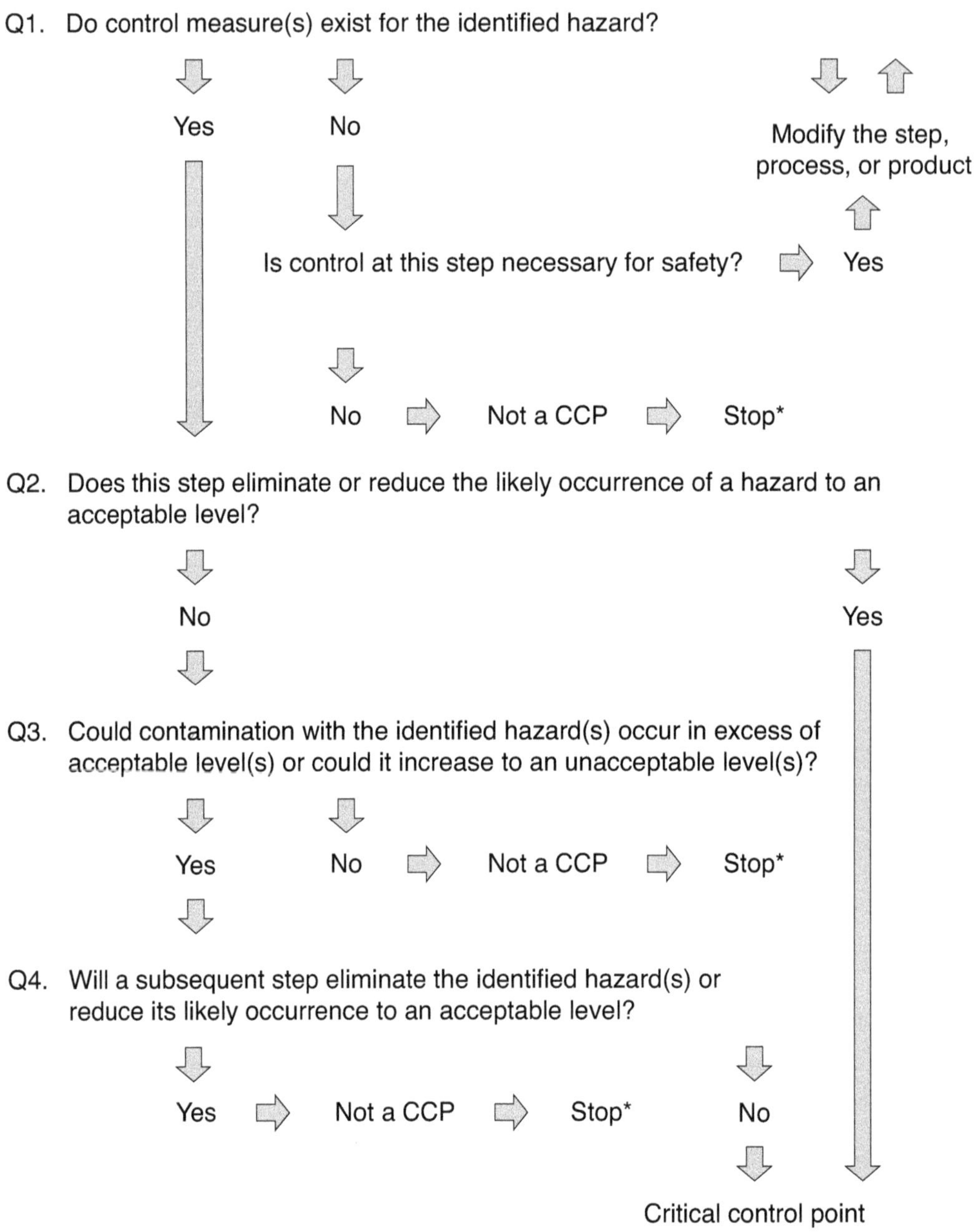

Figure 45.2 Critical control point decision tree example number 2.

end of the manufacturing simulation, the product containers, filled with media, are incubated, and inspected for microbial growth after seven days. A media fill program must be established for any manufacturing facility producing drug products via aseptic process and typical includes topics such as frequency and number of runs; duration and size of runs, line speed, environmental conditions, media selection; incubation and examination of media filled units and interpretation of results. Like process validation, a process simulation program has prerequisite requirements that must be in place prior to executing any media fill. As a rule of thumb, it is recommended that media fills be repeated twice per shift per line per year. All media fill failures must be investigated and root cause determined prior to resuming normal operations.

UNIT OPERATIONS

> Assess unit processes and their validations for deviations requiring investigation. (Analyze)
>
> **Body of Knowledge VI.E.3**

Every lot requires review of supporting documentation and test results prior to releasing a lot for commercial distribution. During this review, all process parameters are reviewed for acceptability and may require investigation to understand the root cause of departure from normal operating ranges and limits. If the parameter is identified as a critical process parameter, investigation is required and even if root cause is identified, the batch may still not be released to the market due to the impact the CPP has on the applicable CQAs.

OPERATING PROCEDURES

> Review qualification and validation results and confirm that validated and qualified parameters are reflected in operating procedures or batch records. (Analyze)
>
> **Body of Knowledge VI.E.4**

cGMP regulations require that lot-specific documentation is produced for every lot of drug product manufactured. This documentation is typically in the form of batch records, which define the validated parameters for the process and provide necessary checks and balances that requirements and limits are met as defined in the process control strategy. Ongoing process monitoring may trigger updates to the control strategy based on actual manufacturing history. Any time a control strategy is updated, the manufacturing records should be evaluated and updated if necessary.

RE-EVALUATION AND REVALIDATION

> Determine appropriate criteria and frequency for re-evaluation and revalidation of unit processes. (Evaluate)
>
> **Body of Knowledge VI.E.5**

Critical unit operations require ongoing monitoring to ensure the manufacturing process remains within the defined specifications and may require periodic re-validation. With the advent of modern approaches to process validation and the move away from a traditional 3-batch model, the need for ongoing monitoring and control of critical operations is crucial to the understanding of the state of control and compliance with validated parameters on a routine basis. This shift in thinking drives the emphasis from single, point-in-time activities to ongoing and constant evaluation, investigation, and adjustment aligning with requirements for establishing design space stemming from *quality by design* (QbD) principles.

One notable exception to this current philosophy relates to those processes specific to ensuring the sterility of products. For terminally sterilized products, periodic validation is required to ensure that the irradiation level adequately sterilizes the load and may be required at specific points in time (that is, when packaging or load for the sterilizer changes). For aseptically produced products, process simulations (a.k.a. media fills) are required to be executed on a semi-annual basis for all filling operations (both manual and automated).

ENVIRONMENTAL MONITORING REQUIREMENTS

> Differentiate between environmental monitoring requirements for different manufacturing area classifications. (Apply)
>
> **Body of Knowledge VI.E.6**

In pharmaceutical manufacturing, cleanroom classifications define the environment in which sterile drugs and biological products should be manufactured, with specific cleanliness levels and particle limits for each grade. Table 45.1 summarizes typical classifications used within pharmaceutical manufacturing.

Environmental monitoring involves regular assessment of the environment to ensure it meets the specified cleanliness requirements which vary depending on the classification of the manufacturing area. Establishing an environmental monitoring program is simple in a facility producing drugs for a single market (that is, US or EU) and gets complicated when that facility now serves multiple markets (that is, US and EU). Two major differences between the US and EU requirements are summarized below:

- The FDA requires only monitoring of ≥ 0.5 µm particles, while the EU requires both ≥ 0.5 µm and ≥ 5.0 µm.
- The FDA specifies requirements for cleanrooms in operation only, while the EU specifies both at rest and in operation conditions for monitoring.

Today, specific sample locations and frequencies are often determined by a risk assessment that is periodically reviewed and updated as ongoing monitoring data is collected.

Table 45.1 Classifications used within pharmaceutical manufacturing.

Classification (EU)	Grade A	Grade B	Grade C	Grade D
Classification (US)	ISO 5	ISO 7	ISO 8	Controlled Not Classified
Typical Unit Operations for Sterile Products	High-risk operations (filling zone, stopper bowls, open ampoules, and vials, making aseptic connections)	Background area of Grade A area	Used for filling products to be terminally sterilized	Less critical operations such as component washing, cleaning, and handling of equipment and accessories, assembly of cleaned components, before sterilization
	Preparation of solutions to be filtered, including weighing operations			

ENVIRONMENTAL MONITORING TOOLS

> Describe and use various monitoring tools to measure viable and nonviable particulates, pressure differentials, temperature, and humidity. (Apply)
>
> **Body of Knowledge VI.E.7**

Pharmaceutical environmental monitoring programs require a number of tools that vary in purpose and usage. Starting with general room conditions, temperature and humidity monitoring can be done with a variety of temperature or relative humidity probes, thermometers, or chart recorders. The device selected depend on the room classification and ongoing monitoring requirements. For manufacturing areas where cleanrooms are used, differential pressure gauges are used to monitor the pressure differences between two areas and can be connected to building automation systems or manually checked before, during, and at the end of manufacturing operations.

Where cleanrooms are used in pharmaceutical manufacturing, additional tools are required to monitor and ensure the cleanliness standards are maintained, including microbial and total particulate counts. Measurement of total particulate counts (formerly referred to as non-viable particulate counts) is done by either an off-line or in-line particulate counter. Most monitoring programs use off-line particulate counters for monitoring general room sampling locations and in-line particulate counters for monitoring critical work areas, especially when using isolated filling lines.

In contrast microbial sampling relies on a combination of tools including active air samplers, surface/touch plates, and settling plates. All of these methods rely on media designed to cultivate one or more microbial types over a period of three to seven days. Active air samplers are designed to intake a specified volume of room and pass it over either a plate or strip containing the selected media. Surface (or touch) plates are designed to come in direct contact with a surface and rely on transfer of any resident microbes to the surface of the media. These same plates are also used in personnel monitoring programs. Settle plates are used to monitor critical work areas and are passive in their monitoring nature. Settle plates are typically placed as close as possible to the work area and allowed to be exposed for a specific duration over the course of aseptic filling operations. For lengthy filling operations, more than one set of settle plates may be required.

Chapter 46

Contamination and Cross-Contamination

INTRODUCTION

It is important to take away the sense of urgency the United States Food and Drug Administration (FDA) exudes when addressing issues of pharmaceutical contamination and cross-contamination. In fact, the Agency refers to the word "contamination" 23 times in 21 CFR 211 (Figure 46.1). That being said, it becomes an immensely important and often daunting task for the manufacturers of pharmaceuticals to implement and employ the appropriate controls to ensure that product contamination and the potential for product cross-contamination never occur during routine processing.

- 21 CFR, Part 211
  - Subpart B—Organization and Personnel; Sec. 211.28 Personnel responsibilities
  - Subpart C—Buildings and Facilities; Sec. 211.42 Design and construction features
- Sec. 211.46 Ventilation, air filtration, air heating and cooling
- Sec. 211.48 Plumbing
- Sec. 211.56 Sanitation
  - Subpart D—Equipment; Sec. 211.67 Equipment cleaning and maintenance
  - Subpart E—Control of Components and Drug Product Containers and Closures
- Sec. 211.80 General requirements
- Sec. 211.82 Receipt and storage of untested components, drug product containers, and closures
- Sec. 211.84 Testing and approval or rejection of components, drug product containers, and closures
- Sec. 211.94 Drug product containers and closures
  - Subpart F—Production and Process Controls; Sec. 211.113 Control of microbiological contamination
  - Subpart G—Packaging and Labeling Control; Sec. 211.130 Packaging and labeling operations

Figure 46.1 "Contamination" cited in 21 CFR 211.

Warning Letter Citations

Failure of pharmaceutical companies to adequately protect production lines from contamination and cross-contamination is a frequently cited form FDA 483 observation noted during establishment inspections. The focus on product contamination has increased significantly since the New England Compounding Center Recall, driven by product contamination resulting in multiple deaths from fungal meningitis.

SOURCES

> Identify potential sources for, and implement controls to minimize, contamination and cross-contamination. (Analyze)
>
> **Body of Knowledge VI.F.1**

Unfortunately, no single cause or event results in the contamination or cross-contamination of pharmaceuticals; otherwise, industry professionals would not need to traverse FDA's enforcement page and read about the ongoing issues impacting the pharmaceutical industry. The causes are as diverse as the industry itself, with many high-profile incidents linked directly to poor *good manufacturing practices* (GMP) for finished pharmaceuticals. The prevention of contamination and cross-contamination rests squarely on the shoulders of manufacturers and the practices they employ, or fail to employ.

One of the primary causes of contamination and cross-contamination is linked to a lack of properly documented and validated cleaning processes used by pharmaceutical manufacturers. Additionally, the improper labeling of drug ingredients and the lack of effective line clearance processes rate high on the list of causes for contamination and cross-contamination of products.

RISK MITIGATION

> Describe and apply various techniques for mitigating the risk of contamination and cross-contamination including cleaning; facility, utility, and equipment design; material and personnel flow; qualified disinfectants; operator training; validation; and monitoring. (Apply)
>
> **Body of Knowledge VI.F.2**

Equipment Cleaning

First and foremost, pharmaceutical manufacturers must establish written procedures that delineate the cleaning requirements for every piece of manufacturing equipment. The procedures must contain sufficient granularity to ensure that each piece of equipment can be adequately cleaned. In some cases it may be necessary to remove equipment from the manufacturing environment to do a proper cleaning. Regardless, the cleaning process must be properly validated as validation is an expectation of 21 CFR 211 and a regulatory requirement for most regulatory bodies outside the United States.

When scripting the procedures for cleaning, multiple parameters require consideration for inclusion into each written procedure. For example, five components of the cleaning process that require consideration are:

- Equipment-related parameters
- Type of residual material (residues) that require cleaning and removal
- Identification of the types of cleaning agents to be used
- Specific type(s) of cleaning processes to be used
- Process variables associated with the actual cleaning of the equipment

Equipment-Related Parameters

From an equipment standpoint, each piece of equipment should have its own equipment file that captures all preventive maintenance activities, including cleaning and, if necessary, calibration. Included in the equipment file should be the *installation qualification* (IQ) for each piece of equipment. Equipment-related parameters that need to be considered include identification of equipment sections that may be a problem for cleaning, identification of the physical/mechanical properties of the equipment (for example, 17-4 stainless steel), and disassembly instructions as appropriate (that is, identification as to whether the equipment can be cleaned in place or requires removal from the production environment).

Residue Removal

Removal of unwanted residue is the underlying reason for the proper cleaning of equipment. Four elements that need to be considered in support of the removal of residue are proper identification of all residues requiring removal as part of the cleaning process, residue reactivity and toxicity, establishment of cleaning limits, and residue solubility.

Cleaning Agents

The selection of cleaning agents and the establishment of parameters for safe and effective use are immensely important. Some of the requirements to be considered when selecting cleaning agents include the chemical composition of the cleaning agent, instructions for properly mixing and using the cleaning agent, establishment of proper handling of cleaning agents, identification of special environmental considerations, and health and safety considerations. A *material safety data sheet* (MSDS) should be available for all cleaning agents selected and employed.

Cleaning Processes

The cleaning process and all parameters relating to the cleaning process must be clearly defined within a written procedure. Elements requiring consideration for inclusion in the procedure include:

- Defining whether the cleaning process is manual or portions of it can be automated
- Identifying whether the equipment can be adequately cleaned in place
- Availability of a user manual to assist with the cleaning process
- Training requirements for personnel performing the cleaning
- Time required to properly clean a piece of equipment
- Frequency of the cleaning process (daily, weekly, and so on)
- Testing or other requirements necessary before placing cleaned equipment back into service

Process Variables

All cleaning processes, regardless of the equipment being cleaned, have variables associated with the process. Process variables requiring consideration for equipment cleaning are the temperature required for the cleaning agent, the duration of each cleaning cycle, including the number of cleaning cycles required, the time frame allocated for each cleaning, operator variability, and the outsourcing of cleaning and maintenance to a third party versus performing the work in-house with the pharmaceutical establishment's personnel.

Proper Identification and Labeling

Upon receipt, all drug ingredients must be properly identified and labeled. It is imperative that product traceability for drug ingredients be maintained to the individual lot/batch level. Additionally, it is incumbent upon pharmaceutical companies to properly store drug ingredients until such time as the ingredients are used. Special consideration should be made for special environmental requirements needed for product storage, as appropriate, for example, storage temperature.

Line Clearance

Pharmaceutical establishments are expected to have appropriate line clearance procedures in place to prevent potential contamination and cross-contamination. The line clearance procedure should clearly define lot/batch controls, documentation practices, product separation, product quarantine (if required), and the overall handling of materials during the manufacturing process.

SUMMARY

FDA is extremely vigilant when it comes to ensuring that pharmaceutical establishments have adequate infrastructure in place, including written procedures, to prevent contamination and cross-contamination. Some of the most effective tools manufacturers can use are validated cleaning procedures that define, in great detail, the cleaning requirements for all pieces of equipment. Additionally, the proper identification and labeling of all drug ingredients and an effective line clearance process are needed as the first line of defense in the prevention of contamination and cross-contamination of pharmaceuticals.

Chapter 47

Reprocessed and Reworked Materials

DISPOSITION PROCESS

> Distinguish reprocessing from reworking and apply appropriate documentation, approval, and disposition methods for these materials. (Apply)
>
> **Body of Knowledge VI.G.1**

REPROCESSING

The reprocessing of any lot or batch of drug is given approval by *quality control* (QC). Approval of a reprocessed lot or batch of a drug by QC is based on documented scientific data, which may include validation. The reprocessing of products that fail to meet their specifications, or rejected product, is generally a rare occurrence. Materials to be reprocessed should be appropriately controlled/quarantined to prevent unauthorized use.

Good manufacturing practices (GMP) state that reprocessing is permitted only when the following conditions are met:

- Reprocessing shall not be performed without the review and approval of the *quality unit* (QU).
- The integrity of the finished product is not affected.
- The reprocessed lot meets specifications.
- The reprocessing is done in accordance with a defined procedure approved by QC.
- All risks have been evaluated, including the consideration of potential formation of by-products and/or overreacted materials.
- Complete records of the reprocessing are kept.
- A new batch number is assigned.
- Validation demonstrates that the quality of the finished product is not affected.

Reprocessing activities including introducing an intermediate or *active pharmaceutical ingredient* (API)—including one that does not conform to standards or specifications—back into the process and reprocessing by repeating a crystallization step or other appropriate chemical or physical manipulation steps (for example, distillation, filtration, chromatography, milling) that are part of the established manufacturing process are generally considered acceptable. Introducing unreacted material back into a process and repeating a chemical reaction is considered to be reprocessing unless it is part of the established process.

If reprocessing of a specific step is used for most batches, this information should be included as part of the standard manufacturing process. Continuation of a process step after an in-process control test has shown that the step is incomplete is considered to be part of the normal process. This action is not considered to be reprocessing.

REWORKING

Reworking is the process of performing additional manufacturing process steps on in-process or final material that does not meet standards or specifications, which is typically performed when there is an unexpected occurrence and this new additional processing step is not preapproved as part of the marketing authorization. The reworked lot may be subject to additional regulatory approvals before release for distribution. Materials to be reworked should be appropriately controlled or quarantined to prevent unauthorized use. The GMP for reworking are:

- Written procedures shall be established and followed, prescribing a system for reprocessing batches that do not conform to standards or specifications and the steps to be taken to ensure that the reprocessed batches will conform with all established standards, specifications, and characteristics.
- The reworking of any lot or batch of drug is given approval by the quality control department.

Approval of a reworked lot or batch of a drug by the quality control department is based on documented scientific data, which may include validation. The reworking of products that fail to meet their specifications is undertaken only in exceptional cases. Reworking is permitted only when:

- An investigation into the reason for nonconformance of batches to established standards or specifications is performed prior to reworking batches.
- All risks have been evaluated to determine their suitability for use, not limited to prolonged storage or exposure to heat- or humidity-sensitive material, and a new batch number is assigned.
- The lot is reworked in accordance with a defined procedure approved by QC, and if warranted, documentation to show that the reworked product is of equivalent quality to that produced by the original process.

- The quality of the finished product is not affected.
- The reworked lot meets specifications. Batches that have been reworked should be subjected to appropriate evaluation, testing, stability testing, if warranted, and documentation to show that the reworked product is of equivalent quality to that produced by the original process.
- Complete records of the reworking are maintained and included in the ongoing stability program.
- The reworked lot is included in the ongoing stability program.

Before acceptance and use, reprocessed material must meet appropriate standards, specifications, and any other relevant criteria. Concurrent validation is often the appropriate validation approach for rework procedures, and this allows a protocol to define the rework procedure, how it will be carried out, and the expected results. If there is only one batch to be reworked, then a report can be written, and the batch released once it is found to be acceptable.

STORAGE

> Describe and apply requirements for segregation and secure storage of these materials. (Apply)
>
> **Body of Knowledge VI.G.2**

Materials are required to be quarantined until they have been sampled, examined, or tested, as appropriate, and released for use. Written procedures are required to be established and followed, prescribing a system for storage of quarantine batches that do not conform to standards or specifications and the necessary steps taken to prevent unauthorized use in manufacturing, to minimize their risk not limited to material segregation, labeling, special storage for rejects. The segregated storage area is assigned for temporary storage until their future use is determined by the Quality Control Department.

Requirements for storage for reworked and reprocessed lots or batches under quarantine include:

- Records of movement are traceable.
- Lots and batches are properly identified and stored in a manner to prevent degradation, contamination, and cross-contamination.
- Correct labeling is verified (not limited to the correlation between the name used by the supplier and the in-house name), container damage, broken seals, and evidence of tampering or contamination.
- Materials are maintained to ensure their integrity, stored appropriately, and suitably spaced to permit cleaning and inspection.

- Materials are stored under appropriate environmental conditions (including temperature, humidity, light exposure, ventilation, and drainage as appropriate) that have no adverse effect on their quality and controlled according to first-in/first-out (FIFO) practices to maintain the material's characteristics.
- Rejected materials are properly identified and controlled under a quarantine system (physical and/or administratively) designed to prevent unauthorized use in manufacturing to minimize their risk (not limited to material segregation, labeling, special storage for rejects, control of material returns).
- Destruction facilities and processes are determined for materials, including suitable facilities and processes. Certificates of destruction should be maintained.

Part VII

Filling, Packaging, and Labeling

Chapter 48

Filling Operations and Controls

Filling operations and controls should be such that the identity, strength, purity, and quality of the product are maintained throughout the process. The requirements for filling and packaging controls are covered in 21 CFR 211, Subpart G. Written procedures that govern the filling processes of each product package, and the execution of these procedures, must be documented on a packaging batch record with appropriate process controls, and any deviations must be documented and reviewed. These procedures help prevent contamination and ensure that filled product is of the strength, identity, purity, and quality it is purported to have.

MATERIALS CONTROL

> Develop and review procedures to ensure the identity, strength, and purity of specified materials (e.g., liquids, powders, ointments, tablets, capsules, and suspensions) and to prevent them from being altered. (Evaluate)
>
> **Body of Knowledge VII.A.1**

Filling operations begin after the bulk manufacturing process has been completed. At this point the blend/bulk has been tested to be uniform, passed all bulk specifications, and been released for filling. Filling processes should be such that the identity, strength, and purity of the bulk are not altered. Clear written filling procedures (filling assembly procedures) specific to each type of product and each packaging type are required in the packaging batch record. *Packaging Batch Records* (PBRs) should be maintained through a change control process to ensure the correct versions are being used in operations. Each PBR is to be verified against the master PBR before a filling operation. The PBR should contain the product name, strength of dosage, dosage form, dosage unit, net contents, and all details and specifications regarding its filling process. The PBR contains a *bill of materials* (BOM) listing the appropriate components, their part numbers, the process of assembly, and specifications of assembling to complete the filling process. The *quality unit* (QU) group will ensure that all components and bulk product needed for filling have been appropriately tested and released for filling operations.

FILLING ASSEMBLY PROCEDURES

PBRs are formally authorized packaging instructions required for each product, pack size, and type. These should normally include:

- The name of the product; a description of its pharmaceutical form, strength, and, where applicable, method of application;
- The pack size expressed in terms of the number, weight, or volume of the product in the final container; a complete list of all the packaging materials required for a standard batch size, including quantities, sizes, and types, with the code or reference number relating to the specifications for each packaging material.

Where appropriate, an example or reproduction of the relevant printed packaging materials and specimens, indicating where the batch number and expiry date of the product have been marked, should be included, as well as any special precautions to be observed. A careful examination of the packaging area and equipment to ascertain the line clearance before and after packaging operations must be documented in the PBR. It needs to include a full description of the packaging operation, including any significant subsidiary operations, the equipment to be used, and the details of in-process controls with instructions for sampling and acceptance limits.

FILLING EQUIPMENT CONTROL

> Analyze the controls needed for various types of production equipment and processes and ensure that the appropriate controls are established to verify filling criteria. (Analyze)
>
> **Body of Knowledge VII.A.2**

Filling equipment should be qualified through valid *installation qualification* (IQ), *operational qualification* (OQ), and *performance qualification* (PQ) protocols to ensure it is installed as per its design specifications and can operate and perform as required to yield desired results within preset specifications. The equipment should be maintained regularly and is to be included in the preventive maintenance program. Any replacement of parts should be like for like and should be managed within a change control process. Each piece of equipment should be assigned a unique identifier, and this number should be noted in filling records of operations that use the equipment.

Filling operations of a pilot or test batch of product should be used to determine the appropriate filling equipment needed in order for the process to yield filled product to preset specifications. This process should be incorporated in the PBR and is to be validated to ensure uniformity in every fill. Further, at predetermined intervals filled goods from every fill should be taken and tested for fill criteria to ensure that filling operations are under control throughout the run. These data are recorded in a filling record that is verified by a second person or QA as part of filled goods release.

CONTAMINATION CONTROLS

> Identify controls to prevent microbial and other contamination at all stages of filling. (Apply)
>
> **Body of Knowledge VII.A.3**

Contamination in filling of a product can occur from other filling processes, from the environment, or from personnel. Filling operations should be such that there is physical or spatial separation between different product filling operations to prevent contamination. Filling areas should be inspected immediately before use to ensure that all materials not required for the next fill have been removed, and this inspection should be documented.

Filling of dry powders requires special precautions to prevent generation and spreading of dust. Proper air control measures are needed for such filling operations, with a regulated supply and extraction of air of suitable quality. Contamination of one product by another product must be avoided. The significance of the contamination risk varies with the type of contaminant and type of product being contaminated. Facilities manufacturing highly sensitive materials such as antibiotics, hormones, biologic preparations, living organisms, cytotoxic substances, or highly active materials should use dedicated equipment in self-contained areas. Products in which contamination is likely to be most significant are products administered by injection, products applied to open wounds, and those given in large doses or for a prolonged time. Several measures can help avoid cross-contamination (Figure 48.1). Measures to prevent cross-contamination and their

- Self-contained areas and dedicated equipment
- Campaign production followed by appropriate cleaning with validated cleaning procedure
- Appropriate air locks, pressure differentials
- Pretreatment of air in the facility
- Appropriate clothing and gowning procedures
- Validated cleaning and decontamination procedures
- Testing for drug residue, allergens, proteins, or micro load

Figure 48.1 Several measures that can help avoid cross-contamination.

effectiveness should be checked periodically according to *standard operating procedures* (SOPs). Production areas where susceptible products are processed should undergo periodic environmental monitoring (for example, for microbiological monitoring and particulate matter where appropriate [ISO 14644]).

STAGED MATERIALS

> Review staged materials and confirm that they are approved for use. (Apply)
>
> **Body of Knowledge VII.A.4**

Staging materials properly is a key process for which to have controls that prevent contamination or substitutions in a filling process. It is essential to have good line-clearance procedures with appropriate checklists and documentation of activities for each fill. The staging area for filling should be clearly marked for each filling process or line. All components needed for a given fill should be brought to the corresponding staging area for the fill, and only components needed for that fill should be there. QA should check and ensure that all components are tested and released before staging for fill. Filled goods should have a location of their own and should not be commingled with components staging. Different products should not be packaged in close proximity unless there is clear physical segregation. The name and batch/lot number of each product being staged for fill, and the fill line, should be clearly displayed in the area. Filled goods should also be labeled immediately with the product name, batch number, and release status. All staged materials should be reconciled at the end of the fill process. Documentation of the quantities of labeling issued, used, destroyed and returned is required, as well as reconciliation against the quantity of drug product finished. Any discrepancies outside narrow preset limits based on historical operating data must be investigated.

STATUS LABELING

> Identify and apply proper status labeling throughout the process. (Apply)
>
> **Body of Knowledge VII.A.5**

The best practice is for products to be labeled immediately after fill, preferably in-line with the fill process to avoid any mistakes. Appropriate procedures for labeling in-line and off-line should be in place in case of any delays. All products at all stages should have a status label. The product should at a minimum have the batch number, product name, and release status. If a particular filled product is removed from the line for any reason, then the status label for that product should contain the reason, for example, QA sample, retain, or nonconforming sample. Any filled goods without a status label are considered contaminated product. Therefore, care must be taken to have status assigned to products at all stages. Once the filling process is completed, all filled goods are removed from the line and located in a designated area with the status of "hold."

For each batch/lot of drug product, appropriate laboratory tests (such as microbiological testing to ensure no contamination) should be used to determine conformity to the finished product specifications. The filled goods taken from the filling process periodically, as per filling procedures for *quality control* (QC) testing, are tested for the criteria specified. Once these tests have been satisfactorily completed, the lot can be released. The products of that lot can be moved to released inventory. Products failing to meet the established specifications must be quarantined, and a nonconformance initiated. The nonconformance should be investigated and products not meeting the specifications must be rejected and disposed of appropriately.

Chapter 49

Environmental Monitoring

> Use various monitoring techniques (e.g., active air sampling, settling plates, swab sampling, nonviable particle counting, and contact plates for surfaces and people) to determine that appropriate environmental conditions are maintained during production operations. (Analyze)
>
> **Body of Knowledge VII.B**

Environmental monitoring is considered by regulators (United States Food and Drug Administration [FDA]) as necessary to ensure that the material/products produced are in accordance with 21 CFR 211.113, "Control of Microbiological Contamination." This section of the Code of Federal Regulations (CFR) specifically states that the "appropriate written procedures . . . shall be established and followed."

A robust environmental monitoring program provides valuable information regarding the point-in-time conditions of the manufacturing environment as well as the ancillary clean areas. It allows for manufacturers to be able to potentially identify areas of contamination and be able to implement *corrective and preventive actions* (CAPAs) accordingly, such as additional cleaning regimens, different cleaning solutions, evaluation of personnel, and optimized material flow. It is important to note that a low level of contamination may not always be detected, and that a trending program may alert manufacturers to future issues.

The specific cleanroom environment that a material/product would be manufactured in would dictate the frequency of monitoring and the specific techniques associated with the environmental monitoring program.

NONVIABLE PARTICLE MONITORING

As part of any controlled environment, nonviable particle monitoring is important both for the identification and control of *nonviable* (nonliving particles, such as dust) and *viable* (living microorganisms) contamination. This monitoring is most critical in an ISO Class 5 (Class 100) environment, typically used for aseptic processing of injectable preparations. Meeting the specifications of the established cleanroom standards is a necessity to perform manufacturing according to the certification specific (and predefined) to the cleanroom.

Typically, the size of most concern is 0.5 µm. This size is representative of the approximate size of a single microorganism. It has been shown that microorganisms can attach themselves to particles and therefore become a biologic contaminant.

Testing may be performed under two different conditions, either dynamic or static. *Dynamic* conditions are those in which activities in the cleanrooms are occurring during normal operations. Testing under these conditions is more desirable when in production to ensure that the cleanrooms are meeting specifications during potentially worst-case operations. *Static* conditions are those in which no operations are occurring and minimal or no personnel are in the cleanroom (with the exception of the analyst), and they do not give a representative picture of cleanliness in the room with personnel flow, components, and active machine operations.

With today's technology, multiple ways exist to perform nonviable monitoring. One method uses a portable particle counting instrument, whereby an analyst performs a system verification of the instrument and, with the instrument's internal vacuum, draws a programmed volume of sample air through the system for analysis (typically through use of a laser). Particle counter sensors may be installed in critical locations in cleanrooms, which are connected to a centralized unit and computer system. They enable alarms to be sent to multiple e-mail addresses or cell phones. These units are typically continuously run and can provide real-time results as well as significant amounts of data for trending purposes.

CONTACT PLATES

Contact, surface, or RODAC (*replicate organism detection and counting*) testing is necessary in an environmental monitoring program. RODAC is a brand name commonly used in the pharmaceutical industry and is a registered trademark of Becton Dickinson, a manufacturer of these agar plates, although several other organizations sell such plates. Contact plates use a slightly convex surface of microbiological media to make direct contact with a flat surface. Predetermined work sites are tested by gently pressing the media surface onto the site and immediately recapping the plate. After testing, the site is wiped down with isopropyl alcohol and a lint-free wipe to ensure that no media residue is left behind. If media residue is left behind, it may become an inadvertent viable area for the growth of microorganisms. Critical areas should be tested at the end of processing to avoid contamination of a sterile processing area.

The contact plates are approximately 65 mm × 15 mm and provide a testing surface area of 25 cm^2. For routine monitoring, *trypticase soy agar* (TSA) with lecithin and polysorbate 80 is the media of choice. The trypticase soy agar is the nutrient media for the microorganisms (specifically bacteria) to grow in. Lecithin and polysorbate 80 are added to deactivate disinfectant residue. TSA contact plates are incubated at 30 °C to 35 °C for 48 to 72 hours.

Another formulation of media, *Sabouraud dextrose agar* (SDA), is used for monitoring of mold and yeast. The main difference between the TSA and SDA agar is that the lower pH of SDA provides for a more favorable environment for growth of fungi. SDA plates are incubated at 20 °C to 25 °C for five to seven days.

Contact plates may be used for personnel monitoring. Four fingers are closed together and the contact plate is lightly applied to the fingertips, and the thumb is lightly pressed on the surface as well. After testing, either the gloves are changed out (in the gowning area for an aseptic area) or disinfected with isopropyl alcohol and wiped with a lint-free wipe to prevent media transfer to work surfaces from gloves. In situations where media fills are occurring, sleeve monitoring may be performed as well. Contact plates should not be used on irregular or textured surfaces, only smooth surfaces. For irregular surfaces, a swab method may be used.

SWAB MONITORING

In areas that can't be accessed and tested with contact plates (for example, machine parts, tubing, or rough or uneven areas), swab monitoring may be performed. A lint-free swab is moistened with a buffer and placed in a sterile, capped tube. When ready for use, which would be at the end of a sterile processing operation, the swab is carefully removed from the tube, using impeccable aseptic technique, and wiped across the surface, rotating the swab to ensure the entire swab is used. The swab is re-submerged in the buffer solution and vortexed (or similar mixing) to ensure that any microbiological contamination is transferred from the swab to the solution. The solution is aseptically poured onto agar plates (TSA or SDA) and incubated. Another swab method may be used: a slightly moistened swab housed in a sterile, capped tube is wiped across the surface and, using the same rotation motion, swabbed directly on the agar plate and incubated.

Regardless of swab method used, a large potential margin for error exists due to the sheer number of steps that may lead to unintentional personnel contamination. The swab monitoring method must undergo method validation to ensure that proper recovery is achieved.

SETTLE PLATES

Settle plates are used as part of an environmental monitoring passive air-sampling program. This type of air testing is considered to be qualitative in that no predetermined quantity of air is being sampled, but rather the quality of the air (types of organisms present) is being tested by plates being placed in critical locations and the agar exposed for a specific amount of time, usually an hour.

Settle plates are composed of TSA or SDA. The plates are exposed and recapped, then incubated. One drawback of settle plates is that they desiccate with extensive exposure time and/or high airflows. In the pharmaceutical industry, this phenomenon displays a "potato chip" appearance when incubated and therefore would inhibit the potential growth of microorganisms. A maximum exposure time should be validated.

ACTIVE (VIABLE) AIR MONITORING

A more representative method of monitoring air quality is *active air monitoring*. Varieties of samplers include impaction, centrifugal, and membrane. These devices allow testing of the number of organisms per volume of air sampled. This method of air testing is considered to be quantitative, that is, the number of organisms can be calculated by the volume of air sampled. Active air samplers draw in a predetermined volume of surrounding air and aim the air stream at the culture medium (agar plate or agar strip) for incubation. These active air samplers are placed in critical areas of an operation, although they should not disrupt the unidirectional airflow.

USING THE DATA

The environmental monitoring data should be recorded over several months to establish the baseline quantity and composition of the microbiologic load in the facility. It can then be used to set alert and action levels during future monitoring. This will help ensure appropriate environmental conditions during pharmaceutical operations.

Chapter 50

In-Process and Finished Goods Inspections

The pharmaceutical industry has recognized the need for better and faster systems for inspection of in-process and finished goods. There are a wide range of inspection solutions, including visual inspection systems, vacuum decay technology, seal quality inspection, blister package inspection, and container leak testing. The implementation of one or more of these inspection methodologies allows drug manufacturers to eliminate catastrophic primary packaging failures and reduce the probability of product recalls while maintaining patient confidence.

The inspection method selected should consider the level of packaging (primary or secondary), process step location (in-process or final inspection), testing costs, and product costs. The method of the inspection is an important consideration when testing or product costs are high, as some tests are destructive. There are two primary types of testing:

- *Destructive testing* (also known as *materials testing*). Accomplished by forcing the part to fail by multiple applications. These tests can include but are not limited to hardness, tensile, compression, impact, fracture toughness, fatigue, and creep.
- *Nondestructive testing* (also known as *noninvasive*). Accomplished by inspecting, testing, or evaluating material without destroying the usefulness of the part or system. These tests can include but are not limited to visual, liquid penetrant, magnetic particle, electromagnetic, ultrasonic, and radiographic. Nondestructive testing can allow a much higher quantity of packaged product to be tested, up to 100% for some types of these tests.

Other testing can include acoustic emission, guided wave testing, ground penetrating radar, laser testing methods, leak testing, magnetic flux leakage, microwave testing, neutron radiographic testing, thermal/infrared testing, and vibration analysis.

In the case of sterile products, inspection methods such as vacuum decay testing, seal force, and dye penetration testing offer an alternative to the required sterility testing to ensure that the primary container maintains the integrity and quality of the drug product throughout its lifespan. According to the United States Food and Drug Administration (FDA) guidance "Container and Closure System Integrity

Testing in Lieu of Sterility Testing as a Component of the Stability Protocol for Sterile Products," some of the advantages of using these inspection methods are:

- A breach in container/closure system integrity can be detected before contamination enters the product.
- Results are timely. Sterility testing typically requires at least seven days of incubation to obtain results.
- The potential for false positive results can be reduced.

INSPECTIONS

> Develop criteria for in-process and finished goods inspections of filled and packaged materials including seal tests, torque testing, and bottle rejection systems. (Create)
>
> **Body of Knowledge VII.C.1**

Most of the companies that provide this type of inspection technology offer nondestructive and noninvasive inspection methods that have the advantage of producing less waste. Some of the current technologies in the marketplace include visual inspection, seal quality inspection, nondestructive leak testing, blister package inspection, and container closure testing. If destructive testing is needed, some additional testing may include burst testing, creep testing, dye penetration, and accelerated aging.

Visual Inspection

Visual inspection of seal characteristics can be utilized in many aspects of the process, equipment, and/or the environment. These characteristics can indicate the lack of seal integrity and/or the possible compromises to the package's durability after performance testing. Generally, qualitative (accept/reject) visual inspection methods for evaluating these characteristics are acceptable.

Seal Quality Inspection (Airborne Ultrasound)

This method is commonly used to inspect the seal quality of medical pouches and other types of flexible packages by transmitting an ultrasonic signal along the seal. The signal creates a high-resolution image of the seal structure and quality. Through the use of this method, container defects such as abrasion, blisters, contamination, incomplete and/or misaligned seal, delamination, wrinkles, flex cracks, crooked seal, and cut seal can be identified. Packaging lines can be automated to reject any units with an incomplete seal.

Nondestructive Leak Testing (Vacuum Decay)

This technology uses the principle of changes in vacuum to determine whether leaks are present in the container, thus giving an indication of the quality and integrity of the primary package container. This inspection methodology is used to inspect filling components such as presterilized stoppers and presterilized caps, and drug product packaged in vials, syringes, and ampules. One of the advantages of this method is that the test is nondestructive and noninvasive; therefore, no sample preparation is required.

Bubble Emission Testing

As is stated in the name, bubble emission testing consists of the visual detection of air (at times, a gas) leaking from a pressurized package or container. Smaller items can be pressurized and immersed in a container of liquid, and larger vessels can be pressurized and inspected by spraying a soap solution that creates fine bubbles on the area being tested (like checking for a leak in a tire). For flat surfaces, a soap solution can be applied to the surface, and a vacuum box can be used to create negative pressure from the inspection side. If there are leaks, bubbles will form, showing the location.

Blister Package Inspection and Container Leak Testing

Blister package inspection (vacuum decay with indirect imaging) is a nondestructive and noninvasive method used for inspection of drugs packaged in blister packages. Leak detection is accomplished by detecting the change in pressure inside the test chamber resulting from gas or vapor egress from a blister package challenged with vacuum. If leaks are present, the inspection diagnostics are capable of identifying the defective blister cavities. The *container leak testing* (pressure decay) method uses changes in pressure to detect the presence of leaks in the container.

Burst Testing

Burst testing allows air into the packaging at a required flow rate and pressure. As air escapes through the walls of the porous (or semi-porous) package, the flow rate must be increased to compensate for the air lost through the walls and create the back-pressure in the porous package. The pressure creates the amount of force needed to cause the seal to fail. A minimum pressure specification is set to qualify as an acceptable seal.

Creep Testing

Creep testing is conducted by performing a slow shear of the adhesive bond. Creep testing steadily inflates the package at a constant rate of pressure and for a certain time period, resulting in a pass/fail result.

Dye Penetration

Dye penetration testing can detect and locate a leak formed between a transparent film and a porous sheet material. The dye penetrant solution is applied to the seal edge to be tested. After contact is made with the dye penetrant for a specified time, the package will then be visually inspected for dye penetration.

Freeze-Dried Product Leak Testing

Freeze-dried (lyophilized) products were previously inspected for vacuum using a spark test, but new methods are being developed and implemented. Leak testing is performed by analyzing the headspace pressure inside the vial using laser absorption spectroscopy (optical measurement of any gas that is present). This method is nondestructive and gives an indication of the actual vacuum level in the vial. Typically, the equipment used for this type of testing has an automatic self-test system that challenges the machine by using calibrated check vials, thus ensuring the reliability of the equipment.

VISION AND DETECTION SYSTEMS

> Ensure that vision and detection systems are qualified, calibrated, and challenged as required for the system. (Apply)
>
> **Body of Knowledge VII.C.2**

Automated Vision Inspection Systems

Automated vision inspection systems help the drug manufacturer analyze the quality of the product, identify product or container defects, and ensure that the data printed on the primary and secondary containers are accurate (for example, drug facts, labels). Most vision inspection system companies agree that the key factors in visual inspection technology are lighting and optics. Lighting in relation to the geometry of the container being inspected should be considered when choosing a visual inspection system, and optics can help when trying to increase contrast between the region of interest (container or drug being inspected) and the background. For example, drug companies use visual inspection systems to inspect blister packages to make sure that they contain the correct color tablets.

Qualification and Calibration

Drug manufacturers must ensure that the inspection technology used is qualified and that any inspection processes used in their operations are validated. The inspection technology selected must be appropriate for the product that will be inspected, and the qualification needs to prove that the equipment is reliable when identifying the characterized product/container defects. Complex detection equipment, especially a vision system, needs to have an *installation qualification* (IQ), *operational qualification* (OQ), and *performance qualification* (PQ) to ensure it is functioning as required. Process validation is then required to ensure that the qualified equipment works consistently on your specific product. This ensures the system has the required detection sensitivity for the expected types of defects with minimal false positive rejections.

This type of equipment should be routinely calibrated, inspected, and checked according to a written preventive maintenance program designed to ensure proper performance. Written procedures for cleaning, operating, and maintaining the automated equipment should be in place. All such programs and procedures must be approved and followed by trained staff.

DEFECT CHARACTERIZATIONS

> Ensure that defect characterizations are identified for each product and can be detected by inspection or test. (Analyze)
>
> **Body of Knowledge VII.C.3**

The inspection methodologies are chosen based on knowledge of your product and the type of defects that are known to occur. For each type of defect, you need to establish the rate that would trigger rejection of the lot, or rework (if possible). The acceptance rate is based on the risk category of the defect, with a very low acceptance rate for critical defects and a much higher rate for minor defects. Each type of inspection is placed in the most appropriate place in the packaging process, such as the vision system placed after seal but before label application, or label barcode reader placed right after the label application.

EQUIPMENT FAILURE DETECTION

> Confirm by inspection or test that equipment failures can be detected. (Apply)
>
> **Body of Knowledge VII.C.4**

Before each packaging/labeling batch is run, the entire line and associated equipment should be checked for cleanliness and absence of packaging components from any prior packaging operation as part of the line clearance.

In addition, during regular operation it is a best practice to challenge the system with a nonconforming product to ensure detection. This should follow a proscribed routine, such as at the start of each new lot or after a certain number of units have been processed. The challenge samples should be representative of the type and severity of defects that the system needs to detect, plus these samples should be properly maintained and replaced on a set schedule. Vision systems can also be challenged by conducting a visual inspection on the same lot or portion of a lot to ensure there is a similar level of detection.

Chapter 51

Product Inspection

STAFF EVALUATION

> Ensure that staff who perform manual and semi-automatic inspections are properly trained and that their inspections meet reproducibility requirements. (Analyze)
>
> **Body of Knowledge VII.D.1**

Each person engaged in manual visual inspections must have the education, training, and experience, or any combination thereof, to perform his or her assigned function, whether that function involves a manual, semiautomated, or automated visual inspection process or operation with reproducibility. Inspectors must be trained in relevant inspection-related procedures, theory discussion, and defect recognition (using examples). Training and testing conditions should be in alignment with those of the actual inspection process and environment. Should trainees inspect product during probationary periods, all units inspected by the trainees should be reinspected by qualified inspectors to assure the quality of the inspection and the development of the trainee.

Likewise, written production and process control procedures must be followed in the execution of the parenteral product inspection, and those inspection activities are documented at the time of their performance. These personnel, procedural, and documentation requirements apply to reinspection activities as well.

INSPECTOR REQUIREMENTS

> Establish requirements for inspectors to have periodic eye examinations. Confirm and document that they take frequent breaks from inspection. (Apply)
>
> **Body of Knowledge VII.D.2**

The qualification process consists of passing an eye examination, successful completion of the training and qualification activities, and a number of successful consecutive passes of the qualification kits as established by the firm's procedures.

Inspectors should have 20-20 vision, or vision correctable to 20-20 vision. Consideration should also be given to color blindness.

The qualification process tests the trainee's ability to identify and segregate defective units from acceptable units by using a test panel (qualification kit) composed of known defective and acceptable containers representative of the process and product, if applicable. Inspectors should be re-qualified at least annually, and the re-qualification process should include at least one successful test on the qualification kit. Quality personnel assigned to perform the verification of inspected lots (AQL inspection) should have, as a minimum, the same training and qualification as would be expected for personnel performing 100% manual visual inspection.

Personnel assigned to perform 100% manual visual inspection must pass eye examinations as part of their initial qualification and on some routine basis thereafter. Vision may be corrected as long as acceptable results are achieved. The examinations should include visual acuity and a color and contrast assessment. Routine annual eye examinations for these parameters are common.

Inspector concentration, ergonomic comfort, and level of fatigue are important considerations for accurate visual inspection results. Inspectors should be provided appropriate relief from the rigor of inspection, which can be accomplished through rest breaks and/or job rotation. A detailed procedure would detail the maximum length of an inspector's shift, followed by a minimum length for a break or other non-inspection activities.

AUTOMATED INSPECTION PROCESSES

> Ensure that automated inspection processes are validated. (Apply)
>
> **Body of Knowledge VII.D.3**

Inspection of finished goods is important for all pharmaceutical products, but 100% inspection of each filled/sealed unit is essential for sterile parentarals. Filled containers of parenteral products must be inspected individually for extraneous contamination and other defects that may adversely affect the integrity of the product. Units found defective in this manner must be rejected. Such inspection is one of many activities that a parenteral manufacturer uses in its effort to assure that its drug product meets the quality and purity characteristics that it is specified to possess. Parenteral product inspection can be accomplished through manual visual inspection or by fully automated methods of inspection.

The inspection process must be designed and qualified to ensure that every lot of all parenteral preparations is essentially free from visible particulates. Note that particulates are one of the main reasons for recalls and for FDA warning letters for sterile drugs. Every container whose contents show evidence of visible particulates must be rejected. Likewise, evidence of any other serious (critical or major) product/container defect provides grounds for rejecting that final container. Defect rates should be calculated at the end of each inspection process. Maximum allowable reject rates must be established.

Trends in types and size of particulates should be monitored and evaluated (for example, as part of an overall quality system). Intrinsic particulates (from the manufacturing process) are to be treated separately from extrinsic particulates (from environmental sources external to the manufacturing process itself). Three classifications of common defects have been defined (Table 51.1). Container or closure defects include cracked containers, broken finish (vial), damaged stopper or seal that compromises integrity (vial), or wrong or missing component (for example, a vial stopper or seal) or container. Examples of product defects include foreign material or particulate matter, precipitates and agglomerates, underfills and overfills, and melted cake (or a collapsed cake in some cases) for lyophilized products.

Parenteral product inspection activities must be performed in accordance with established specifications, standards, sampling plans, and test procedures that are scientifically sound and appropriate to ensure that the drug product conforms to appropriate standards of identity, strength, quality, and purity. The inspection process can occur as a separate process or in-line with an associated process such as filling, labeling, or packaging. Drug product containers undergoing inspection are fully assembled (container/closure assembly) and as such can be inspected in an unclassified area. Adequate facilities, equipment, and materials must be available and used to execute the inspection process regardless of the methodology and technology being used. Special considerations must be given to drug product solutions that are prone to forming product-related particulates (for example, protein formulations). The consideration should include specific product-related training and inspection techniques, and/or characterization data for product storage time and conditions.

Reinspections may be permitted, but the number of reinspections should be limited, specified, and justified, and any special characteristics of the product should be taken into consideration, if applicable (for example, temperature and light exposure and sensitivity). The need for any reinspection event should be evaluated and authorized by the appropriate quality function. Additional, and oftentimes specialized, training and/or qualification for a reinspection activity is customary and expected.

Quality control (QC) acceptance criteria must be applied to the parenteral product inspection process, and the results must be considered when assessing a drug

Table 51.1 Three typical classifications of common defects.

Critical	Affects the safety, identity, strength, quality, or purity of the product and could cause injury to the patient
Major	Potential to affect the quality of the product, or can lead to serious container impairments of a lesser degree (a functional risk, whereby the product is impossible or difficult to use)
Minor	Affects the cosmetic appearance only (no potential to affect the quality of the product)

product batch's condition for approval and release. Each appropriate specification and appropriate statistical quality control criterion must be met. The statistical QC criteria must include appropriate acceptance levels and/or appropriate rejection levels. Statistically sound sampling plans for *acceptable quality level* (AQL) inspection must be used. ISO 2859.1 or ANSI/ASQ Z1.4 level II sampling commonly are used for the quality assurance acceptance sampling, after the inspection event, which serves as a *quality assurance* (QA) audit of the entire process (the manufacturing process, as well as the 100% inspection process). AQLs for particulate and other serious defect attributes are generally within the range of 0.065% to 1.0%. Reinspections are also subject to additional AQL sampling and inspection.

MANUAL VISUAL INSPECTION

The basic manual visual inspection process typically involves the individual viewing of 100% of the filled containers of a parenteral drug product lot. Viewing distance, particle motion in the liquid, and container volume are important factors to be considered in the design of any manual visual inspection program. The viewing must be performed against a nonglare black-and-white background under a specified nonglare light intensity. The viewing is performed by a person who has been trained and qualified to inspect for specified defects.

The purpose of the inspection is to detect container or drug product solution defects according to the established training and qualification procedures, under a controlled setting and specified inspection parameters. Containers considered defective are segregated from containers considered acceptable. Further handling of the segregated containers must be delineated in the firm's procedures.

Lighting may be fluorescent, incandescent, spot, and/or polarized, and must be nonglaring. The light intensity at the container viewing area must be specified and of a strength that facilitates defect detection. Common acceptable illumination intensities are between 2000 and 3750 lux. Higher intensities can be used for colored glass and plastic containers, and for certain types of products (for example, colored solutions). Higher intensities should be justified and characterized by the firm. Light intensity must be monitored to ensure it is within the parameters specified in the inspection procedures. Procedures should be in place to ensure sufficient relamping frequency.

The rate of the manual visual inspection process must be specified. This rate is oftentimes referred to as the "pace." A common pace, per container, is five seconds of viewing against the black background and five seconds of viewing against the white background. Otherwise, the pace or rate must be defined and characterized. Rate consistency is customarily accomplished with the aid of a pacing mechanism (for example, an LED or other type of light prompting). Pacing is the methodology used for achieving a uniform and effective rate of inspection within the defined and acceptable quality standards and levels.

Another form of manual visual inspection of parenteral products typically involves a *presentation machine*. These machines typically use many of the same features of the standard manual inspection process (that is, lighting, inspection zone, container movement to create particle movement) yet handle and present the vial to the operator in a controlled and repeatable manner. Although this process is referred to as *semiautomated* visual inspection of parenteral product, it has many of the same requirements for operator training and qualification. The machine manipulates the container during viewing, yet the operator performs the inspection.

Qualification

As an alternate method to the 100% visual inspection, fully automatic inspection machines exist for particle, cosmetic defect, and leak detection (container/closure defect) inspection. When an automated method is used, it must provide at least the same level of inspection security as the manual visual inspection method. Any automated method used must provide detection rates at confidence levels that are equal to or greater than those achieved by manual inspection. Automated inspection machines should be challenged using defects obtained from the production operation. It is also possible to have a combination system, such as a manual visual inspection with an automated leak detection system.

As with any other process, equipment, or system used in the manufacture of finished pharmaceuticals, automated inspection machines and processes must be validated and documented under the company's overall validation policy. Typical performance qualification for automated inspection equipment would involve the inspection of three consecutive lots to assure reproducibility of results. Likewise, the equipment must be routinely calibrated, inspected, and checked according to a written program designed to assure proper performance. Written procedures for cleaning, operating, and maintaining the automated equipment are required. All such programs and procedures must be approved and followed.

Defect libraries, or sample sets (also referred to as *qualification/test panels* or *kits*), should be representative of those particles and defects found in the operation. Particle sizes and the types of defects present in the sets should represent the range of sizes and defects that have been, or could be, found in the production environment. When automated inspection equipment is used, it can be calibrated using the standard sets. Defect libraries must be reviewed at some appropriate frequency and modified, augmented, or revised to assure that the kits are maintained in a state that supports the inspection program's needs.

Chapter 52

Packaging Operations and Controls

One of the most important functions that the product packaging provides is the protection of its contents. The identity, strength, and purity of the drug product can become compromised if the packaging component that is in direct contact with the dosage form contaminates the product and/or fails to protect it from environmental conditions such as temperature, contaminants, and moisture (for example, humidity).

In addition to the evaluation and qualification of the primary packaging components (those that come in direct contact with the product), drug manufacturers need to evaluate the suitability of the secondary packaging components, as a failure in packaging performance may translate to potential quality problems. For example, secondary packaging components such as cartons that are intended for additional light-exposure protection need to be controlled to ensure that the cartons are not removed while the product is in storage.

The regulations for manufacturing of container closure systems detail the qualification and controls regarding the primary packaging components, as well as the variables to consider when selecting a primary packaging component (for example, vial, syringe, ampule, stopper) for the dosage form. This chapter, however, only addresses the aspects of operations and controls related to the secondary packaging components (for example, cartons, pallets) and how the secondary packaging plays a role in the protection of the product.

CONTENT PROTECTION

> Develop and apply procedures to prevent the environment or events from altering the identity, strength, purity, and quality of the package content. (Create)
>
> **Body of Knowledge VII.E.1**

According to the United States Food and Drug Administration (FDA) guidance for industry "Container Closure Systems for Packaging Human Drugs and Biologics," the secondary packaging components are used for several purposes:

- To provide protection from excessive transmission of moisture or solvents into or out of the packaging system
- To provide light protection for the packaging system
- To provide additional microbiological protection by protecting the packaging system from microbial intrusion
- To provide protection from excessive transmission of reactive gases (atmospheric oxygen, inert headspace filler gas, or other organic vapors) into or out of the packaging system

The secondary packaging offers a means for product identification and presentation. For example, cartons and packaging inserts commonly are used to display information related to drug facts, dosage, and expiration dates, and the information presented on them is governed by *good manufacturing practices* (GMP). On this subject, the Federal Food, Drug, and Cosmetic Act (FD&C) states that a drug or device shall be deemed to be misbranded if packaging and labeling are inappropriate, unrecognizable, misleading, or allow inks or adhesives in contact with the product. A particular type of packaging that grants additional security to the final drug product is the tamper-evident package, which can be designed as the primary packaging, secondary packaging, or a combination of both packaging systems. Currently, the Agency requires this type of package for most *over-the-counter* (OTC) drug products. The definition as stated in 21 CFR 211.132 is:

> *A tamper-evident package is one having one or more indicators or barriers to entry which, if breached or missing, can reasonably be expected to provide visible evidence to consumers that tampering has occurred. To reduce the likelihood of successful tampering and to increase the likelihood that consumers will discover if a product has been tampered with, the package is required to be distinctive by design or by the use of one or more indicators or barriers to entry that employ an identifying characteristic (e.g., a pattern, name, registered trademark, logo, or picture).*

In summary, drug manufacturers are facing numerous challenges when it comes to ensuring the compliance and security of the final package while maintaining brand recognition and shelf presence (especially for OTC drugs). As a result, manufacturers need to balance cost efficiency, regulations, and the new trends pertaining to child-resistant and senior-friendly packaging.

TYPES OF PLASTICS USED FOR PHARMACEUTICAL PACKAGING

Plastics are synthetic polymers that are sensitive to heat. Plastic containers are light in weight. Generally, they are resistant to inorganic chemicals but are often attacked by organic chemicals and by organic solvents and oils. Plastics contain some additives (for example, antioxidants, lubricants, plasticizers, stabilizers, filler, pigments) that may contaminate the drug product. Very few types of plastics completely prevent the entry of water vapor, and some are permeable to gases like oxygen and carbon dioxide.

There are two general types of plastics: thermoplastic and thermosetting. *Thermoplastics* soften on heating to a viscous fluid, which hardens on cooling.

Table 52.1 Common plastics and their properties.

Plastic	Properties
High-density polyethylene (HDPE)	Stiffness, strength, toughness, resistance to moisture, resistance to permeability to gas, ease of processing
Low-density polyethylene (LDPE)	Strength, toughness, flexibility, resistance to moisture, ease of sealing, ease of processing
Polystyrene (PS)	Versatility, clarity, easily formed
Rigid polyvinyl chloride (PVC)	Versatility, ease of blending, strength, toughness, clarity, transparency
Polypropylene (PP)	Strength, toughness, resistance to heat, chemicals, grease and oil, barrier to moisture
Polyamide (Nylon 6,10)	Superior color fastness, excellent abrasion resistance, durability
Polycarbonate	Durable, high impact resistance
Acrylic polymers	Transparency, resistance to breakage, elasticity
Polyethylene terephthalate (PET)	Clarity, lightness, strength, toughness, barrier to liquid and gas

Thermosetting plastics when heated become flexible, but they do not become liquid. Because of this, they are usually hard and brittle at room temperature. Table 52.1 lists several common plastics and their properties.

TYPES OF GLASS USED FOR PHARMACEUTICAL PACKAGING

Glass containers provide an excellent barrier with a proper closure system. Tinted glass can provide protection for photosensitive drugs. There are four types of glass used for pharmaceutical packaging:

- Type I—borosilicate glass
- Type II—treated soda-lime glass
- Type III—regular soda-lime glass
- Type IV (NP)—general purpose soda-lime glass

Type I borosilicate glass has a high melting point, so it can withstand high temperatures, is resistant to chemical substances, provides reduced leaching action, and can withstand sterilization. Type I borosilicate glass is used for laboratory glass apparatus and water for injection.

Type II treated soda-lime glass is fairly resistant to attack by water for a period of time. Sulfur treatment neutralizes the alkaline oxides on the surface, thereby rendering the glass more chemically resistant. Type II treated soda-lime glass is used for alkali-sensitive products, infusion fluids, blood and plasma, and large-volume containers.

Type III regular soda-lime glass contains high concentrations of alkaline oxides and imparts alkalinity to aqueous substances, flakes easily, and may crack due to sudden change of temperature. Type III regular soda-lime glass is used for all solid dosage forms (tablets, powders, and so on) and oily injections.

Type IV (NP) general purpose soda-lime glass is non-parenteral glass. Type NP general purpose soda-lime glass is only used for oral and topical purposes.

Additionally, the personnel involved in the packaging and labeling operation must have the required training and experience to perform their assigned duties. Personnel qualifications must be current in order to support the different inspection activities performed throughout the packaging and labeling operations. Equipment changeovers and setup are critical when the production lines are used for packaging and labeling of different products. FDA has issued Form 483 observations for drug manufacturers failing to maintain control of the packaging line as a result of deficiencies in personnel training. An FDA 483 observation regarding this subject cited the manufacturer for failure to adequately train employees engaged in the manufacture, processing, or holding of a drug product in GMP or in the particular operations performed by the employees [21 CFR 211.25(a)]. For example, employees engaged in the production of 260 mg/26 mL liquid oral product, lot #XYZ, which was mislabeled as 20 mg/2 mL, did not follow the company's written procedures pertaining to product changeover on the labeling system or those requiring documentation of filler settings.

QUALIFICATION AND MAINTENANCE OF EQUIPMENT

> Ensure that equipment used in packaging operations is qualified and maintained. (Apply)
>
> **Body of Knowledge VII.E.2**

Packaging Equipment: Only qualified equipment should be used on the packaging line, so the Quality Unit signs off on change control forms before any significant changes are made in the packaging area. Proper maintenance and calibration of the packaging and labeling equipment must be performed and documented.

LINE CLEARANCE OPERATIONS

> Determine that line clearance is performed and documented. (Analyze)
>
> **Body of Knowledge VII.E.3**

Line clearance involves the inspection of production facilities before the packaging/labeling operation to ensure that packaging and labeling materials from other lots are not present. An additional inspection of the facilities should be conducted at the end of the packaging operation. This ensures "line clearance" to prevent mix up of prior labeling on the new product lot. The quality unit (QU) must document these in the packaging batch record.

QUALITY CHECK CRITERIA

> Identify and apply specified criteria when quality checks are performed. (Analyze)
>
> **Body of Knowledge VII.E.4**

Figure 52.1 depicts the steps in a typical final packaging process, performed and recorded in the packaging batch record by Packaging Operations. The Quality Unit must perform the appropriate quality checks at critical steps to ensure the finished good meets specifications. These critical steps are designated with a star in the following diagram:

- Inspection of packaging and labeling materials to determine suitability for use;
- Inspection and verification of the packaging line (equipment) to ensure that the equipment is ready for use;
- Other preparation activities, which include verification of batch records being present in order to document the packaging/labeling operations;
- Label application to primary packaging (labeling equipment). Ensure the correct label is attached to the container and legible. Ensure that the lot code and expiration date is on the primary container (unless exempted).

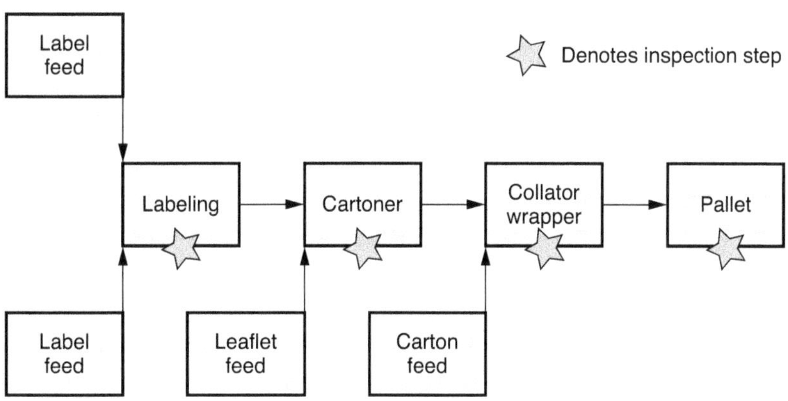

Figure 52.1 Typical final packaging process. Inspection of the product as it goes from the labeling operation to the palletizing operation occurs at every step to assure that containers and packages have the correct label. At the end of the process, the quality unit performs a visual examination of a representative sample of the packaged product.

Source: Figure used with permission of Dixie A. Dean, Roy Evans, and Ian Hall, eds. 2000. *Pharmaceutical Packaging Technology.* London and New York: Taylor & Francis.

- Packaging inserts application (cartoner).
- Product in its primary packaging is placed in cartons (cartoner), that is, single dose per cartoner or multiple doses per cartoner. Ensure that the expiration date is visible on the outer package and legible.
- Product is collated and arranged on pallets for storage or distribution. Ensure shipping containers are properly labeled, and pallet tags are correct.

CUT-LABEL PROCEDURES

> Apply appropriate procedures for cut labels and splices. Apply)
>
> **Body of Knowledge VII.E.5**

Most manufacturers are changing from cut labels to roll labels due to the increased risk of mix-up with individual labels. Where cut labels are used, drug manufacturers need to apply at least one of the following special control procedures: (1) dedication of labeling/packaging lines to each different strength of each different drug product, (2) use of automated or semiautomated equipment to conduct a 100% inspection for the correct labeling, and (3) use any automated technique (including size/shape differentiation) that prevents incorrect labeling from being processed by the equipment. Visual inspection should be verified by a second individual who did not perform the initial 100% inspection.

HAND-APPLIED LABEL PROCEDURES

> Ensure that hand-applied labels are 100% inspected. (Apply)
>
> **Body of Knowledge VII.E.6**

Typically, low-volume or investigational products may be hand labeled. During hand-labeling operations, conduct 100% visual inspection for correct labeling for hand-applied labeling that includes the label placement and orientation. Visual inspection should be verified by a second individual who did not perform the initial 100% inspection.

PACKAGING CONTROLS

> Distinguish between controls needed for different types of packaging processes. (Apply)
>
> **Body of Knowledge VII.E.7**

To ensure adequate control of the packaging and labeling operation, drug manufacturers must exercise, at minimum, several GMPs:

- Packaging line identification, indicating the type of product and lot number being packaged, should be present from the start of the packaging process until completion and cleanup;
- Examination of packaging and labeling materials before use to determine their suitability for the packaging operation. Incoming packaging components are sampled and inspected against their specifications before being released by the QU for future use. They must be stored in a manner to prevent environmental and microbial contamination;
- Label control, including packaging insert controls. This involves reconciliation of the number of labels sent to the packaging line against the number used, destroyed, and returned to the labeling storage area;
- Bulk product must be stored in a manner to prevent deterioration and environmental/microbial contamination until it is packaged and protected in its primary container;
- Clear and detailed standard operating procedures and batch production records that allow for documentation of the key steps of the packaging operation (lot number, line-clearance procedures, label reconciliation, number of units packed, and results of the work-in-progress [WIP] inspection and packaged product inspection);
- Segregation of packaged product that needs to undergo inspection by the quality unit;
- Prevention of cross-contamination or errors by physical separation of different packaging/labeling operations;
- Controls over contract manufacturers that perform the secondary packaging and labeling operations.

CONTAMINATION CONTROLS

> Identify controls to prevent microbial and other contamination at all stages of packaging. (Analyze)
>
> **Body of Knowledge VII.E.8**

To prevent microbial, chemical, and physical contamination during packaging, manufacturers should establish risk-based preventative measures. One way to establish a risk-based program is through the use of a Hazard analysis and critical control points (HACCP) program.

HACCP is used to manage risks with microbial, chemical, and physical contamination. HACCP can is used in many industries including food, drugs, cosmetics, medical devices, and biologics. HACCP has seven steps:

1. Conduct a hazard analysis and identify preventive measures for each step of the packaging process.
2. Determine the critical control points.
3. Establish critical limits.
4. Establish a system to monitor the critical control points.
5. Establish the corrective actions to be taken when monitoring indicates that the critical control points are not in a state of control.
6. Establish a system to verify that the HACCP system is working effectively.
7. Establish a record-keeping system.

Potential areas of use for HACCP might be to identify and manage risks associated with microbial, chemical, and physical contamination hazards. HACCP is useful when product and process understanding is sufficiently comprehensive to support identification of critical control points. The output of an HACCP analysis is risk management information that facilitates monitoring of critical points not only in the manufacturing process but also in other life cycle phases.

The foundation of an effective HACCP program is a rigorous prerequisite program. Prerequisite programs generally consist of the following elements:

- *Facilities:* The establishment should be located, constructed, and maintained according to sanitary design principles. There should be linear product flow and traffic control to minimize cross-contamination from raw to cooked materials.
- *Supplier control:* Each facility should assure that its suppliers have in place effective GMP and food safety programs. These may be the subject of continuing supplier guarantee and supplier HACCP system verification.
- *Specifications:* There should be written specifications for all raw materials, products, and packaging materials.
- *Production equipment:* All equipment should be constructed and installed according to sanitary design principles. Preventive maintenance and calibration schedules should be established and documented.
- *Cleaning and sanitation:* All procedures for cleaning and sanitation of the equipment and the facility should be written and followed. A master sanitation schedule should be in place.
- *Personal hygiene:* All employees and other persons who enter the manufacturing plant should follow the requirements for personal hygiene.
- *Training:* All employees should receive documented training in personal hygiene, GMP, cleaning and sanitation procedures, personal safety, and their role in the HACCP program.

- *Chemical control:* Documented procedures must be in place to assure the segregation and proper use chemicals in the plant. These include cleaning chemicals, fumigants, and pesticides or baits used in or around the plant.
- *Receiving, storage, and shipping:* All raw materials and products should be stored under sanitary conditions and the proper environmental conditions such as temperature and humidity.
- *Traceability and recall:* All raw materials and products should be lot-coded and a recall system in place so that rapid and complete traces and recalls can be done when a product retrieval is necessary.
- *Pest control:* Effective pest control programs should be in place.

For more information about HACCP refer to 21 CFR Part 120 Hazard Analysis and Critical Control Point (HACCP) Systems or ISO 22000 Food Safety Management Systems—Requirements for Any Organization in the Food Chain.

TAMPER-EVIDENT PACKAGING

> Ensure that tamper-evident and child-proof packaging requirements are established for required products. (Apply)
>
> **Body of Knowledge VII.E.9**

OTC human drugs for retail sale, with some limited exceptions, are required to have tamper-evident packaging if the product will be accessible to the public (for example, on a store shelf). It must have at least one indicator or barrier to entry that would be evident to the consumer if the package has been opened. The packaging must have a distinctive design or identifying characteristic (for example, pattern, trademark, logo or picture) that cannot be easily duplicated or commonly available. The tamper-evident feature can be on the primary container, secondary package (for example, carton), or a combination of systems and should remain intact with normal shipping and handling. In addition, all two-piece hard gelatin capsule products must be sealed with a tamper-evident technology such as a "locking" cap or banded seal.

OTC drugs that require tamper-evident packaging must identify these features in prominent text on the package, including a description of their identifying characteristic (for example, a logo-imprinted safety seal). An exemption is possible, but must be justified using a citizen petition.

Chapter 53
Labeling Operations and Controls

Labels are an integral part of the drug product packaging process and the marketing of the final drug product. They provide essential information regarding the contents of the drug product, such as strength, dosage, medical claims, and drug warnings. According to the United States Food and Drug Administration (FDA), drug products that are mislabeled can pose a threat to public health, cause product recalls, and seriously damage the image of the drug manufacturer.

As a result, labels are designed, produced, and controlled to meet *good manufacturing practices* (GMP) that ensure the identity and quality of the product. Additionally, it is ultimately the responsibility of the drug manufacturer to ensure that label integrity (print quality, label adhesion, and so on) is never compromised during the life span of the drug product, from distribution to its intended use. The requirements for label control and packaging components are becoming increasingly stringent. To address these requirements, drug manufacturers, as well as label manufacturers, are moving toward electronic verification of the printed labels, and improved inventory control systems that allow tracking using bar codes printed on each label.

In February 2004, FDA released a rule requiring bar codes on human drugs and biologic products. By implementing bar code tracking, FDA hopes to reduce the number of medication errors, thus ensuring that health professionals provide patients with the right drug and proper dosage. The rule also applies to establishments that handle blood components that will be used for transfusions. According to FDA, this new rule will help prevent nearly 500,000 adverse events and transfusion errors and concurrently will help save US$93 billion in healthcare costs over 20 years.

FDA's final rule requires the use of a linear bar code to indicate the *National Drug Code* (NDC) on most prescription drug products and certain *over-the-counter* (OTC) drug products.

LABEL PRINTING IN PACKAGING

> Confirm and document that any printing done separately or during packaging is performed correctly. (Analyze)
>
> **Body of Knowledge VII.F.1**

LABEL RECEIPT

In general, the labels destined for the packaging operation of the drug product are printed by a vendor that has been qualified by the drug manufacturer. To ensure that the correct labels are received, each shipment of labels that arrives at the manufacturing site is subjected to several steps:

- The receiving department at the drug manufacturer receives the labels. Records must be maintained for each shipment of labels received from the vendor (printer).
- A quarantine area or controlled area is designated in which to store the received labels while incoming inspection is done by the *quality unit* (QU).
- The labels are sampled for inspection against defined reject criteria (*acceptable quality level* [AQL] inspection method).
- The QU releases or rejects the labels. If the label is released, an identification or lot number typically is assigned.
- The labels are maintained in a secured and controlled area while awaiting release for packaging use. Access to this controlled area should be granted only to a few individuals. Additionally, labels destined for different products or different strengths of the same product need to be stored separately to avoid potential errors.

LABELING ISSUANCE TO MANUFACTURING

GMP guidelines are in place to ensure that not only the type of label but also its quantity is strictly controlled when being issued for packaging of the drug product. The manufacturers of the drug must have *standard operating procedures* (SOPs) that detail the steps to follow when labels are issued to manufacturing.

Label specimen verification is required to ensure that the current version of the label is being documented on the drug product batch record. Label verification is required to guarantee that all labels to be used match the identity of the label template (specimen) specified in the drug product batch record. The label verification step implies a 100 percent inspection of all labels that will be installed in the labeling equipment. Each label is counted and verified either by manual methods or by electronic verification. Several methods of electronic verification currently are used by drug manufacturers, and they typically use an electronic visual inspection of each label by means of a camera that compares a photograph of each label with the approved label specimen.

SOPs specify how many labels will be issued against the planned number of containers to be labeled. Furthermore, the completed batch records must show comparisons between the number of labels that were issued to manufacturing, the number of labels used, and the number of labels not used during the operation (destroyed or returned). An investigation is initiated and completed by the QU if any discrepancies are found during the label reconciliation process.

Unused labels can be returned to the controlled storage area if they have not been printed with the drug product batch number; however, most companies have procedures requiring that unused labels be destroyed in an effort to prevent labeling errors.

QUALITY OF PRINT USED

> Ensure that any type of print information (engraved and embossed) on packaging materials is clear and resistant to fading, smudging, or erasure. (Apply)
>
> **Body of Knowledge VII.F.2**

Label control is extremely important for all drug products, as labeling mix-ups and printing errors continue to be a major source of product recalls. Drug manufacturers need to ensure that the information printed on the labels does not degrade over time. Thus, it becomes fundamental that the inks used for label manufacturing are light resistant, do not smear, and do not leave traces of odor. Additionally, the printing must conform to the approved label specimen (for example, uniformity, art specification).

Another factor that is controlled by drug manufacturers is the labeling operation. Just as batch records are required to document label issuance operations, they are also required for recording key aspects of the actual labeling operation (Figure 53.1).

EXPIRATION DATING

The guidelines for GMP require that the drug expiration date is noted on the label. Moreover, if the drug product is to be reconstituted at the time of use, the label must include the expiration dates for the reconstituted and un-reconstituted drug product.

LABEL CHANGES

> Determine whether regulatory notification and approval is required for proposed label changes. (Evaluate)
>
> **Body of Knowledge VII.F.3**

The development of labeling artwork is a collaborative process involving both the commercial side of the organization (for example, marketing) and the medical and regulatory side of the organization (for example, medical affairs, regulatory affairs). Often, competing demands for space occur as the groups negotiate how to incorporate all of the required safety and regulatory information while maintaining the look and feel or the corporate branding that is important to the company. This negotiation process ultimately results in the artwork (graphical representation of the proposed labeling) that is used by printing firms to produce the final printed labeling.

Like the drug product itself, the labeling that accompanies the product has a life cycle of its own. The collaborative development of labeling begins early in the product development phase and evolves as more information is learned about the drug, specifically its utility and safety profile. The labeling evolves after approval as more is learned about the drug product through its use in the general population. Product labeling needs to be current with all new information to ensure that

- Labeling equipment identification, and confirmation of calibration status
- Names of personnel involved in labeling operations
- Line clearance verification before and after label operations
- Label segregation to avoid mix-ups
- Label reconciliation numbers
- In-process examination of label quality (that is, label adherence, location on container)

Figure 53.1 Labeling operations.

medical practitioners and patients have the information that they need to make the best possible healthcare decisions.

FDA has detailed documents on the requirements for labeling changes for approved drug products, biological products, and devices describing the types of changes that (1) require prior approval, (2) require submission but may proceed to distribution, or (3) are notified in the annual report. The regulatory or quality group must ensure that these requirements are met before the new label is released for use.

A graphical representation of the product labeling is developed and is typically what is reviewed and approved internally before being sent to the printers for conversion to the actual labeling materials. An SOP must be in place describing the label development, approval and output process. These graphical files need to be controlled to ensure that what is submitted to the printers is current and contains the most accurate information regarding the safety and efficacy of the product. Version control is critical to this process, and procedures should be in place, both within the organization and at the labeling printers, to control current and obsolete versions of the graphical files.

LABEL RECONCILIATION

Confirm that label reconciliation is performed and documented, and discrepancies are investigated. (Analyze)

Body of Knowledge VII.F.4

Labels and labeling should be formally issued to the production order and recorded therein. Best practice is to maintain a copy of the first and last label in the production record. Label and labeling reconciliation is simply accounting for the labels that were issued to the production order, consumed, returned to inventory, and/or destroyed. If there is a discrepancy (the number of labels cannot be reconciled), an investigation should be initiated and documented.

UNUSED LABELS

> Confirm that procedures are established and used for controlled, unused batch-coded labels and labeling materials. (Analyze)
>
> **Body of Knowledge VII.F.5**

The manufacturer must establish and maintain procedures to address any labeling that is not used, and that has been printed with lot (batch) number and expiration date (or any other batch-specific information) to ensure they are destroyed. No matter what destruction process is used, however, it should render the labeling materials unusable to prevent their unauthorized or mistaken use. A register or log of labels that were destroyed should be maintained. Additionally, the destroyed labeled should be documented and reconciled within the production batch record.

LABEL PRODUCTION

> Define terms related to offline printing, roll label splicing, gang printing, secure storage, and destruction. (Understand)
>
> **Body of Knowledge VII.F.6**

Regulatory agencies expect that there will be appropriate rigor in the printing process to ensure that labeling is accurate and that errors do not occur. Pharmaceutical companies should treat labeling vendors as they would any other vendor; that is, labeling vendors should be evaluated or audited and qualified in a manner similar to any other raw material or packaging component vendor.

Vendor Proof

Once the graphical representations are reviewed and approved internally, the files are forwarded to the labeling vendors to be printed. The labeling vendor should be expected to take the files and generate a vendor proof that is reviewed by the pharmaceutical company to ensure that the new or revised labeling is accurate and matches the internally approved artwork. If issues are noted, procedures should be in place to ensure that corrections are made and that preventive measures are implemented to preclude future problems.

Gang Printing

Gang printing is a practice that involves placing many different print projects on the same printing sheet in an effort to reduce printing costs and paper waste. FDA *good manufacturing practices* (GMP) regulations explicitly prohibit the use of gang printing of labeling for different drug products or different strengths or net contents of the same drug product unless the labeling from the gang-printed sheets is adequately differentiated by size, shape, or color. Because of the high potential for errors, pharmaceutical companies typically do not allow gang printing of drug product labeling.

Roll Label Splicing

Flexographic printing has become commonplace in the production of pharmaceutical labeling. After the labels are printed, they are typically die cut and, if necessary, wound onto master rolls. These master rolls are then slit into individual label widths and counted to provide the required number of labels per roll. It is not uncommon for the roll stock to break during the various steps of the process or for the process to require multiple rolls of label stock to complete the printing run. When this occurs, it may be necessary to splice the roll stock together. It is critical for the label vendor to clearly identify the presence of splices in each roll of labeling. The vendor should provide a report that demonstrates that they have verified that the labeling on each side of the splice is correct and that there are no issues with the printed copy on either side of the splice. Upon receipt of the labeling at the pharmaceutical company, verification of the spliced rolls should take place.

Off-Line Printing

The packaging and labeling of pharmaceutical products typically occur together as a single operation. Labeling is applied to the packaged product, and a unique batch number and expiration date is printed on the package (for example, bottom of the bottle) or on the labeling as part of this operation. In label print cases, where the batch and expiration date are printed online, it is possible to return unused/unprinted labeling to the secured storage area for use on another packaging run.

There are instances, however, where the labeling must be printed off-line with the batch-specific information. Equipment on the packaging lines may not be capable of printing the batch-specific information, or the type of product may preclude the ability to print this information online. In these instances, very tight controls over the use of this batch-specific labeling need to be in place. Any labeling that is printed off-line with batch-specific information should be stored separately from other labeling, and there should be very tight controls on reconciling the quantity used during the packaging run. Any labeling that is printed off-line with batch-specific information cannot be returned to inventory and must be discarded and destroyed after completion of the packaging operation.

SECURE STORAGE OF LABELING

GMP regulations require that printed materials and labeling be stored in an area with limited access and in a manner that will prevent errors. According to FDA GMP regulations, the labels and labeling materials for each different drug product, strength, dosage form, or quantity of contents should be stored separately (that is, adequately segregated, either spatially or in separate containers) and with suitable identification to preclude errors.

Similar requirements are identified in the Canadian GMP regulations. Printed packaging materials need to be stored in an area that is restricted to designated personnel who are supervised by individuals with training and experience and who are directly responsible to the person in charge of the manufacturing or a person having the same qualification.

DESTRUCTION OF LABELING

On a global basis, the GMP regulations are consistent when it comes to destruction of labeling. Any labeling that fails to meet specifications upon incoming receipt must be destroyed. Labeling that is superseded by new labeling, or is considered outdated, should be destroyed. The destruction must be appropriately documented.

CONTROL OF GRAPHICAL FILES AND PRINTING PLATES

One of the major challenges in the area of label printing is the frequency with which product labeling changes in today's regulatory environment. Regulatory agencies want to ensure that healthcare providers and patients have access to current information related to the safety of pharmaceutical products, and, as a result, updates to the labeling are common. Version control and the handling of current and obsolete graphical files and printing plates are of critical importance to ensure that this current information is incorporated correctly and in a timely fashion.

Labeling vendors, like the pharmaceutical company itself, should have procedures in place to control current and obsolete versions. Current versions should be accessible for use, while obsolete versions should be archived or removed from an active file location. Printing plates for obsolete versions should be destroyed and documentation of their destruction provided to the pharmaceutical company for verification of the destruction process.

ACCESS CONTROL

> Ensure that controls are established for the creation, storage, and issuance of labeling such as product labels, package inserts, and printed cartons. (Analyze)
>
> **Body of Knowledge VII.F.7**

GMP regulations state that access to the labeling storage area should be limited to authorized personnel that have the proper training to follow the applicable SOPs and to understand the importance of preventing labeling errors. Similarly, labeling that is returned to the labeling storage area (for example, after completion of a packaging operation) should be identified appropriately and stored in such a manner as to prevent errors.

The QU must ensure that the correct lot number and expiration date are used and legible at the start of the printing process, and by inspection throughout processing of the lot.

EXAMPLES OF FDA WARNING LETTERS

Failure to ensure that strict control is exercised over labeling issued for use in drug product labeling operations as required by 21 CFR § 211.125(a). Your firm does not exercise adequate controls over the labeling issuance and labeling operation of the drug products being handled at your facility. Your firm's personnel were not aware of the labeling requirements set forth by the abovementioned regulation.

Failure to ensure that written procedures are designed and followed to assure that correct labels, labeling, and packaging materials are used for drug products as required by 21 CFR § 211.130. Your firm had not established or implemented procedures to assure that correct labels were applied to your . . . products.

Specifically, your primary label includes an UDI, yet your firm has not established and maintained procedures to control labeling activities to include performing a labeling inspection to ensure the correct UDI is present. Your Quality Supervisor told our investigator that you do not have any UDI procedures and your SOP-01201 Labeling and Packaging Controls, Rev D, eff 12/21/2020, does not include UDI.

Other Examples of FDA Warning Letters

Your product labeling specifies a 10-day use limit under the "Warnings" heading in the "Drug Facts" panel (that is, "Stop use and ask a doctor if . . . pain gets worse or lasts more than 10 days") but this warning appears only within the Drug Facts panel of the retail package. Its placement within the perimeter of the Drug Facts panel suggests that it applies exclusively to the analgesic use discussed in that panel. It is contradicted by the statement elsewhere on the labeling "For calcium, adults and children 12 years and over, take up to 4 caplets per day . . . ," without any limitation on duration indicated.

Your product labeling warns against use by children and teens, to prevent Reye's syndrome (that is, "Warning—Reye's syndrome: Children and teenagers should not use this medicine for chicken pox or flu symptoms before a doctor is consulted about Reye's syndrome, a rare but serious illness reported to be associated with aspirin."), but this warning appears only within the "Drug Facts" panel of the retail package. Its placement within the perimeter of the Drug Facts panel suggests that it applies exclusively to the analgesic use discussed in that panel. It is contradicted by the statement elsewhere on the labeling "For calcium, adults and children 12 years and over, take up to 4 caplets per day . . . ," without any warning that there are circumstances where it is not appropriate for children and teenagers to take the product.

Chapter 54

Filling and Packaging Records

TERMS

> Define terms related to these records including evidence of line clearance, printed material reconciliation, and yields. (Understand)
>
> **Body of Knowledge VII.G.1**

Filing and packaging records generally contain evidence of the following activities:

- Line clearance is the process that ensures equipment and work areas are free of products, materials, packaging, labeling, and procedures, work instructions, and forms from the previous job.
- Theoretical yield is the quantity that would be produced at any appropriate phase of manufacture, processing, or packing of a particular drug product, based on the quantity of components to be used, in the absence of any loss or error in actual production.
- Reconciliation is accounting for the labels that were issued to the production order, consumed, returned to inventory, and/or destroyed.

REGULATIONS AND GUIDANCE DOCUMENTS

An essential aspect of a robust documentation program is the existence of records that provide a detailed history of the manufacturing stages, testing, and any relevant event that was part of the execution of a batch of drug product. As a result, it is expected that the regulations for *good manufacturing practices* (GMP) of finished pharmaceuticals (21 CFR 211, Eudralex Volume 3, Chapter 4, among others) require companies to establish not only written procedures for the production of the finished drug products but also batch production and process control records to assure product identity, strength, quality, and purity. The written procedures, as well as the batch records, provide clear instructions on how to execute a batch of drug product per approved procedures, and at the same time they guarantee uniformity from batch to batch.

SETUP INSTRUCTIONS

> Ensure that packaging line setup instructions are appropriate for all components. (Apply)
>
> **Body of Knowledge VII.G.2**

Filling and Packaging Records

Separate batch production record(s) are prepared for each batch of drug product to cover the different manufacturing steps from filling to packaging. The batch production records should be designed in such a way that they provide clear instructions on how to manufacture the batch and ensure consistency in execution from batch to batch even when differently qualified individuals execute the manufacturing operation. The batch production record section on filling should include:

- Initials or signature of the staff performing each significant step in the packaging and labeling the operation, initials/signature of the person checking the step was performed, and the dates completed.
- Identification of the major equipment and lines used.
- Component information including, but not limited to, material description, material number, quantity used, and expiration dating.
- Checks on the work area, such as line clearance procedures before the filling operation begins (area free of items not intended for use in the operation) and results of the inspection after use of the area.
- In-process (for example, in-line weight checker, vision system) and laboratory control results, including any sampling performed.
- Yield calculations at the end of each manufacturing step; these calculations are performed by a qualified individual and independently verified by another organizational unit staff member. For example, the amount of bulk product using in filling (# containers × avg weight) plus waste can be reconciled against the total bulk product brought to the filling line.
- Documentation on any major equipment used during the production of the batch, and documentation of equipment suitability verification (for example, checks on calibration due dates, cleaning status).
- Manufacturing time constraints. When appropriate, time limits should be established for the completion of the filling step to guarantee the quality of the drug product; when deviation from the production time limit occurs, the event should be documented with a justification on the batch record.

- Fill weight calculation/verification.
- Interventions and line stoppages on the filling batch record listing the time of the event and the duration of the intervention or line stoppage.
- Any sampling performed.

The batch production record section on final packaging/labeling should include:

- Initials or signature of the staff performing the operation; initials/ signature of the person checking the step was performed, with the dates completed.
- Identification of the major equipment and lines used.
- Product description (name, type, strength, pack size).
- Product specifications.
- Inventory record of each component used during the packaging operation (cartons and labels, among others). Where possible a sample of the labeling used as part of the batch record should be included.
- Review of the label information against the established label specifications, including a copy of each label signed and dated by the staff responsible for label approval.
- Line clearance operation of the packaging area before and after use.
- Special precautions to be observed during the packaging process.
- Final product inspection by the *quality unit* (QU) to ensure the correct labeling has been applied, and the lot number and expiration dating are correct.
- Any sampling performed, including reserve samples.

Production, Process Control, and Laboratory Record Review

The filling, packaging, and labeling records must be reviewed by the corresponding organizational unit and the QU to ensure that the finished drug product was manufactured according to the established and approved procedures. The review of the production and process control records must include an assessment of any event failures or deviations that could potentially affect the quality of the product. These can include investigations into product yields or label reconciliation outside of the minimum or maximum limits. Where deviations from the written procedures exist, the appropriate organizational units must document the changes and their corresponding justification. For sterile products, the production record review should include, but it is not limited to:

- Events that could have impacted the environment in the *critical zone* (an area designed to maintain sterility of sterile materials). Sterilized product, containers, closures, and equipment may be exposed in the critical zone.
- Review of maintenance activities that could have impacted the execution of the batch.

- Batch and trending data for utility and support systems (*heating, ventilation, and air conditioning* [HVAC], particle monitoring system, and so on).
- Documentation concerning adherence to critical process parameters such as fill time limits and sterilization expiration dates.

The laboratory records are reviewed to ensure compliance with specifications and standards, usually summarized on *certificate of analysis* (COA) that becomes part of the product batch record. The information additionally documented by the laboratory includes descriptions of the samples received for testing, batch number, quantity, the date the samples were taken, the date the samples were received for testing and tested, calculations, and the initials or signatures of the staff that performs and verifies the test results.

Part VIII

Product Development and Technology Transfer

Chapter 55

Quality by Design Concepts

To understand the most current thinking behind product development and regulatory expectations, it is important to understand the principles built into the International Conference on Harmonization (ICH) *Pharmaceutical Development* (ICH Q8) guidance. The concepts in ICH Q8 came about as a compromise between members of the ICH, which included the United States Food and Drug Administration (FDA) and Pharmaceutical Research and Manufacturers of America (PhRMA), and their counterparts from Europe and Japan. ICH Q8 describes the suggested contents for the 3.2.P.2 Pharmaceutical Development section of a regulatory submission in the ICH M4 *common technical document* (CTD) format. ICH Q8 was published in the Federal Register, May 22, 2006, as guidance for pharmaceutical product development.

Before the ICH Q8 guidance, in the United States the design and development information submitted in an application was variable; some information was submitted in the Investigational New Drug (IND) application, some information was submitted in European Union (EU) reports, and some information was distributed in New Drug Applications (NDAs) in an inconsistent format. Europe had incorporated key elements of the ICH Q8 guidance in that submissions described formulation development, the critical product attributes, and design of the manufacturing process. Japan had limited expectations, with more information for complex dosage forms. In general, though, there was limited (regulatory) incentive to truly understand processes and products, and to optimize them.

In these situations, opportunities exist to develop more flexible regulatory approaches, for example, to facilitate: risk-based regulatory decisions (reviews and inspections); manufacturing process improvements, within the approved design space described in the dossier, without further regulatory review; reduction of post-approval submissions; real-time quality control, leading to a reduction of end-product release testing.

After implementation of the ICH Q8 guidance, product quality and performance may be achieved and assured by design of effective and efficient manufacturing processes, product specifications are based on mechanistic—mathematical relationship between input and output—understanding of how formulation and process factors impact product performance, and there is the ability to effect continuous improvement and continuous real-time assurance of quality.

In the United States, the ICH guideline is discretionary and nonbinding. It represents FDA's thinking on NDA submissions, and companies can offer alternative approaches. ICH, however, provides the preferred FDA format for submissions.

CRITICAL QUALITY ATTRIBUTES (CQAS) AND CRITICAL PROCESS PARAMETERS (CPPS)

> Identify critical quality attributes (CQAs) for products and critical process parameters (CPPs) for processes. (Evaluate)
>
> **Body of Knowledge VIII.A.1**

QUALITY BY DESIGN

Quality by design (QbD) designs quality into the product using knowledge of the process. With QbD, a relationship exists between quality attributes and product efficacy and performance; with increased knowledge and control, decreased regulatory oversight is justified because all critical sources of variability are identified and explained, variability is controlled by the process, and product quality attributes can be accurately and reliably predicted over the design space established for materials used, process parameters, and environmental and other conditions.

ICH Q8 shows that pharmaceutical development is a learning process. Understanding is gained, for example, by knowledge, formal experimental designs, *process analytical technology* (PAT), and life cycle knowledge. Formal experimental design (*design of experiments* [DOE]) is a structured, organized method for determining the relationship between factors affecting a process and the output of that process. PAT is a system for designing, analyzing, and controlling manufacture through timely measurements (that is, during processing) of critical quality and performance attributes of raw and in-process materials and processes with the goal of ensuring final product quality. Both successes and failures are described as part of the life cycle story that demonstrates QbD.

CRITICAL PROCESS PARAMETER

The *critical process parameters* (CPPs) are the measured variables that are known to have an effect on one or more product quality attributes. All CPPs must be identified and controls established within a defined acceptable range. The ranges for these parameters may be typically obtained in later development studies (that is, full-scale engineering studies or technology transfer studies) aimed to optimize the process for commercialization, conducted in parallel with the human clinical trials. The process optimization efforts must ensure that the enhanced process remains equivalent to that used for the making of the clinical trials product.

CRITICAL QUALITY ATTRIBUTES

The *critical quality attributes* (CQAs) are intrinsic quality characteristics that are desired or needed to ensure patient safety and benefit of the drug product. In ICH

Q6A, some *active pharmaceutical ingredient* (API) attributes should be considered critical, regardless of the drug product end use. Identification, physicochemical properties, appearance, assay, and purity are applicable to all drug products. Particle size, microbial purity, and polymorphism depend on the drug product. The ICH Q6A decision trees can be used to determine the criticality of these quality attributes for solids, solutions, or sterile products. All CQA must be identified and have established acceptance criteria.

DESIGN SPACE

> Define the concept of design space as it is used throughout the product lifecycle. (Understand)
>
> **Body of Knowledge VIII.A.2**

Information from pharmaceutical development studies is a basis for risk management (using ICH Q9). Critical parameters carry the risks, critical formulation and process parameters generally are identified through an assessment of the extent to which their variation can have an impact on the quality of the drug product, and this assessment helps define design space. *Design space* is therefore defined as the multidimensional combination and interaction of input variables (for example, material attributes) and process parameters that have been demonstrated to provide assurance of quality.

Within the regulatory framework, working within the design space submitted in an application is not considered a change. Movement out of the design space is considered to be a change and would normally initiate a regulatory post-approval change process requiring regulatory notification. Therefore, ICH Q8 Section 2.4, "Design Space," facilitates regulatory flexibility. In the traditional submission process, there is limited knowledge, and any change needs new data and new approval. With the Q8 model implementation, the influence of factors is explored, creating knowledge, and risk analysis of the impact of change is possible; there is the ability to move within a defined area post-approval that gives flexibility for continuous improvement without the need for further approval. In addition, expanded design space facilitates flexible regulatory approaches. An increased understanding of the product life cycle, including material attributes—manufacturing process and process controls—facilitates the approval and submissions process.

VALIDATION PRINCIPLES WITH QUALITY BY DESIGN

Validation is evidence of a high degree of assurance that a specific process will consistently produce a product meeting its predetermined specifications and quality attributes. In ICH Q8, *continuous process verification* is defined as an alternative approach to process validation in which manufacturing process performance is continuously monitored and evaluated. Product quality and performance are achieved and assured by design of effective and efficient manufacturing processes. Traditionally, validation has focused on system operations and testing the output of the system. In the ICH Q8 validation approach, validation focuses on system design and on testing the quality that is being built into the process. The Q8 validation approach focuses on process design space and ongoing assessments. It uses DOE to define CQA, and design space to document the process used to verify CQA. The CQA and their interrelation are documented, and the CPP for each CQA is identified. The effect that CQA variability has on the process is identified. The design space for the process is formed based on findings.

The CQA and CPP are identified through experiments, risk analysis, and analyzer assessments. Validation is based on the range of control of the process within the design space rather than map testing to the frozen specification parameters.

PROCESS ANALYTICAL TECHNOLOGY (PAT) TOOLS

> Identify process analytical technology (PAT) tools including multivariate data analysis, process analyzers, and process and endpoint controls. Describe their use in supporting the manufacture of quality products. (Understand)
>
> **Body of Knowledge VIII.A.3**

FDA realized that changes were needed in its inspection approach; the perception was that the existing regulatory system was rigid and did not favor innovation. To encourage innovation, the FDA launched a new initiative in August 2002—Pharmaceutical cGMPs for the 21st Century: A Risk-Based Approach. This program included up-to-date concepts of risk management and quality systems, use of the latest scientific advances in technology, risk-based approaches that encouraged innovation, and a move to make FDA submissions and inspections more consistent. The Agency's resources addressed the most significant health risks. The desired state of pharmaceutical process is achieved when science and engineering principles are used effectively for assessing and mitigating risks related to poor product and process quality. This process should be accomplished with design of effective and efficient manufacturing processes, specifications that are based on understanding of how formulation and process factors affect product performance, continuous real-time quality assurance, and regulatory policies and procedures that accommodate the most current level of scientific knowledge. Together with the design space concept in QbD, the FDA envisioned *process analytical technology* (PAT) as part of the new initiatives that would reduce the risks associated with manufacturing uncertainty; risks would be fully contained and controlled.

PAT is a system for designing, analyzing, and controlling manufacturing through timely measurements during processing of critical quality and performance attributes of raw and in-process materials and processes, to ensure final product quality. Analytical requirements of PAT include chemical, physical, microbiological, mathematical, and risk analysis. PAT is not simply online/off-line, real- time analytical measurements. PAT combines multivariate tools for design, data acquisition, and analysis, process analyzers, process control tools, and continuous improvement and knowledge management tools.

MULTIVARIATE TOOLS

Multivariate tools include mathematical approaches, such as statistical DOE, response surface methodologies, process simulation, and pattern recognition tools, to increase understanding of the relevant multifactorial relationships (for example, between formulation, process, and quality attributes) and applicability of this knowledge in different scenarios (that is, generalization). These tools enable the identification and evaluation of product and process variables that may be critical to product quality and performance. They may identify potential failure modes and mechanisms and quantify their effects on product quality.

PROCESS ANALYTICAL TECHNOLOGY

PAT adds process analyzers for in-process multivariate (multiple variable) measurements, which are taken in one of three ways:

- *At-line,* where the sample is removed and analyzed close to the process stream
- *Online,* where the sample is diverted from the manufacturing process to an analyzer, and possibly returned to the stream
- *In-line,* which may be an invasive or noninvasive process that analyzes the sample while it is part of the process stream

PROCESS CONTROL TOOLS

We are most familiar with traditional in-process control tools used for monitoring such things as temperature, pH, pressure, flows, and other physical parameters (that is, univariate measurements). These measurements provide enough information to adjust equipment to more optimal settings for each variable as required, but do not typically provide enough information to learn more about the process as a whole. With PAT, rapid, dedicated in-/online testing is added to these traditional tools to monitor samples periodically and from targeted/strategic locations to actively manipulate the process to maintain a desired state. In the PAT framework, validation can be demonstrated through this continuous quality assurance, where a process is continuously monitored, evaluated, and adjusted using validated in-process measurements, tests, controls, and process end points.

CONTROL CHARTS

Control charts are decision-making tools that provide information for timely decisions concerning production processes, environmental monitoring, continuous process monitoring for validated processes, and issue trending. Control charts are also problem-solving tools that help locate and identify the causes of poor or marginal quality.

Table 55.1 Variables and attributes control chart selection.

Type	Distribution	Sample	Application
Variables control charts			
Type	**Distribution**	**Sample**	**Application**
X and R	Normal	$2 \leq 10$	Measurement subgroups
X and s	Normal	> 10	Measurement subgroups
Attributes control charts			
Type	**Distribution**	**Sample**	**Application**
c	Poisson	Constant	Count number of defects per item
u	Poisson	Varies	Count number of defects per item
np	Binomial	Constant	Count of defective items
p	Binomial	Varies	Count of defective items
g	Binomial	Individual	Interval between rare events
Variables or attributes control charts			
Type	**Distribution**	**Sample**	**Application**
X and mR	Normal	1	Individual counts or measurements

Source: Adapted from Mark A. Durivage. 2015. *Practical Engineering, Process, and Reliability Statistics.* Milwaukee: ASQ Quality Press.

Control charts contain a centerline—usually the mathematical average of the samples plotted—upper and lower statistical control limits that define the constraints of common cause variation, and performance data plotted over time. The type of control used is dependent on the amount and type of data available. Table 55.1 provides guidance on variables and attributes control chart selection.

CONTINUOUS IMPROVEMENT AND KNOWLEDGE MANAGEMENT TOOLS

Continuous learning through data collection and analysis over the life cycle of a product is important to increase the process knowledge base. A knowledge base can be of most benefit when it consists of scientific understanding of the relevant multifactorial relationships, for example, between formulation, process, and quality attributes. The desired state of pharmaceutical process gives the ability to effect continuous improvement and continuous real-time assurance of quality—continuously optimize the process within the design space—so that each batch is a new validation batch, and each batch provides the opportunity for a better process model.

Chapter 56

Phase-Appropriate Good Manufacturing Practices (GMP) Requirements

PRODUCT LIFE CYCLE DEVELOPMENT

> Apply phase appropriate Good Manufacturing Practices (GMPs) throughout the product life cycle. (Apply)
>
> **Body of Knowledge VIII.B.1**

The United States Food and Drug Administration (FDA) published a final rule in the Federal Register on "Current Good Manufacturing Practice and Investigational New Drugs Intended for Use in Clinical Trials," effective September 15, 2008. This guidance takes into consideration the nature of phase 1 clinical studies. It recognizes that the studies are of short duration in well- controlled clinical settings with healthy subjects; therefore, it exempts most phase 1 investigational drugs and the *active pharmaceutical ingredients* (APIs) for the Investigational New Drug (IND) from compliance with 21 CFR 210 and 211. FDA is considering guidelines on investigational drugs used for phase 2 and phase 3 clinical trials; however, until they are published, 21 CFR 210 and 211 continue to apply.

In Europe, the Clinical Trials Application (CTA) requires *good manufacturing practices* (GMP) for all clinical trial phases and focus on the quality of the end product, and places emphasis on the *qualified person* (QP) for oversight of product quality.

DEVELOPMENT PHASES

> Identify recommendations and requirements in relation to phases of development including method qualification/validation, comparability protocols, and adoption of critical process parameters and specifications. (Understand)
>
> **Body of Knowledge VIII.B.2**

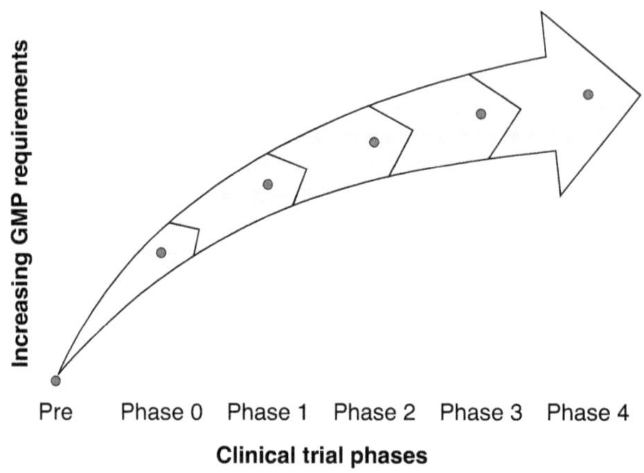

Figure 56.1 Clinical trial phases depicting increasing GMP requirements.

FDA recognizes five categories for describing the clinical trial of a drug based on the study's characteristics, such as the objective and number of participants (see Figure 56.1). The five phases are:

- *Phase 0.* Exploratory study involving very limited human exposure to the drug, with no therapeutic or diagnostic goals (for example, screening studies, microdose studies).
- *Phase 1.* A phase 1 clinical trial includes the initial introduction of an investigational new drug product, including biological drug products, into humans. It is appropriate that the process by which a drug product is manufactured in the development phase be well documented and controlled in order to assure the reproducibility of the product for further testing and for ultimate commercial production.
- Such studies are conducted to establish the basic safety of the drug, and are designed to determine the metabolism and pharmacologic actions of the drug in humans. The total number of subjects in a phase 1 clinical trial is generally limited to no more than 80 subjects. This is in contrast to phase 2 and phase 3 clinical trials when a substantially greater number of subjects are involved, more subjects are exposed to the drug product, and the effectiveness of the drug product is also tested in addition to safety. In phase 1 trials, researchers closely monitor the participants and adjust the dose a little at a time until they find the amount that works the best with acceptable side effects. This dose is usually the one used for further testing.
- *Phase 2.* Phase 2 trials look at how effective the new treatment is. Phase 2 trials are slightly larger than phase 1 trials and usually involve 25 to 100 people. Researchers start with the dose and method of giving the new treatment that were found to be best in phase 1. The phase 2 participants are given the new treatment, and the researchers watch to see if the treatment has some benefit. The benefits the researchers look for may be different depending on the goals of the research. For

example, whether the cancer tumor gets smaller or stops growing, a longer time before the cancer comes back, a longer survival time, or better quality of life. If a certain percentage of participants benefit from the treatment and the side effects are still acceptable, the new treatment will probably go on to a phase 3 trial.

- *Phase 3.* These are studies that gather more information about safety and effectiveness by studying different populations (1000 to 3000 people) and different dosages, and by using the drug in combination with other drugs. However, once an investigational drug product has been manufactured by, or for, a sponsor and is available for use in a phase 2 or phase 3 study, thus demonstrating an intent to expose more subjects to the investigational drug and requiring that the regulations' CGMP requirements be met, the same investigational drug product used in any subsequent phase 1 study by the same sponsor must be manufactured in compliance with part 211. In addition to drug products that, if eventually approved, would be approved under Section 505 of the Act (21 USC 355), this rule applies to investigational biological products that are subject to the CGMP requirements. Examples of such products include recombinant and nonrecombinant therapeutic products, vaccine products, allergenic products, in vivo diagnostics, plasma derivative products, blood and blood products, gene therapy products, and somatic cellular therapy products (including xenotransplantation products) that are subject to the CGMP requirements.
- *Phase 4.* Phase 4 trials usually look at whether the treatment offers benefits or produces long-term side effects that weren't studied or seen in the phase 2 or phase 3 trials. Phase 4 trials usually are done after a treatment has been approved for use by FDA. Phase 4 trials are less common than phase 1, 2, or 3 trials and may involve hundreds of thousands of people. These include post-market requirement and commitment studies that are required of or agreed to by the sponsor. These studies gather additional information about a drug's safety, efficacy, or both. For example, one phase 4 trial is interested in participation from women diagnosed with breast cancer who are taking an aromatase inhibitor. Aromatase inhibitors are known to cause joint pain and bone loss. This phase 4 trial is looking at whether the joint pain some women have while taking an aromatase inhibitor is linked to more defects in their cartilage compared to women not taking an aromatase inhibitor.

COMBINATION PRODUCTS

> Identify Good Manufacturing Practices (GMP) requirements and various studies required for combination drug-device or drug-delivery products. (Understand)
>
> **Body of Knowledge VIII.B.3**

Under 21 CFR Part 3, a *combination product* is a product composed of any combination of a drug, device, or biological product. The drugs, devices, and biological products included in combination products are referred to as "constituent parts" of the combination product.

Under 21 CFR 3.2(e), a combination product includes:

- A product comprised of two or more regulated components, that is, drug/device, biologic/device, drug/biologic, or drug/device/biologic, that are physically, chemically, or otherwise combined or mixed and produced as a single entity (a "single entity" combination product, such as a prefilled syringe or drug-eluting stent)
- Two or more separate products packaged together in a single package or as a unit and comprised of drug and device products, device and biological products, or biological and drug products (a "co-packaged" combination product, such as a surgical or first-aid kit)
- A drug, device, or biological product packaged separately that according to its investigational plan or proposed labeling is intended for use only with an approved, individually specified drug, device, or biological product where both are required to achieve the intended use, indication, or effect and where upon approval of the proposed product the labeling of the approved product would need to be changed, for example, to reflect a change in intended use, dosage form, strength, route of administration, or significant change in dose (a "cross-labeled" combination product, as might be the case for a light-emitting device and a light-activated drug)
- Any investigational drug, device, or biological product packaged separately that according to its proposed labeling is for use only with another individually specified investigational drug, device, or biological product where both are required to achieve the intended use, indication, or effect (another type of cross-labeled combination product)

As a combination product, the constituent parts of a combination product retain their regulatory status (as a drug or device, for example) after they are combined.

The final rule on cGMP requirements for combination products applies to all combination products. The cGMP requirements for constituent parts of cross-labeled combination products that are manufactured separately and not co- packaged are the same as those that would apply if these constituent parts were not part of a combination product (for example, for a drug/device combination product, 21 CFR parts 210 and 211 would apply to the manufacture of the drug constituent part[s] of the cross-labeled combination product, and 21 CFR part 820 would apply to the device constituent part[s]).

If a streamlined approach is used, the manufacturer must demonstrate compliance with all of the relevant provisions of either the drug cGMPs or quality system regulation, and the provisions specified in 21 CFR 4.4(b) for the other set(s) of cGMP requirements applicable to its product.

CLINICAL TRIALS MATERIAL

> Describe and apply requirements for production and packaging of clinical trials material and investigational medicinal products (IMPs). (Apply)
>
> **Body of Knowledge VIII.B.4**

A clinical study involves research using human volunteers (also called *participants*) that is intended to add to medical knowledge. There are two main types of clinical studies: clinical trials (also called *interventional studies*) and observational studies.

In a *clinical trial*, participants receive specific interventions according to the research plan or protocol created by the investigators. These interventions may be medical products, such as drugs or devices; procedures; or changes to participants' behavior, such as diet. Clinical trials may compare a new medical approach to a standard one that is already available, to a placebo that contains no active ingredients, or to no intervention. Some clinical trials compare interventions that are already available to each other. When a new product or approach is being studied, it is not usually known whether it will be helpful, harmful, or no different than available alternatives (including no intervention). The investigators try to determine the safety and efficacy of the intervention by measuring certain outcomes in the participants. For example, investigators may give a drug or treatment to participants who have high blood pressure to see whether their blood pressure decreases.

OBSERVATIONAL STUDIES

In an *observational study*, investigators assess health outcomes in groups of participants according to a research plan or protocol. Participants may receive interventions (which can include medical products such as drugs or devices) or procedures as part of their routine medical care, but participants are not assigned to specific interventions by the investigator (as in a clinical trial). For example, investigators may observe a group of older adults to learn more about the effects of different lifestyles on cardiac health.

PARTICIPATING IN A CLINICAL STUDY

Every clinical study is led by a principal investigator, who is often a medical doctor. Clinical studies also have a research team that may include doctors, nurses, social workers, and other healthcare professionals. Clinical studies can be sponsored, or funded, by pharmaceutical companies, academic medical centers, voluntary groups, and other organizations such as National Institutes of Health.

Clinical studies can take place in many locations, including hospitals, universities, doctors' offices, and community clinics. The location depends on who is conducting the study. The length of a clinical study varies depending on what is being studied. Participants are told how long the study will last before they enroll.

INFORMED CONSENT

Informed consent is a process used by researchers to provide potential and enrolled participants with information about a clinical study. This information helps people decide whether they want to enroll or continue to participate in the study. The informed consent process is intended to protect participants and should provide enough information for a person to understand the risks of, potential benefits of, and alternatives to the study.

INSTITUTIONAL REVIEW BOARDS

Each federally supported or conducted clinical study and each study of a drug, biological product, or medical device regulated by FDA must be reviewed, approved, and monitored by an *institutional review board* (IRB). An IRB is made up of doctors, researchers, and members of the community. Its role is to make sure that the study is ethical and that the rights and welfare of participants are protected. The IRB also reviews the informed consent document.

In addition to being monitored by an IRB, some clinical studies are also monitored by data monitoring committees (also called *data safety and monitoring boards*).

INVESTIGATIONAL PRODUCTS

21 CFR Part 4 does not alter the scope of the underlying cGMP regulations for drugs, devices, biological products, and human cellular and tissue products (HCT/Ps). In particular, Part 4 does not alter the applicability of these cGMP regulations to investigational products.

Under 21 CFR 812.1, investigational devices are exempt from Part 820 except for design control requirements under 21 CFR 820.30. This exemption also applies to investigational combination products that include an investigational device constituent part.

Measures the sponsor might take to ensure cGMP compliance at all manufacturing facilities for the combination product include auditing and other oversight activities. For example, when multiple facilities participate in the manufacturing process, the sponsor may facilitate cGMP compliance by coordinating interactions among the facilities. A sponsor can enter into comprehensive quality agreements with the various facilities and suppliers involved and coordinate development of such agreements among these entities. These quality agreements may, for instance, specify expectations as to which facility will perform what activities and develop and maintain the documentation needed to demonstrate compliance with which cGMP requirements (based, for example, on which aspects of manufacture each facility conducts).

CHANGE CONTROL

Coordination of changes between manufacturers participating in the manufacture of a combination product is an important CGMP issue. The manufacturer of a co-packaged or single entity combination product should ensure appropriate consideration of any implications for the safety or effectiveness of its combination product that might arise from changes to it, including to any constituent part of it. Therefore, the manufacturer should establish arrangements with its suppliers to receive notice of changes.

Chapter 57

Raw Materials, Packaging, and Infrastructure for Product Development

> Select appropriate development studies for raw material selection and evaluate the results to determine their critical quality attributes. (Analyze)
>
> **Body of Knowledge VIII.C**

The United States Food and Drug Administration (FDA) provides guidance for raw materials, packaging, and infrastructure for product development within the guidance for industry *Process Analytical Technology—A Framework for Innovative Pharmaceutical Development, Manufacturing, and Quality Assurance.* The premise of this document is to ensure that the current concepts of risk management and quality systems approaches are incorporated into the design, development, and manufacture of pharmaceuticals while maintaining product quality.

Effective innovation in development, manufacturing, and quality assurance would be aided by answering several questions:

- What are the mechanisms of degradation, drug release, and absorption?
- What are the effects of product components on quality?
- What sources of variability are critical?
- How does the process manage variability?
- What are the packaging considerations?

The goal of *process analytical technology* (PAT) is to enhance understanding of and control the manufacturing process, which is consistent with current drug quality system thought: quality cannot be tested into products; it should be built in or should be by design.

Quality is built into pharmaceutical products through a comprehensive understanding of:

- The intended therapeutic objectives, patient population, route of administration, and pharmacological, toxicological, and pharmacokinetic characteristics of a drug

- The chemical, physical, and biopharmaceutic characteristics of a drug
- Design of a product and selection of product components and packaging based on drug attributes
- The design of manufacturing processes using principles of engineering, material science, and *quality assurance* (QA) to ensure acceptable and reproducible product quality and performance throughout a product's shelf life

Many tools are available that enable process understanding for scientific, risk-managed pharmaceutical development, manufacture, and QA. These tools, when used within a system, can provide effective and efficient means for acquiring information to facilitate process understanding, continuous improvement, and development of risk mitigation strategies. In the PAT framework, these tools can be categorized as:

- Multivariate tools for design, data acquisition, and analysis
- Process analyzers
- Process control
- Continuous improvement and knowledge management

Multivariate tools for design, data acquisition, and analysis from a physical, chemical, or biological perspective, and pharmaceutical products and processes, are complex multifactorial systems. Many development strategies can be used to identify optimal formulations and processes. The knowledge acquired in these development programs is the foundation for product and process design. Methodological experiments based on statistical principles of orthogonality, reference distribution, and randomization provide effective means for identifying and studying the effect and interaction of product and process variables. Traditional one-factor-at-a-time (OFAT) experiments do not address interactions between product and process variables. When used appropriately, these tools enable the identification and evaluation of product and process variables that may be critical to product quality and performance. The tools may identify potential failure modes and mechanisms and quantify their effects on product quality.

Some process analyzers provide nondestructive measurements that contain information related to biologic, physical, and chemical attributes of the materials being processed. These measurements can be:

- *At-line,* where the sample is removed and analyzed close to the process stream;
- *Online,* where the sample is diverted from the manufacturing process to an analyzer, and possibly returned to the stream;
- *In-line,* which may be an invasive or noninvasive process that analyzes the sample while it is part of the process stream.

Process control tools used for the design and optimization of drug formulations and manufacturing processes within the PAT framework can include the following steps, although the sequence of steps can vary:

- Identify and measure critical material and process attributes relating to product quality.
- Design a process measurement system to allow real-time or near real-time (for example, on-, in-, or at-line) monitoring of all critical attributes.
- Design process controls that provide adjustments to ensure control of all critical attributes.
- Develop mathematical relationships between product quality attributes and measurements of critical material and process attributes.

Continuous improvement and knowledge management through data collection and analysis over the life cycle of a product are important. These data can contribute to justifying proposals for post-approval changes. Approaches and information technology systems that support knowledge acquisition from such databases are valuable for the manufacturers and can facilitate scientific communication.

Guidance is provided with regard to acceptance criteria that should be established for all new drug substances and new drug products, that is, universal acceptance criteria, as well as those that are considered specific to individual drug substances and /or dosage forms.

Generally, proposals to implement these concepts should be justified by the applicant and approved by the appropriate regulatory authority before being put into effect.

- Periodic or skip testing is the performance of specified tests at release on pre-selected batches and /or at predetermined intervals, rather than on a batch-to-batch basis with the understanding that those batches not being tested still must meet all acceptance criteria established for that product. When tested, any failure to meet acceptance criteria established for the periodic test should be handled by proper notification of the appropriate regulatory authority(ies). If these data demonstrate a need to restore routine testing, then batch by batch release testing should be reinstated.
- The concept of different acceptance criteria for release versus shelf-life specifications applies to drug products only; it pertains to the establishment of more restrictive criteria for the release of a drug product than are applied to the shelf-life. In Japan and the United States, this concept may only be applicable to in-house criteria, and not to the regulatory release criteria. In the European Union (EU) there is a regulatory requirement for distinct specifications for release and for shelf-life where different.
- In-process tests are tests which may be performed during the manufacture of either the drug substance or drug product, rather than as part of the formal battery of tests which are conducted prior to release. In-process tests that are only used for the purpose of adjusting process parameters within an operating range. For example, hardness and friability of tablet cores that will be coated and individual tablet

weights, are not included in the specification. However, this approach should be validated to show that test results or product performance characteristics do not change from the in-process stage to finished product.

- The experience and data accumulated during the development of a new drug substance or product should form the basis for the setting of specifications. Some examples are microbiological testing for drug substances and solid dosage forms, extractables from product containers, and particle size testing. These tests may be performed as an in-process test. Dissolution testing for immediate release solid oral drug products made from highly water-soluble drug substances may be replaced by disintegration testing.

It is recognized that only a limited amount of data may be available at the time of filing, which can influence the process of setting acceptance criteria. As a result it may be necessary to propose revised acceptance criteria as additional experience is gained with the manufacture of a particular drug substance or drug product (example: acceptance limits for a specific impurity). The basis for the acceptance criteria at the time of filing should necessarily focus on safety and efficacy.

Chapter 58

New Product Development Studies and Reports

Analyze studies and reports including stability reports, material compatibility, method development, and development reports to support product development and submissions. (Evaluate)

Body of Knowledge VIII.D

Drug development is the process of taking a new chemical lead through the stages necessary to allow it to be tested in human clinical trials, and of identifying the process boundaries for realization of safe and effective products. Drug discovery is the first stage of drug development. At this stage, *new chemical entities* (NCEs) that have promising activity against a particular biologic target important in disease are discovered.

After discovery, the safety, toxicity, pharmacokinetics, and metabolism in humans, including dose and schedule, are assessed before human clinical trials. Studies are conducted to determine the major toxicities of the NCE, including an assessment of major organ toxicity (effects on the heart and lungs, brain, kidney, liver, and digestive system), and effects on other parts of the body that might be affected by the drug (for example, the skin if the new drug is to be delivered through the skin). These studies can be made either in vitro or by using experimental animals. The physicochemical properties of the NCE are established (that is, the chemical makeup, stability, and solubility, determination of the starting materials and their reaction paths, and basic specifications). The process to make the chemical is developed for suitability (for example, as capsules, tablets, aerosol, intramuscular injectable, subcutaneous injectable, or intravenous formulations). With the process of gathering human clinical trials data, long-term or chronic toxicities are determined, as well as effects on systems not previously monitored (that is, fertility, reproduction, immune system).

The process for realization of safe and effective products is identified after the discovery phase, and confirmed with the clinical study. The process may have started as small-scale experiments, and may have been increasingly scaled-up, with appropriate *quality control* (QC) and manufacturing methods identified to ensure consistent product quality. The finalized process is summarized in a comprehensive development report, which should be approved by high-level company management. Technology transfer and other pre-validation activities should be considered part of the overall development report. The process of transfer includes verification of the parameters, controls, and associated supporting systems through formal studies, with implementation of any new required upgrades to ensure consistent quality product before moving into the process validation. The development report should include historical data, dosage form and formulation designs, design of manufacturing methods, specification and test method, and other information:

- *Historical data.* Information on the pharmaceutical development of new drug substances and drug products at stages from early development phase to final steps before process validation.
- *Dosage form and formula designs; design of manufacturing methods.* Successes and failures that contributed to learning more about the process, change histories of important processes and control parameters, and quality profiles of batches (including stability data).
- *Specifications and test methods.* Drug substances, intermediates, drug products, raw materials, and components, and their rationale (validity of specification range of critical tests such as assay, impurities, and dissolution; rationale for selection of test methods, reagents, and columns; and traceability of raw data).
- *Laboratory method development and validation.* Testing methods (for example, chemistry and microbial) are developed and validated before the human clinical trials, including, if applicable, successful transfer to the testing site.
- *Process model.* Brief description of the process, typically accompanied by a process flow diagram; each major step and equipment in the process are indicated, and, ideally, the batch record with all appropriate procedures and specifications should be finalized when the development report is finalized.
- *Active pharmaceutical* ingredient *(API),* raw *materials, and components.* APIs must be appropriately characterized, validated (manufacturing process), and identified as to source, and must satisfy quality specifications for manufacturing of drug product; critical excipients must be defined and well controlled.
- *Critical process* parameter *(CPP).* CPPs are the measured variables that are known to have an effect on one or more product quality attributes; all CPPs must be identified and controls established within a defined acceptable range. The ranges for these parameters may be typically obtained in later development studies (that is, full-scale engineering studies, technology transfer studies aimed to optimize the process for commercialization, conducted in parallel with the human clinical trials); process optimization efforts must ensure that the enhanced process remains equivalent to that used for the making of the clinical trials product.

- *Critical quality* attribute *(CQA)*. CQAs are intrinsic quality characteristics that are desired or needed to ensure patient safety and benefit of the drug product; in ICH Q6A, some API attributes should be considered critical, regardless of the drug product end use. Identification, physicochemical properties, appearance, assay, and purity are applicable to all drug products; particle size, microbial purity, and polymorphism depend on the drug product. ICH Q6A decision trees can be used to determine the criticality of these quality attributes for solids, solutions, or sterile products. All CQAs must be identified and acceptance criteria established.

REPORTS

Method development is a crucial process that supports product development, particularly in industries such as pharmaceuticals, chemistry, and materials science. It involves the design, optimization, and validation of methods and procedures for various aspects of product development, including quality control, testing, and analysis. Key steps for method development to support product development include:

- *Define objectives and scope.* Clearly define the objectives of method development, such as what specific parameters or attributes you need to measure or analyze. Determine the scope, including the types of samples, instruments, and techniques to be used.
- *Literature review.* Conduct a comprehensive literature review to identify existing methods or approaches that may be applicable to your product or analysis needs. Analyze the strengths and weaknesses of existing methods and consider how they align with your objectives.
- *Method selection.* Decide whether to adapt an existing method, modify it, or develop a completely new method. Consider factors such as accuracy, precision, sensitivity, and specificity when selecting or designing the method.
- *Validation studies.* Validate the method to ensure it meets predefined criteria for accuracy, precision, specificity, linearity, and robustness. Document validation studies thoroughly, including protocols, data, and results.
- *Method documentation.* Prepare a detailed method development report that includes a step-by-step description of the method, instrumentation used, and validation procedures. Provide guidelines for sample preparation, data analysis, and troubleshooting.
- *Quality control (QC) procedures.* Develop QC procedures to ensure the ongoing reliability of the method. Establish control charts, reference standards, and periodic system suitability tests.
- *Method transfer and scaling.* If applicable, develop procedures for transferring the method from the development phase to production

or other laboratories. Consider scalability and adaptability to different equipment or environments.

- *Training and implementation.* Train personnel on the developed method, ensuring they understand the procedures and quality standards. Implement the method in the relevant areas of product development or quality control. Ensure all documentation related to method development and validation is compliant with regulatory requirements if applicable.

Method development is a dynamic process that requires careful planning, experimentation, and documentation. It plays a pivotal role in ensuring the quality, safety, and consistency of products during their development and throughout their life cycle. Effective method development contributes to informed decision-making and successful product development.

Stability reports are a critical component of the *new product development* (NPD) process, especially for industries like pharmaceuticals and chemicals where product stability is paramount. These reports document the product's physical, chemical, and microbiological stability over time, ensuring that it meets quality and safety standards throughout its shelf life. Some of the key elements typically included in stability reports for new product development are:

- *Product description.* Begin with a detailed description of the new product, including its intended use, composition, and any unique characteristics.

- *Stability testing protocol.* Define the stability testing protocol, including the storage conditions (for example, temperature, humidity, light), testing intervals (for example, monthly, quarterly), and duration (example, six months, 12 months, 24 months).

- *Sample selection.* Explain how representative samples are selected for stability testing. These samples should accurately reflect the product's formulation, packaging, and manufacturing process.

- *Storage conditions.* Specify the storage conditions to which the samples will be subjected. These conditions should mimic the expected storage environment for the product.

- *Test methods.* Describe the analytical methods used to assess stability, including physical, chemical, and microbiological testing. These methods should be validated and well-documented.

- *Testing parameters.* Outline the specific parameters and attributes being monitored. This may include color, odor, pH, viscosity, chemical composition, potency, and microbiological counts.

- *Acceptance criteria.* Define the criteria for determining product stability. What changes in attributes or parameters are considered acceptable or unacceptable? This may involve setting specific limits or ranges for each attribute.

- *Testing schedule.* Provide a schedule of when stability testing will be conducted, and specify the frequency of testing at different time points (for example, 0 months, 3 months, 6 months).
- *Sample handling and storage.* Detail how samples are handled, labeled, and stored throughout the testing period. Ensure that storage conditions are consistently maintained.
- *Data collection.* Document the results of stability testing at each time point. Include any deviations from the acceptance criteria and corrective actions taken, if necessary.
- *Data analysis.* Analyze the stability data to assess trends, degradation kinetics, and potential factors influencing stability.
- *Summary of findings.* Summarize the key findings of the stability testing, including any changes observed in product attributes, and provide an overall assessment of product stability.
- *Recommendations.* Based on the stability data, make recommendations regarding the product's shelf life, storage instructions, and any necessary formulation or packaging adjustments.
- *Conclusion.* Conclude the stability report with a clear statement about whether the product meets stability requirements and is suitable for commercialization.

Stability reports are crucial not only for ensuring product quality and safety but also for regulatory compliance. They provide a robust foundation for decision-making throughout the NPD process and help companies make informed choices about product formulation, packaging, and storage requirements.

DEVELOPMENT REPORTS

Development reports are essential documents that support and document various aspects of the product development process. These reports provide valuable information, data, and insights to guide decision-making and ensure that the development process is efficient and effective. Some examples of reports can include:

- Project kick-off report. This report marks the start of the development process. It outlines the project's goals, objectives, scope, and initial timelines. It may also include an overview of the team members, roles, and responsibilities.
- *Requirements analysis report.* This report documents the detailed requirements for the new product. It includes functional, technical, and design specifications, often with input from stakeholders and end-users.
- *Design and prototyping report.* Documents the design process and prototypes created during development. It includes design decisions, iterations, and rationale.

- *Risk assessment and mitigation report.* Identifies potential risks associated with the development process and outlines strategies to mitigate them. This report helps in proactive risk management.
- *Progress toward milestones and deliverables.* Periodic reports that track progress toward key project milestones and deliverables, ensuring that the project stays on schedule.
- *Lessons learned report.* After the project's completion, this report reflects on what went well, what could have been improved, and key takeaways for future development efforts.

These development reports play a critical role in keeping stakeholders informed, managing project risks, ensuring product quality, and documenting the evolution of the product from concept to market. They also serve as valuable references for post-development analysis and continuous improvement.

Chapter 59

Scale-up and Transfer Activities

DEVELOPMENT AND VALIDATION PRINCIPLES

> Identify and distinguish between development and validation studies. (Understand)
>
> **Body of Knowledge VIII.E.1**

According to International Conference on Harmonization (ICH) Q8 (R2 *Pharmaceutical Development*), drug development studies contribute to scientific understanding that supports the establishment of product specifications, manufacturing *critical process parameters* (CPPs), and in-process *critical quality attributes* (CQAs). ICH Q8 (R2) provides guidance for the development studies that would be included in dossiers or New Drug Application (NDA) submissions to competent authorities. It describes a concept, known as *design space*, whereby and within which changes to a process may proceed with no regulatory resubmission of a dossier or NDA. Design space may be applicable to scale-up activities or to site changes.

Development studies and reports support specifications for the excipients, *active pharmaceutical ingredients* (APIs), intermediates, and drug products, and manufacturing process parameters. Process validation exercises and reports confirm the manufacturability at commercial scale of the drug substance or drug product in conjunction with transfer activities. Although the United States Food and Drug Administration (FDA) drug product submission process permits regulatory submissions before executing process validation, FDA expects that process validation will occur and be internally approved before any product is released to the market after an NDA or Biological License Application (BLA) approval.

The distinction between development studies and validation studies (or exercises) is that development studies may, in fact, result in unexpected negative results, disproven hypotheses, or novel responses, hence, a new shape of the design space. Those development studies may be performed on any scale of equipment from laboratory, to pilot, to commercial scale of simulated manufacturing. Validation exercises, by definition, occur on commercial-scale equipment. Executed process validation protocols must neither subtract from nor add to anything in preapproved/signed-off protocol.

TECHNOLOGY TRANSFER TYPES

> Define different types of technology transfer including manufacturing site change and analytical laboratory site change. Analyze inter-site comparison of results. (Analyze)
>
> **Body of Knowledge VIII.E.2**

The only official regulatory guidance mentioning the term *technology transfer* is ICH Q10 *Pharmaceutical Quality System,* and then only in relation to manufacturing processes and not analytical methods. Within a product's manufacturing life cycle, there exists drug development, technology transfer, commercial manufacturing, and product discontinuation. In Section 3.2.1, Table 1, technology transfer is described: "Monitoring during scale-up activities can provide a preliminary indication of process performance and the successful integration into manufacturing. Knowledge obtained during transfer and scale-up activities can be useful in further developing the control strategy."

For manufacturing, *technology transfer* serves as progressive scale-up at the final manufacturing site and by the actual personnel of that site. Included in the process transfer are relevant packaging and cleaning (residue) validation activities. No competent authority regulation mandates that *research and development* (R&D) or the transferring site assist in this transfer, but such collaboration improves the early success of *conformance batches,* as FDA terms commercial-scale pre-validation batches. Industry generally refers to these batches as *demonstration batches.*

Analytical technology transfer pertains to the transfer of analytical methods between laboratories for the testing of raw materials, finished product, or cleaning validation samples. It is a final location-specific exercise to acclimate the *quality control* (QC) personnel at the transfer receiving site in the applicable assays. Protocols are often written that contain statistical acceptance criteria for similarity of assay results between the transferring laboratory and the receiving laboratory (by blinded samples). The analytical test means and standard deviations must usually be compared or combined to arrive at an acceptably low difference (of means) and tight precision. A formal report is typically prepared to summarize the transfer activities. Although no competent authority regulation defines universally applicable or expected statistics, The United States Pharmacopeia (USP) Chapter <1224> is concerned with general requirements for the transfer of validated analytical methods.

SUCCESSFUL TECHNOLOGY TRANSFER

> Define various studies including ranging, capability, in-process control, hold times, and shipping to ensure successful transfer between development and commercial processes. (Evaluate)
>
> **Body of Knowledge VIII.E.3**

To enable a successful manufacturing process transition and scale-up between R&D bench scale, R&D pilot scale, and commercial scale, numerous process development studies are undertaken. The term *unit operations*, coined by Arthur D. Little in 1915, is generally used to refer to distinct physical changes or unit actions (for example, pulverizing, mixing, and drying); unit operations involving chemical changes are sometimes referred to as *unit processes*. The physical changes comprising unit operations primarily involve contact, transfer of a physical property, and separation between phases or streams.

These studies do not require preapproved protocols as with formal validation; however, for process development study results to be able to contribute to a development report either included in a marketing application (for example, NDA, BLA) or available for competent authority inspection review, its raw data must adhere to record-keeping practices as rigorous as contained in 21 CFR 58, "Good Laboratory Practices."

Key Elements of Technology Transfer

The change management system should provide management and documentation of adjustments made to the process during technology transfer activities.

Key elements of Technology transfer should include:
Documentation/ Information

- The documentation should be consistent and there should be controlled procedures for the technology transfer, including an understanding of prior knowledge from similar products.

Personnel

- Personnel roles and responsibilities should be clearly defined, including the cross-functional experts.

Technology Evaluation and Development

- Utilize principles in ICH Q8, Q9, and Q10
- Assure that there are robust processes and corresponding analytical methods and that they are well-understood.
- Include Uni/Multivarient Design of experiments
- Identification/verification of CPPs and CQA including important parameters.

Execution

- This includes the successful manufacture of demonstration batches
- Understand the validation requirements and strategies
- Continuous monitoring such as (PAT and proactive process analysis)

Monitoring during scale-up activities can provide a preliminary indication of process performance and the successful integration into manufacturing. Knowledge obtained during transfer and scale up activities can be useful in further developing the control strategy.*

Process Ranging Studies

Process ranging studies are experiments that, to achieve known CQAs of the intermediate or finished drug (or even the API and its precursors), establish allowable values of CPPs, which are equipment controls or controllable environmental conditions that influence in-process or finished product (or API) test results. In the language of statistical *design of experiments* (DOE), "factors" are the CPPs (for example, rotary tablet press speed), and "levels" are any allowable settings of those equipment controls or factors (for example, 100 rpm). Statistics books and readily purchasable software provide many choices in DOE methodology. DOE is the alternative to highly inefficient one-factor-at-a-time (OFAT) experimentation. DOE indicates which factors have the most impact on resultant CQAs and specifies how much variation around a nominal (target) value of a CPP is allowable. Any software packages used during process development studies contributing to a formal development report must be validated by the software provider or the drug or API manufacturer. See FDA guidance *General Principles of Software Validation; Final Guidance for Industry and FDA Staff.*

Process Capability Studies

Process capability studies are experiments intended to determine, for a particular process, how close to target specification a product output is running, as well as how widely dispersed that output is. The *process capability index* is a measurement of the process's ability to create product within specification limits. Figure 59.1 is a graphical representation of process capability ratios compared to the statistical control limits and part specification limits.

In-Process Control Studies

In-process control (IPC) studies are experiments that determine and develop online or off-line measurement(s) that help predict acceptable API or finished drug release test results and enable tighter control over an adjustable process. Regardless of the dosage form (or API), IPC measurements, tests/inspections, or go/ no-go attribute test results can be plotted/trended/interpreted via run charts or *statistical process control* (SPC) charts. In SPC, control limits for *central tendency* (that is, means or medians) and for *variance* (that is, range or standard deviation) can be calculated manually or using validated software. The main purpose of SPC is the use of unbiased statistics to decide when and when not to adjust a process parameter to maintain process control. Individual results or means of three or more consecutive points (subgroups) may be plotted versus time or versus lot number. A loss

* ICH Q10.

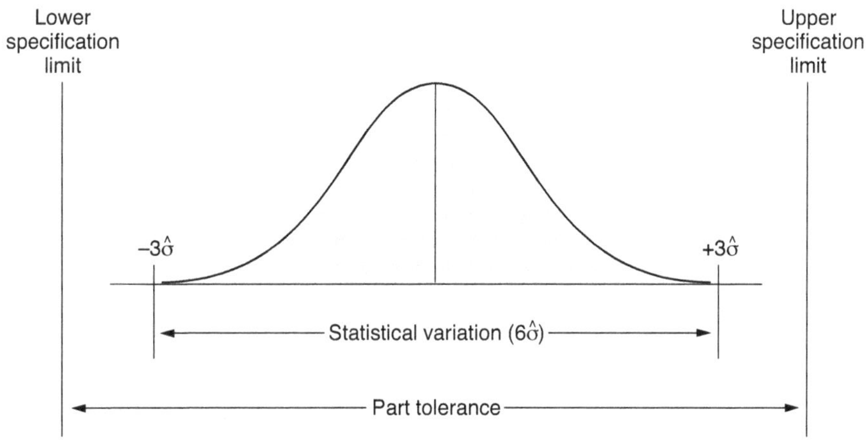

Short-term capability	**Desired ratio**
$C_r = \frac{6s}{USL-LSL}$	< 1
$C_p = \frac{USL-LSL}{6s}$	> 1
$C_{pk} = \text{Minimum of } \frac{\bar{X}-LSL}{3s}, \frac{USL-\bar{X}}{3s}$	> 1
Long-term capability	**Desired ratio**
$P_r = \frac{6\sigma}{USL-LSL}$	< 1
$P_p = \frac{USL-LSL}{6\sigma}$	> 1
$P_{pk} = \text{Minimum of } \frac{\bar{X}-LSL}{3\sigma}, \frac{USL-\bar{X}}{3\sigma}$	> 1

Figure 59.1 Relationship between statistical control limits and product specifications.
Source: Adapted from Mark A. Durivage. 2015. *Practical Engineering, Process, and Reliability Statistics.* Milwaukee: ASQ Quality Press.

of statistical control (and a need for process adjustment) is indicated on a control chart by several factors (Figure 59.2).

Hold-Time Studies

As the name implies, a study must exist for every time lapse (that is, hold time) anticipated or intended to occur between any two API and drug manufacturing process steps. Tests used to judge the quality of the material in question after any hold time must be scientifically reliable and, ideally (but not necessarily), validated. Product and process design risk identification is expected to identify all hold times in API and drug manufacturing processes, and those risks assessed through hold-time studies. Cleaning processes have hold times that must be characterized and the maximum duration established. Bulk hold time represents the

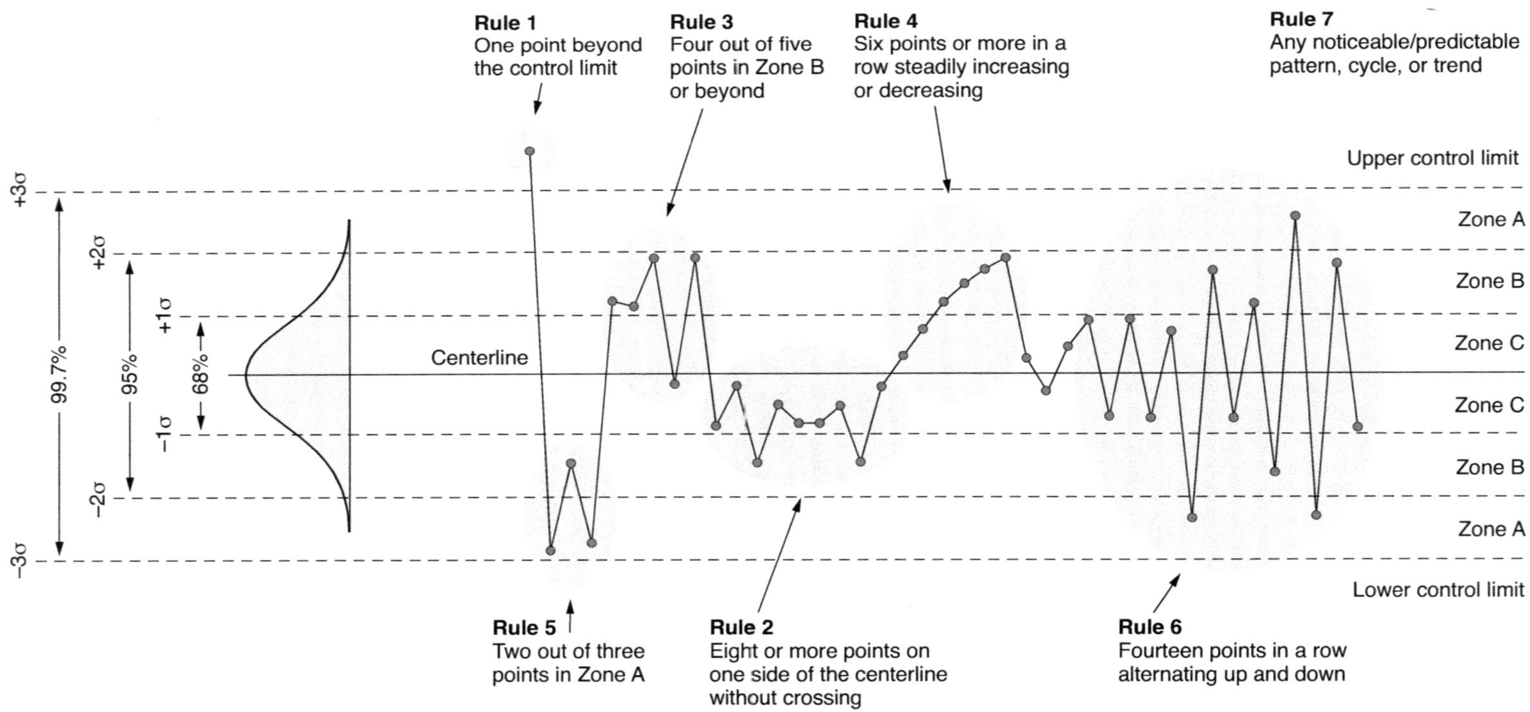

Figure 59.2 Control chart interpretation rules.

Source: Adapted from Stephen A. Wise and Douglas C. Fair. 1997. *Innovative Control Charting: Practical SPC Solutions for Today's Manufacturing Environment.* Milwaukee: ASQ Quality Press. Used with permission.

time lapse between completion of most of an API or drug manufacturing process and the primary packaging. For materials consisting of an emulsion, a suspension, or other multiple-phase combination, hold-time studies are most critical. Hold times are critical for cold-chain products (for example, proteins, vaccines, sensitive antibiotics) that undergo a primary or secondary packaging operation at ambient temperature.

Physicomechanical Simulations and Shipping Studies

In addition to ICH and any competent authority–mandated stability (and photostability) studies minimally required to be approved to ship and market a new drug product through and into various climatic zones, several studies are usually expected or mandated: physicomechanical simulations of vibration, shaking, package impact, and pressure changes, and shipping studies. The best reference for these physicomechanical and shipping studies is USP General Chapter <1079> "Good Storage and Shipping Practices." According to <1079>, a drug can take any number of routes to the ultimate consumer or patient, which may include a complex series of transfers outside the marketing owner's control.

It is usually expected in shipping studies for APIs and finished drugs that within the secondary or tertiary packaging will be judiciously placed (and calibrated) temperature and other indicators that may be inspected or downloaded at the receiving site. The API or drug product must undergo complete release testing at its destination. APIs and drug products with temperature sensitivities reflected in the labeling must use temperature cycling and shipping studies that greatly stress the insulating or refrigerating properties of the tertiary package. PDA Technical Report 39 provides strategies for cold-chain dominated shipping studies. Stability, physicomechanical, shipping, and cold-chain studies enable the product development team to conclude what excursions in labeled storage conditions are permitted and which necessitate material destruction or return. PDA Technical Report 53 provides guidance on the requirements for stability testing of new drug products to support distribution conditions.

Appendix A
At-a-Glance Comparisons

Table A.1 US, EU/PIC-S, Canadian, and WHO GMPs.

Item	US cGMP	EU/PIC-S GMP	Canada	WHO
Subpart A—General Provisions				
Scope	211.1	Introduction 1.1, 1.2, 1.3	Introduction Division 1A and 2 of Part C	General Consideration
Minimum requirements for preparation of drug products for humans or animals	211.1a			
Correspondence with other FDA regulations for biological drugs and for HCT/Ps	211.1b			
Exclude OTC products if the products and all their ingredients are ordinarily marketed and consumed as human foods	211.1c			
Definitions	211.3	Glossary	Glossary Appendix B of GUI-0001	Glossary
Subpart B—Organization and Personnel				
Responsibilities of Quality Control Unit (QCU)	211.22			
Approval/rejection of components, materials, products; review of batch records; and products made by contractors	211.22a	1.4, 1.5, 1.8, 1.9, 2.4, 2.6, 2.7, 6.1, 6.5, 6.6, 7.5	C.02.013, C.02.014, C.02.015	9.6, 9.8, 9.10, 17.4
Adequate laboratory facilities	211.22b	1.3, 1.8, 1.9, 3.26 to 3.29, 6.5, 6.6	C.02.013, C.02.015	17.3
Approval/rejection of procedures and specifications	211.22c	1.4, 1.9, 2.6	C.02.015	9.10

Source: José Rodríguez-Pérez. 2014. *The FDA and Worldwide Current Good Manufacturing Practices and Quality System Requirements Guidebook for Finished Pharmaceuticals*. Milwaukee: ASQ Quality Press.

Continued

Table A.1 *Continued.*

Item	US cGMP	EU/PIC-S GMP	Canada	WHO
Written procedures and responsibilities of QCU	211.22d	1.4, 1.7, 1.9, 2.3, 2.4, 2.6, 4.24, 6.2	C.02.013, C.02.015	9.10
Personnel Qualification	211.25			
Education, training, and experience of employees in assigned task and in GMP	211.25a	1.8, 2.1, 2.8 to 2.12	C.02.006	9.2, 9.4, 10.1 to 10.5
Education, training, and experience of supervisory personnel	211.25b	1.8, 1.9, 2.1, 2.3 to 2.12	C.02.006	9.2, 9.4, 9.7, 10.1 to 10.5
Sufficient number of personnel	211.25c	1.5, 1.8, 1.9, 2.1, 6.1	C.02.006	9.2
Personnel Responsibilities	211.28			
Clean clothing and protective apparel	211.28a	2.16	C.02.008	11.6, 11.8
Good sanitation and health habits	211.28b	2.13 to 2.20	C.02.008	9.4, 11.1 to 11.3, 11.5, 11.7
Access to limited-access areas	211.28c	3.5, 3.21, 5.16, 6.4	C.02.004, C.02.011, C.02.013	9.5, 16.7
Report to supervisors health conditions that may affect products being manufactured	211.28d	2.15	C.02.008	11.4
Consultants	211.34	2.11, 7, 9.2	C.02.006, C.02.024	7, 8.3, 10.6
Subpart C—Buildings and Facilities				
Design and Construction Features	211.42			
Suitable size, construction, and location	211.42 (a)	3.8, 3.9, F.3	C.02.004	12.1, 12.4, 12.5
Adequate space and flow of operations designed to prevent contamination	211.42 (b)	3.7, 3.8, F.3	C.02.004	12.2, 12.10, 12.25, 12.26

Continued

Table A.1 *Continued.*

Item	US cGMP	EU/PIC-S GMP	Canada	WHO
Adequate and defined areas to prevent contamination	211.42 (c)	3.6, 3.13, 3.22, 3.23, 3.26, 3.33	C.02.004	12.14, 12.17, 12.19, 12.22 to 12.24, 12.33
Penicillin operations maintained separately	211.42 (d)	3.6	C.02.004	12.24
Lighting—adequate lighting	211.44	3.3, 3.16	C.02.004	12.8, 12.32
Ventilation, Air Filtration, Air Heating and Cooling	211.46		C.02.004	
Adequate ventilation	211.46 (a)	3.3, 3.12	C.02.004	12.8, 12.30
Adequate environmental controls	211.46 (b)	3.12, 3.14	C.02.004	12.30, 16.16
Air filtration systems	211.46 (c)	3.12	C.02.004	12.30
Separate air-handling system for penicillin	211.46 (d)	3.6	C.02.004	12.24
Plumbing	211.48			
Potable water	211.48 (a)	3.10	C.02.005, C.02.008, C.02.029	12.28, 14.6, Annex 3 of TRS 929
Adequate drain system	211.48 (b)	3.11	C.02.004	12.29
Sewage and Refuse	211.50	No	C.02.007	14.44 to 14.45
Washing and Toilet Facilities (Including Hot and Cold Water)	211.52	3.31	C.02.004	12.12
Sanitation	211.56			
Building maintained in a clean and sanitary condition	211.56 (a)	3.2, 3.4	C.02.004	3.1, 12.7
Written procedures for sanitation and cleaning, including sufficient details	211.56 (b)	3.2, 3.43, 4.29	C.02.007	14.44 to 14.45, 15.46, 15.48

Continued

Table A.1 *Continued.*

Item	US cGMP	EU/PIC-S GMP	Canada	WHO
Written procedure for pest control, cleaning, and sanitizing agents	211.56 (c)	4.29	C.02.007	12.9
Sanitation procedures apply to company employees as well as to contractors and temporary sanitation workers	211.56 (d)	No, but implicit	C.02.007	15.31, 15.48
Maintenance—Good State of Repair	211.58	3.2	C.02.004	12.6
Subpart D—Equipment				
Equipment Design, Size, and Location	211.63	3.34	C.02.005	13.1, 13.2
Equipment Construction	211.65			
Product-contact surfaces cannot interact with product	211.65 (a)	3.9, 3.39	C.02.005	13.9
Lubricant and coolant not in contact with product	211.65 (b)	3.38	C.02.005	14.3
Equipment Cleaning and Maintenance	211.67			
Equipment and utensils cleaned, maintained, and sanitized periodically	211.67 (a)	3.36, 3.37	C.02.005, C.02.007	13.6, 13.8
Detailed written procedures for cleaning and maintenance of equipment	211.67 (b)	3.36, 3.43	C.02.005, C.02.007	13.12, 16.17, 16.18, 16.22
Records kept of maintenance, cleaning, sanitation, and inspection	211.67 (c)	4.29	C.02.020	15.31, 16.2
Automatic, Mechanical, and Electronic Equipment	211.68			
Use of automatic, mechanical, or electronic equipment	211.68 (a)	3.41	C.02.005	16.23

Continued

Table A.1 *Continued.*

Item	US cGMP	EU/PIC-S GMP	Canada	WHO
Control of computerized systems	211.68 (b)	Annex 11	C.02.005	15.9
Verification of automated equipment performance	211.68 (c)	5.33	C.02.005, C.02.020	4.11
Filters—Use of Filter in Manufacturing	211.72	Annex 1 84–87	No	Annex 6 of TRS 902, 7.6 to 7.9
Subpart E—Control of Components and Drug Product Containers and Closures				
General Requirements	211.80			
Detailed written procedures for handling materials	211.80 (a)	5.2	C.02.011, C.02.013	14.5, 16.2
Handling to prevent contamination	211.80 (b)	5.10	C.02.011	16.10 to 16.14
Stored off the floor to permit cleaning and inspection	211.80 (c)	5.7	C.02.011	14.5
Identification of materials	211.80 (d)	5.13, 5.29	C.02.011	14.13
Receipt and Storage of Untested Components, Drug Product Containers, and Closures	211.82			
Visual examination of materials upon receipt	211.82 (a)	5.3, 5.4, 5.27	C.02.011	14.9 to 14.11
Quarantine materials before release	211.82 (b)	5.5	C.02.011	14.4
Testing and Approval or Rejection of Components, Drug Product Containers, and Closures	211.84	Annex 8		
Materials withheld from use until properly released	211.84 (a)	5.31	C.02.009, C.02.010, C.02.016, C.02.017	14.15, 17.13

Continued

Table A.1 *Continued.*

Item	US cGMP	EU/PIC-S GMP	Canada	WHO
Representative samples of materials collected	211.84 (b)	6.12	C.02.009, C.02.010, C.02.016, C.02.017	14.12, 17.7
Adequate process to collect samples	211.84 (c)	6.11, 6.13	C.02.009, C.02.010, C.02.013, C.02.016, C.02.017	17.8 to 17.11
Examination of samples	211.84 (d)	5.30, 6.15 to 6.22	C.02.009, C.02.010, C.02.016, C.02.017	14.14, 17.14
Approval or rejection of materials	211.84 (c)	6.17	C.02.009, C.02.010, C.02.016, C.02.017, C.02.019	14.15, 17.5
Use of Approved Components, Drug Product Containers, and Closures	211.86	5.7	No	14.5
Retesting of Approved Components, Drug Product Containers, and Closures	211.87	5.29	C.02.009	14.13
Rejected Components, Drug Product Containers, and Closures	211.89	5.61	C.02.011, C.02.014	14.28
Drug Product Containers and Closures	211.94			
Containers and closures cannot interact with products	211.94 (a)	No	No	2.1
Container closure system to protect product	211.94 (b)	No	No	2.1
Use clean containers and closures	211.94 (c)	5.48	No	16.19
Detailed written standard and specification for containers and closures	211.94 (d)	No	C.02.011	No

Continued

Table A.1 *Continued.*

Item	**US cGMP**	**EU/PIC-S GMP**	**Canada**	**WHO**
Subpart F—Production and Process Control				
Written Procedures; Deviations	**211.100**			
Detailed written procedures for production and process control	211.100(a)	5.2	C.02.011, C.02.013	16.1, 16.2
Follow procedures and documentation of manufacturing operations and deviations	211.100 (b)	5.15	C.02.011	16.3
Charge-in of components	211.101			
Batch formulated to not less than 100%	211.101 (a)	No	C.02.011	
Adequate measure and identification of components	211.101 (b)	5.32	C.02.011	14.12 to 14.18
Supervise measuring of components	211.101 (c)	5.33	C.02.011	14.17
Supervise addition of components to the batch	211.101 (d)	No	C.02.011	No
Calculation of Yield	211.103	5.8, 5.39	C.02.011	16.4, 16.20
Equipment Identification	211.105			
Storage container and processing lines	211.105 (a)	3.42, 5.12	C.02.005, C.02.011	13.3, 13.4
Major equipment	211.105 (b)	5.12	C.02.011	16.6
Sampling and Testing of In-Process Materials and Drug Products	211.110			
Detailed written procedures	211.110 (a)	1.8, 3.17, 4.18, 5.21 to 5.24, 5.38, 6.18	C.02.011, C.02.013	16.9, 16.16, 17.6 to 17.12
Validated specifications	211.110 (b)	4.15	C.02.011	15.13 to 15.17

Continued

Table A.1 *Continued.*

Item	US cGMP	EU/PIC-S GMP	Canada	WHO
Testing of in-process materials	211.110 (c)	4.15	C.02.013	15.20, 17.6 to 17.12, 17.17
Quarantine rejected in-process materials	211.110 (d)	5.61	C.02.011, C.02.014	14.28
Time Limitations on Production	211.111	4.15	C.02.011	15.23
Control of Microbiological Contamination	211.113	Annex 1		
Handling of nonsterile products	211.113 (a)	5.10, Annex 1	C.02.007	16.10 to 16.14
Written procedures for sterile products, including validation of processes	211.113 (b)	5.10, Annex 1	C.02.029	Annex 6 of TR 902
Reprocessing	211.115			
Written procedures	211.115 (a)	5.62 to 5.64	C.02.014	14.29, 14.31, 15.40
Quality unit preapproval	211.115 (b)	5.63	C.02.014	14.29 implicit
Subpart G—Packaging and Labeling Control				
Materials examination and usage criteria	211.122			
Detailed written procedures including evaluation of labeling and packaging materials before use	211.122 (a)	4.14, 5.40, 5.42	C.02.013, C.02.016, C.02.017	12.21, 14.19 to 14.21, 14.23, 16.2, 17.15, 17.16
Approval or rejection of materials	211.122 (b)	6.17	C.02.014, C.02.016, C.02.017	15.18
Keep records of each shipment or materials received	211.122 (c)	4.22	C.02.017	14.21, 16.2
Store materials separately under controlled access	211.122 (d)	5.41	C.02.011	14.20
Obsolete materials	211.122 (c)	5.43	C.02.011, C.02.016	14.22

Continued

Table A.1 *Continued.*

Item	US cGMP	EU/PIC-S GMP	Canada	WHO
Gang-printed labels	211.122 (f)	No	C.02.011	No
Special control procedures when using cut labeling	211.122 (g)	5.51	C.02.011, C.02.016	14.20, 16.30
Monitoring of printing devices	211.122 (h)	5.50, 5.52	C.02.011	16.29
Labeling Issuance	211.125			
Strict control over labeling materials	211.125 (a)	5.2	C.02.011, C.02.017	16.25
Examination of labeling materials	211.125 (b)	5.2	C.02.011, C.02.017	16.26
Reconciliation of labeling materials	211.125 (c)	5.56	C.02.011, C.02.017	16.34
Destruction of excess imprinted labeling	211.125 (d)	5.57	C.02.011, C.02.017	16.35
Control of returned labeling	211.125 (c)	5.57	C.02.011, C.02.017	16.35
Detailed written procedures for issuance of labeling	211.125 (f)	5.2	C.02.011, C.02.017	16.2
Packaging and Labeling Operations	211.130			
Prevention of mix-ups during operations	211.130 (a)	3.15, 5.44	C.02.011	16.25, 16.26
Identification and handling of unlabeled drug product containers	211.130 (b)	5.49	C.02.011	16.28
Identification of drug product with lot or control number	211.130 (c)	4.19	C.02.011	15.39, 16.27
Examination of materials before packaging operations	211.130 (d)	4.19, 5.47	C.02.011	16.26
Inspection of facilities immediately before use	211.130 (c)	4.19, 5.45	C.02.011	16.26

Continued

Table A.1 *Continued.*

Item	US cGMP	EU/PIC-S GMP	Canada	WHO
Tamper-evident packaging requirements for OTC human drug products	211.132	No		No
General authority	211.132 (a)	No	No	No
Requirements for tamper-evident packaging	211.132 (b)	No	No	No
Labeling	211.132 (c)	No	No	No
Request for exemptions from requirements	211.132 (d)	No	No	No
OTC products subject to approved NDA	211.132 (c)	No	No	No
Poison Prevention Packaging Act	211.132 (f)	No	No	No
Drug Product Inspection	211.134			
Inspection during finishing operations	211.134 (a)	5.54	C.02.019	16.32
Inspection after finishing operations	211.134 (b)	5.54	C.02.019	16.32
Recording of finished product inspections	211.134 (c)	4.21	C.02.019	16.32
Expiration Dating	211.137	4.16 not explicit		
Requirement to bear an expiration date	211.137 (a)		C.02.027	15.11
Storage conditions	211.137 (b)		C.02.027	17.23
Reconstituted products	211.137 (c)		C.02.027	15.11
Expiration date as part of labeling	211.137 (d)		C.02.027	15.11
Homeopathic products	211.137 (c)	No	No	No
Allergenic extracts	211.137 (f)	No	No	No

Continued

Table A.1 *Continued.*

Item	**US cGMP**	**EU/PIC-S GMP**	**Canada**	**WHO**
New drug product for investigational use	211.137 (g)	No	No	No
OTC drug exemptions	211.137 (h)	No	No	No
Subpart H—Holding and Distribution				
Warehousing Procedures	211.142			
Quarantine of products before release	211.142 (a)	3.21, 5.2	C.02.011	14.4, 14.26
Storage of products	211.142 (b)	3.18, 3.19, 5.58, 5.60	C.02.011, C.02.015	15.21, 15.23, 16.2
Distribution Procedures	211.150	5.2	C.02.011	16.2
Oldest products are distributed first	211.150 (a)	No	No	14.5
Tracking of distributed lots	211.150 (b)	4.28, 8.13	C.02.011, C.02.012	6.6, 15.45
Subpart I—Laboratory Controls				
General Requirements	211.160			
Laboratory controls approved by the QCU. Deviations documented	211.160 (a)	1.4, 1.9, 4.2, 4.3, 4.13 to 4.16, 4.25, 4.26	C.02.009, C.02.013, C.02.015, C.02.016, C.02.017, C.02.018	15.14 to 15.15, 16.3
Validated laboratory controls	211.160 (b)	4.25, 4.26, 6.7, 6.11, 6.15 to 6.21	C.02.009, C.02.018	17.6 to 17.12, 17.14, 17.17, 17.18
Testing and Release for Distribution	211.165			
Testing to each lot	211.165 (a)	1.9, 4.26, 6.3	C.02.014, C.02.015, C.02.018	15.42, 17.18
Microbial test	211.165 (b)	4.26	C.02.018, C.02.029	17.18
Detailed written sampling plans	211.165 (c)	4.25, 6.11 to 6.13	C.02.018	15.37, 17.7 to 17.11
Acceptance criteria	211.165 (d)	6.17	C.02.014, C.02.018	15.21

Continued

Table A.1 *Continued.*

Item	US cGMP	EU/PIC-S GMP	Canada	WHO
Validated test methods	211.165 (c)	1.9, 6.15	C.02.018	15.13
Reprocessed material testing	211.165 (f)	5.61 to 5.64	C.02.014	14.29
Stability Testing	211.166	6.23 to 6.33	C.02.027, C.02.028	
Written stability program	211.166 (a)	Not explicit	C.02.027, C.02.028	17.22 to 17.25
Determination of expiration date	211.166 (b)	Not explicit	C.02.020, C.02.027	17.23
Homeopathic products	211.166 (c)	No	No	No
Allergenic extracts	211.166 (d)	No	No	No
Special testing requirements	211.167			
Sterile and pyrogen-free products	211.167 (a)	Annex 1	C.02.029	Annex 6 of TRS 902
Ophthalmic ointments	211.167 (b)	Annex 9	No	No
Controlled-release dosage form	211.167(c)	No	No	No
Reserve samples	211.170			
API	211.170 (a)	6.12, Annex 19	C.02.020, C.02.025, C.02.026	17.3, 17.21
Finished drug product	211.170 (b)	6.12, Annex 19	C.02.025, C.02.026	17.3, 17.21
Laboratory Animals	211.173	3.33, 6.22	C.02.004	No
Penicillin Contamination	211.176	No	No	No
Subpart J—Records and Reports				
General requirements	211.180			
Retention time for finished drug product records	211.180 (a)	4.10 to 4.12, 6.8 to 6.10	C.02.021	15.8

Continued

Table A.1 *Continued.*

Item	US cGMP	EU/PIC-S GMP	Canada	WHO
Retention time for components and materials	211.180 (b)	4.10 to 4.12, 6.8 to 6.10	C.02.021	15.8
Record availability for inspections	211.180 (c)	No	No	No
Retention methods	211.180 (d)	No	No	No
Annual product review	211.180 (c)	1.10, 1.11, 4.29	C.02.011	1.6
Procedure for escalation of issues to management	211.180 (f)	1.6	Sec. 4.2.1	1.2
Equipment Cleaning and Use Log	211.182	4.31	C.02.005, C.02.020, C.02.024	15.46, 15.47
Component, Drug Product Container, Closure, and Labeling Records	211.184	1.9	C.02.020	
Components and materials detailed information	211.184 (a)	4.22, 4.23	C.02.020	15.33
Test and examination results of components and materials	211.184 (b)	4.26	C.02.010, C.02.017, C.02.020	15.43
Detailed inventory of components and materials used	211.184 (c)	4.20	C.02.017, C.02.020	15.32, 15.33
Examination results of labels and labeling	211.184 (d)	4.21	C.02.017, C.02.020	15.33
Disposition of rejected components, materials, and labeling	211.184 (c)	5.61	C.02.017, C.02.020	15.33
Master Production and Control Records	211.186			
Issuance	211.186 (a)	4.1 to 4.6, 4.17 to 4.19	C.02.011, C.02.020	
Detailed information to be included	211.186 (b)	4.1 to 4.6, 4.17 to 4.19	C.02.011, C.02.020	15.22, 15.23
Batch Production and Control Records	211.188			

Continued

Table A.1 *Continued.*

Item	US cGMP	EU/PIC-S GMP	Canada	WHO
Verified accurate reproduction of master production and control records	211.188 (a)	4.20 to 4.21	C.02.011, C.02.020	15.25
Documentation of each significant step in the process	211.188 (b)	4.20 to 4.21	C.02.011, C.02.020	15.26, 15.27, 15.28, 15.30
Production Record Review—Prior Release, Investigation of Unexplained Discrepancies	211.192	2.4, 4.20, 4.21, 4.29, 5.8, 5.39, 5.55, 5.56	C.02.014, C.02.015, C.02.019	16.4, 16.20, 16.36, 17.20
Laboratory Records	211.194			
Detailed laboratory records	211.194 (a)	4.26, 6.7, 6.17	C.02.015, C.02.020	15.42, 15.43, 17.3
Changes to test methods	211.194 (b)	4.29	C.02.015	4.11
Reference standards	211.194 (c)	6.21	C.02.015	14.34, 14.35, 14.41, 15.12
Calibration of laboratory equipment	211.194 (d)	6.7	C.02.015	13.5, 16.23
Stability testing	211.194 (c)	6.17	C.02.015	15.43
Distribution Records	211.196	4.28	C.02.012, C.02.022	15.45
Complaint Files	211.198			
Detailed written procedures for handling complaints	211.198 (a)	4.29, 8.1 to 8.8	C.02.015	5.1 to 5.10
Complaint record information and retention period	211.198 (b)	4.11	C.02.023	5.1 to 5.10
Subpart K—Returned and Salvaged Drug Products				
Returned drug products	211.204	4.29, 5.65	C.02.014	14.33
Salvaged drug products	211.208	No	No	No

Notes: No = No correspondence with the specific US cGMP section. US cGMP updated as of April 1, 2014. EU EudraLex Vol. 4, Canada, and WHO cGMP used for this comparison are the respective current versions as of January 1, 2014.

Table A.2 FDA cGMP and quality system elements.

Quality system element	US cGMP citation
Management responsibilities	
Leadership	
Structure	• Establish quality function: § 211.22(a) • Definition: § 210.3(b)(15) • Notification: § 211.180(f)
Build the quality system	• Quality Unit procedures: § 211.22(d) • Quality Unit procedures, specifications: § 211.22(c) with reinforcement in § 211.100(a) and § 211.160(a) • Quality Unit control steps: § 211.22(a), with reinforcement in § 211.42(c), § 211.84(a), § 211.87, § 211.101(c)(1), § 211.110(c), § 211.115(b), § 211.142, § 211.165(d), § 211.192 • Quality Unit quality assurance; review and investigate: § 211.22(a), § 211.100(a–b), § 211.180(f), § 211.192, § 211.198(a) • Record control: § 211.180(a–d), § 211.180(c), § 211.180(d), § 211.180(c), § 211.186, § 211.192, § 211.194, § 211.198(b)
Establish policies, objectives, and plans	• Procedures: § 211.22(c–d), § 211.100(a)
Management review	• Record review: § 211.100, § 211.180, § 211.192, § 211.198(b)(2)
Resources	
General arrangements	
Develop personnel	• Qualifications: § 211.25(a) • Staff member: § 211.25(c) • Staff training: § 211.25(a–b)
Facilities and equipment	• Buildings and facilities: § 211.22(b), § 211.28(c), § 211.42–211.58, § 211.173 • Equipment: § 211.63–211.72, § 211.105, § 211.160(b)(40), § 211.182 • Laboratory facilities: § 211.22(b)
Manufacturing operations	
Design and develop product and processes	• Production: § 211.100(a)
Examine inputs	• Materials: § 210.3(b), § 211.80–211.94, § 211.101, § 211.122, § 211.125

Continued

Table A.2 *Continued.*

Quality system element	US cGMP citation
Perform and monitor operations	• Production: § 211.100, § 211.103, § 211.110, § 211.111, § 211.113 • QC criteria: § 211.22(a–c), § 211.115(b), § 211.160(a), § 211.165(d), § 211.188 • QC checkpoints: § 211.22(a), § 211.84(a), § 211.87, § 211.110(c)
Address nonconformities	• Discrepancy investigation: § 211.22(a), § 211.100, § 211.175, § 211.192, § 211.198 • Recalls: 21 CFR Part 7
Evaluation activities	
Analyze data for trends	• Annual review: § 211.180(c)
Conduct internal audits	Not covered in cGMP
Risk assessment	Not covered in cGMP-—covered in guidance ICH Q9
Corrective action	• Discrepancy investigation: § 211.22(a), §211.192
Preventive action	Not covered in cGMP
Promote improvement	• § 211.110

Note: US cGMP updated as of April 1, 2014.

Appendix B

21 CFR 4 Regulation of Combination Products Overview

SUBPART A—CURRENT GOOD MANUFACTURING PRACTICE REQUIREMENTS FOR COMBINATION PRODUCTS

SUBPART B [RESERVED]

Appendix C

21 CFR 211 Pharmaceuticals GMP Overview

SUBPART A—GENERAL PROVISIONS

Sec. 211.1 Scope.

Sec. 211.3 Definitions.

SUBPART B—ORGANIZATION AND PERSONNEL

Sec. 211.22 Responsibilities of quality control unit.

Sec. 211.25 Personnel qualifications.

Sec. 211.28 Personnel responsibilities.

Sec. 211.34 Consultants.

SUBPART C—BUILDINGS AND FACILITIES

Sec. 211.42 Design and construction features.

Sec. 211.44 Lighting.

Sec. 211.46 Ventilation, air filtration, air heating and cooling.

Sec. 211.48 Plumbing.

Sec. 211.50 Sewage and refuse.

Sec. 211.52 Washing and toilet facilities.

Sec. 211.56 Sanitation.

Sec. 211.58 Maintenance.

SUBPART D—EQUIPMENT

Sec. 211.63 Equipment design, size, and location.

Sec. 211.65 Equipment construction.

Sec. 211.67 Equipment cleaning and maintenance.

Sec. 211.68 Automatic, mechanical, and electronic equipment.

Sec. 211.72 Filters.

SUBPART E—CONTROL OF COMPONENTS AND DRUG PRODUCT CONTAINERS AND CLOSURES

Sec. 211.80 General requirements.

Sec. 211.82 Receipt and storage of untested components, drug product containers, and closures.

Sec. 211.84 Testing and approval or rejection of components, drug product containers, and closures.

Sec. 211.86 Use of approved components, drug product containers, and closures.

Sec. 211.87 Retesting of approved components, drug product containers, and closures.

Sec. 211.89 Rejected components, drug product containers, and closures.

Sec. 211.94 Drug product containers and closures.

SUBPART F—PRODUCTION AND PROCESS CONTROLS

Sec. 211.100 Written procedures; deviations.

Sec. 211.101 Charge-in of components.

Sec. 211.103 Calculation of yield.

Sec. 211.105 Equipment identification.

Sec. 211.110 Sampling and testing of in-process materials and drug products.

Sec. 211.111 Time limitations on production.

Sec. 211.113 Control of microbiological contamination.

Sec. 211.115 Reprocessing.

SUBPART G—PACKAGING AND LABELING CONTROL

Sec. 211.122 Materials examination and usage criteria.

Sec. 211.125 Labeling issuance.

Sec. 211.130 Packaging and labeling operations.

Sec. 211.132 Tamper-evident packaging requirements for over-the-counter (OTC) human drug products.

Sec. 211.134 Drug product inspection.

Sec. 211.137 Expiration dating.

SUBPART H—HOLDING AND DISTRIBUTION

Sec. 211.142 Warehousing procedures.

Sec. 211.150 Distribution procedures.

SUBPART I—LABORATORY CONTROLS

Sec. 211.160 General requirements.

Sec. 211.165 Testing and release for distribution.

Sec. 211.166 Stability testing.

Sec. 211.167 Special testing requirements.

Sec. 211.170 Reserve samples.

Sec. 211.173 Laboratory animals.

Sec. 211.176 Penicillin contamination.

SUBPART J—RECORDS AND REPORTS

Sec. 211.180 General requirements.

Sec. 211.182 Equipment cleaning and use log.

Sec. 211.184 Component, drug product container, closure, and labeling records.

Sec. 211.186 Master production and control records.

Sec. 211.188 Batch production and control records.

Sec. 211.192 Production record review.

Sec. 211.194 Laboratory records.

Sec. 211.196 Distribution records.

Sec. 211.198 Complaint files.

SUBPART K—RETURNED AND SALVAGED DRUG PRODUCTS

Sec. 211.204 Returned drug products.

Sec. 211.208 Drug product salvaging.

Appendix D

21 CFR 820 Medical Device GMP Overview

SUBPART A—GENERAL PROVISIONS

Sec. 820.1 Scope.

Sec. 820.3 Definitions.

Sec. 820.5 Quality system.

SUBPART B—QUALITY SYSTEM REQUIREMENTS

Sec. 820.20 Management responsibility.

Sec. 820.22 Quality audit.

Sec. 820.25 Personnel.

SUBPART C—DESIGN CONTROLS

Sec. 820.30 Design controls.

SUBPART D—DOCUMENT CONTROLS

Sec. 820.40 Document controls.

SUBPART E—PURCHASING CONTROLS

Sec. 820.50 Purchasing controls.

SUBPART F—IDENTIFICATION AND TRACEABILITY

Sec. 820.60 Identification.

Sec. 820.65 Traceability.

SUBPART G—PRODUCTION AND PROCESS CONTROLS

Sec. 820.70 Production and process controls.

Sec. 820.72 Inspection, measuring, and test equipment.

Sec. 820.75 Process validation.

SUBPART H—ACCEPTANCE ACTIVITIES

Sec. 820.80 Receiving, in-process, and finished device acceptance.

Sec. 820.86 Acceptance status.

SUBPART I—NONCONFORMING PRODUCT

Sec. 820.90 Nonconforming product.

SUBPART J—CORRECTIVE AND PREVENTIVE ACTION

Sec. 820.100 Corrective and preventive action.

SUBPART K—LABELING AND PACKAGING CONTROL

Sec. 820.120 Device labeling.

Sec. 820.130 Device packaging.

SUBPART L—HANDLING, STORAGE, DISTRIBUTION, AND INSTALLATION

Sec. 820.140 Handling.

Sec. 820.150 Storage.

Sec. 820.160 Distribution.

Sec. 820.170 Installation.

SUBPART M—RECORDS

Sec. 820.180 General requirements.

Sec. 820.181 Device master record.

Sec. 820.184 Device history record.

Sec. 820.186 Quality system record.

Sec. 820.198 Complaint files.

SUBPART N—SERVICING

Sec. 820.200 Servicing.

SUBPART O—STATISTICAL TECHNIQUES

Sec. 820.250 Statistical techniques.

Appendix E

21 CFR 1271 Human Cells, Tissues, and Cellular and Tissue-Based Products Overview

SUBPART A—GENERAL PROVISIONS

SUBPART B—PROCEDURES FOR REGISTRATION AND LISTING

SUBPART C—DONOR ELIGIBILITY

SUBPART D—CURRENT GOOD TISSUE PRACTICE

SUBPART E—ADDITIONAL REQUIREMENTS FOR ESTABLISHMENTS DESCRIBED IN § 1271.10

SUBPART F—INSPECTION AND ENFORCEMENT OF ESTABLISHMENTS DESCRIBED IN § 1271.10

Appendix F

Body of Knowledge—Pharmaceutical GMP Professional Certification (CPGP)

This body of knowledge (BoK) covers compliance with good manufacturing practices (GMPs), as regulated and guided by national and international agencies for the pharmaceutical industry. It covers finished human and veterinary drugs and biologics, combination devices including their component raw materials (e.g., active pharmaceutical ingredients (APIs) and excipients), and packaging and labeling operations.

This BoK includes topics and subtopics, and each topic or subtopic has a subtext explanation and assigned cognitive level. The cognitive level, in parentheses at the end of each subtext entry, refers to the highest cognitive level at which the topic will be tested. A more complete description of cognitive levels is provided at the end of this document. These details will be used by the Examination Development Committee as guidelines for writing test questions and are designed to help candidates prepare for the exam by identifying specific content that can be tested. The subtext is not intended to limit the subject matter or be all-inclusive of what might be covered in an exam but is intended to clarify how topics relate to the role of the Certified Pharmaceutical GMP Professional (CPGP).

I. Regulatory Agency Governance (17 Questions)

A. Global regulatory framework
Identify the acts, statutes, and directives that apply to pharmaceuticals. (Understand)

B. Regulations and guidances
Interpret frequently used regulations and guidelines/guidances/drafts including those published or administered by the Pharmaceutical Inspection Convention and Pharmaceutical Inspection Cooperation Scheme (PIC/S), the International Conference on Harmonization (ICH), the World Health Organization (WHO), the European Medicines Agency (EMA), the Food & Drug Administration (FDA), Health Canada, USDA 9CFR, and other national regulatory agencies and international standards (e.g. ISOs). (Understand)

C. Mutual recognition agreements
Interpret requirements that govern product registration, import or export of raw material or finished product, and the sharing of inspection findings. (Understand)

D. Regulatory inspections
Define and describe various types of inspections including pre-approval, system-based, for-cause, and license renewal, and describe the frequency for each. Describe how to prepare for and host in-person and remote evaluations and how to accommodate record requests and digital reporting. (Understand)

E. Enforcement actions
Define and describe various global enforcement actions and consequences (e.g., warning letters, consent decree, license withdrawals, product seizure, and import alerts). (Understand)

F. Regulatory agency reporting

1. **Post-marketing changes** Describe how post-marketing changes to specifications, processes, and methods are assessed for impact to determine the appropriate reporting method [e.g., scale up and post-approval changes (SUPAC)]. (Understand)

2. **Regulatory reporting requirements** Describe global reporting requirements including supplements, field alerts, biological product deviation reports, adverse events, product recalls, annual reports, and variations to dossiers and applications. (Understand)

3. **Product Surveillance** Describe monitoring requirements for risk evaluation and mitigation strategy (REMS) and pharmacovigilance. (Understand)

G. Site Master File (SMF), Validation Master Plan (VMP), Drug Master File (DMF), and Site Reference File (SRF)
Describe the purpose and content of Site Master Files (SMFs), Validation Master Plans (VMPs), Drug Master Files (DMFs), and Site Reference Files (SRFs). (Understand)

II. Quality Systems (26 Questions)

A. Quality management system (QMS)
Describe key elements of the structure of a Quality Management System (QMS). Outline the requirements for the development, operation, and management review for suitability and effectiveness as defined in ICH Q7, ICH Q10, EU GMP, and other guidances. (Evaluate)

B. Quality unit (site) management
Describe quality management elements for individual sites or units including responsibilities for the quality unit management such as Qualified Person, Management Representative, batch release (disposition) requirements for investigational and commercial products, and the need for quality units to be independent from operations. (Understand)

C. **Risk management and assessment**

1. **Risk management**
   Use various methods to apply risk management principles as described in ICH Q9, ICH Q12, and other guidance and regulatory documents. (Apply)

2. **Risk assessment**
   Use the Quality Management Maturity (QMM) assessment program to assess the effectiveness of the Quality Management System to ensure process and product quality. (Apply)

D. **Training and personnel qualification**

1. **Needs analysis**
   Identify the requirements for determining the type of training, qualification, and experience needed by quality staff members, operations personnel, and related functions. (Apply)

2. **Staff development requirements**
   Determine proof of proficiency based on regulations, guidances, and directives including documented evidence and periodic reassessment. (Apply)

3. **Training effectiveness and role of supervisor**
   Apply various methods for testing and evaluating training effectiveness. Identify the role and responsibilities of supervisors including ensuring staff are adequately trained to perform their assigned functions. (Apply)

E. **Change control and management**

1. **Pre-change analysis** Assess the impact that proposed changes will have on products, processes, facilities, utilities, and systems to minimize risk and ensure regulatory compliance. (Analyze)

2. **Change implementation**
   Implement and document the change following a change implementation plan. (Create)

3. **Post-change analysis** Analyze data and other inputs to determine the results of a change and evaluate any new risk factors created by the change. (Analyze)

F. **Investigations and corrective and preventive action (CAPA)**

1. **Trigger events**
   Identify events that require investigation, root cause analysis, and impact assessment both directly and indirectly related to the event. (Evaluate)

2. **Response actions**
   Define immediate action, corrective action, preventive action, management responsibility, and methods of implementing them. (Evaluate)

3. **CAPA feedback and trending**
   Describe how CAPA trending is used to modify appropriate quality system elements. (Create)

G. **Audits and self-inspections**

1. **Audits processes and results** Develop audit schedules, differentiate between various audit types (systems, product, and process) conducted either remotely or on-site, document evidence of audit completion, and analyze audit results to assess conformance to requirements. (Evaluate)

2. **Audit follow-up** Use various methods to evaluate and verify the effectiveness of corrective actions taken. (Evaluate)

H. **Documents and records management**

1. **GMP document system** Describe the GMP document system to determine compliance to regulatory requirements including corporate standards, master plans, procedures, manufacturing, and test instructions. (Analyze)

2. **Records**
   Review various records (e.g., logbooks, tags, and training evidence) to confirm compliance to requirements such as Attributable, Legible, Contemporaneous, Original, Accurate (ALCOA) and PIC/S guidelines for data integrity. (Analyze)

3. **Record retention**
   Identify regulatory requirements for record retention. (Understand)

I. **Product complaints and adverse event reports**

1. **Product complaints** Describe and distinguish between product complaints and adverse events. Evaluate complaint-handling processes. (Evaluate)

2. **Adverse events** Describe regulatory requirements for the reporting of adverse events (e.g. counterfeit product, fraud). (Analyze)

3. **Event response** Evaluate the level of action that needs to be taken in response to adverse events such as corrections, and product removal. (Evaluate)

J. **Product trend requirements**
   Describe and distinguish between components of periodic product assessment, such as the U.S. annual product review (APR) and the European product quality review (PQR), with regard to data trends and other required elements. (Understand)

K. **Supplier and contractor quality management**

1. **Supplier quality systems** Identify and interpret standards and regulations related to monitoring supplier, vendor, and contractor quality management systems. (Understand)

2. **Supplier controls** Assess the adequacy of controls over supplier, vendor, and contractor selection and procurement and receipt of raw materials, components, and contract services. Determine the need for formal contracts/quality agreements. (Evaluate)

3. **Supplier evaluation** Assess the quality systems of suppliers, vendors, and contractors using various methodologies including supplier qualification, certification, evaluation, audit, and supplied product or service performance trending. (Evaluate)

4. **Quality agreements**
Explain the purpose and describe the main elements of a quality agreement. (Understand)

5. **Outsource processes**
Use various methods for the management of contract manufacturing, drug development, testing laboratories, and other outsourcing activities including defining the roles and responsibilities of each party. (Apply)

III. Laboratory Systems (20 Questions)

A. Compendia (United States, Europe, and Japan)

1. **Required vs. informational compendia** Describe and distinguish between required and informational ("general") compendial chapters. (Apply)

2. **Marketing requirements vs. compendia** Distinguish among the U.S. Pharmacopoeia (USP), European Pharmacopoeia (PhEur or EP), and Japanese Pharmacopoeia (JP) in terms of requirements for marketing authorization. (Understand)

3. **Compendial methods review** Review compendial methods to ensure they are verified as suitable for use in the testing lab. (Analyze)

4. **Compendial or non-compendial requirements review** Review test methods, qualifications, validation, and verification against required compendial chapters (general and informational, as needed). (Analyze)

5. **Biological, microbiological, chemical, and physical test methods** Identify and interpret results from compendia identification tests, quantitative analysis, qualitative analysis, and other tests or studies for biological, microbiological, chemical, and physical tests. (Apply)

B. Laboratory investigations of atypical results

1. **Test data** Describe and develop procedures for investigating each type of test data including biological, microbiological, chemical test, and unknowns. (Analyze)

2. **Atypical results** Identify, analyze, and interpret data on processes or products that are out-of-specification or out-of-trend. Determine

the outcome of the laboratory portion of the investigation and the criteria for further investigation. (Evaluate)

C. **Instrument management**

1. **Instrument controls** Apply operating procedures for instrument identification, classification, qualification, calibration, and preventive maintenance. (Apply)

2. **Instrument calibration** Determine that instruments are calibrated within the specified range of operation and that they are accurate and precise. (Apply)

D. **Specifications**

1. **Types of specifications** Determine whether approved specifications exist for raw materials, intermediates, packaging components, labels, and finished products. (Analyze)

2. **Test data and specifications** Compare test data with specifications to determine whether raw materials, intermediates, packaging, labels, and finished products meet requirements. (Analyze)

3. **Specifications revision** Review and update specifications when methods are revised, or compendia are changed. (Evaluate)

E. **Laboratory record-keeping and data requirements**

1. **Record-keeping requirements** Identify and review record-keeping requirements for data acquisition systems to ensure data integrity. (Apply)

2. **Record review** Ensure data integrity and prevent loss of data by reviewing laboratory records including audit trails of electronic data to detect errors, falsification, and fraud. (Evaluate)

3. **Certificates of analysis (COAs)** Review Certificates of Analysis (COAs) to ensure they are complete, internally reviewed, and appropriately retained. (Apply)

F. **Laboratory handling controls**

1. **Sample handling** Determine whether samples are identified and handled in accordance with requirements including name, sample identification, and chain of custody. (Apply)

2. **Reagents, solutions, and standards identification** Determine whether reagents, solutions, and standards are identified and labeled, and quantities are traceable in accordance with requirements including opened-on, expiry, (validated) use-by, or recertify-by dates. (Apply)

3. **Storage requirements** Describe and use procedures to store samples, reagents, solutions, solvents, and standards in appropriate environmental conditions (e.g., temperature, humidity, light

exposure, and absence of oxygen) to maintain the material's characteristics for testing. (Apply)

G. **Stability programs**

1. **Release tests vs. stability-indicating tests** Define and distinguish between release and stability-indicating tests. (Analyze)
2. **Stability test data** Review stability data against specifications and identify trends that can establish, support, or challenge an expiry date. (Evaluate)
3. **Stability-point failure** Identify the stability-point failure of a product or material and evaluate the implications for regulatory compliance. (Evaluate)

H. **Reserve samples and retains**
Describe the various regulatory requirements for retains and reserve samples. (Apply)

IV. **Infrastructure: Facilities, Utilities, and Equipment (17 Questions)**

A. **Facilities**

1. **Buildings** Determine and document the requirements for appropriate size and construction of buildings and areas, and the location of control systems. Ensure that construction and location facilitate proper operation and minimize the risk of error and cross contamination including meeting requirements that specify separation of antibiotics, hormones, and toxins. (Apply)
2. **Manufacture and storage environment** Identify requirements for appropriate lighting, ventilation, and drainage to avoid adversely affecting product (either directly or indirectly) during manufacturing and storage. (Apply)
3. **Facilities change control** Use various methods to verify that change control practices are in use to maintain the qualified state of the facilities. (Apply)

B. **Utilities**

1. **Water supply systems** Identify and interpret regulatory requirements for the design of water supply systems including various unit operations (e.g., dechlorination, reverse osmosis, deionization, and distillation), delivery lines, back-flow or back-siphonage prevention, and drainage systems as appropriate for the type of water (potable, purified, and water for injection) needed in various processing steps. (Apply)
2. **Compressed air and gas systems** Identify and apply regulatory requirements related to compressed air and gas systems including storage, flow regulation, filtration, venting, and purging. (Apply)

3. **Utility design for production** Identify and select utility designs related to production steps (e.g., washing, sterilizing, and depyrogenation) for use with specific materials and processes. (Apply)

4. **Utilities design specifications** Review operations of utilities to ensure they meet design specifications. (Apply)

5. **Utilities change control** Verify that change control practices are used to maintain the qualified state of affected utilities. (Apply)

C. **Equipment**

1. **Equipment planning** Review equipment location, design, construction, installation, and maintenance based on the operations to be conducted. (Apply)

2. **Equipment layout** Determine the layout of equipment to minimize the risk of errors, to facilitate effective cleaning and maintenance, and to avoid contamination or any other undesired effect on product quality. (Apply)

3. **Equipment cleaning and maintenance** Review procedures and schedules for equipment cleaning, maintenance, and sanitization (where necessary) to ensure that they meet requirements. (Apply)

4. **Equipment cleaning validation or verification** Evaluate the need and methodology for product-contact cleaning validation and verification. Evaluate the difference between single-use, multi-use, and dedicated equipment. (Evaluate)

5. **Equipment change control** Verify that change control has maintained the qualified state of equipment. (Apply)

D. **Qualification and validation**
Verify that the qualifications and validations of facilities, equipment, and utilities are conducted in accordance with various requirements including factory and site acceptance testing (FAT/SAT), design, installation, operational, and performance qualification (DQ/IQ/OQ/PQ) prior to process validation. (Analyze)

E. **Maintenance and metrology systems**

1. **Maintenance procedures** Verify that procedures are used for routine and non-routine maintenance of heating, ventilation, air conditioning (HVAC) systems, air and water filters, and other equipment and utilities. (Analyze)

2. **Metrology change control** Verify that appropriate calibration and engineering/equipment change control procedures are used and that a metrology program exists for the calibration of instruments that control manufacturing facilities, utilities, and equipment. (Analyze)

F. **General cleaning, sanitization, and sterilization systems**

1. **Cleaning procedures**
Review cleaning procedures in accordance with cleaning validation whenever validation is required and performed. (Apply)

2. **Sanitization procedures** Review sanitization procedures for facilities and equipment and ensure all are in accordance with any required validation studies including details on cleaning schedules, methods, equipment, materials, sanitizers, disinfectants, sporicides, and sterilants. (Apply)

3. **Pest control** Review and verify that a pest control program is established and that it uses authorized rodenticides, insecticides, fungicides, fumigating agents, and appropriate traps for pest elimination. (Apply)

4. **Sterilization processes** Verify that appropriate sterilization processes are established and validated. (Apply)

G. **Automated or computerized systems**

1. **Validation procedures** Review procedures for implementation and validation of automated or computerized systems. Verify that critical parameters for their operation and maintenance are controlled and monitored. (Evaluate)

2. **Open and closed computerized systems** Distinguish between open and closed computerized systems. (Apply)

3. **Configuration control** Verify that version control and configuration are maintained and monitored. (Evaluate)

4. **Security requirements** Evaluate on-site, multi-site, and cloud-based computerized systems to ensure they meet regulatory and guidance requirements for key elements such as access control, data protection, electronic signature, change control, data archiving, maintenance, transcription, audit trail, and periodic system monitoring. (Evaluate)

H. **Business continuity and disaster recovery planning**

1. **Supply chain impact** Review plans and verify procedures for disaster recovery, record recovery, and business continuity that will guard operations from interruption to the supply chain and ensure data integrity. (Evaluate)

2. **Contingency plan** Verify the testing and effectiveness of contingency plans as required or proceduralized. (Apply)

V. **Materials and Supply Chain Management (17 Questions)**

A. **Receipt of materials**

1. **Incoming materials** Describe and use processes to receive, label, and store incoming materials (raw materials, bulk chemicals,

components, and product labeling) and take appropriate action on deviations (damaged materials, materials from unapproved suppliers, and missing documentation). (Apply)

2. **Inventory controls** Describe and use procedures for documenting inventory transactions, updating changes in material status, allocation, and "stop shipments" for quality holds. (Apply)

B. **Sampling processes**

1. **Sampling plans** Review sampling plans for representative sampling, appropriate sample size, and test or inspection criteria. (Apply)
2. **Sampling environment** Differentiate and apply the requirements for sampling environment and utensils to the type of the material being sampled. (Apply)
3. **Cleaning** Ensure the sampling environment is appropriately cleaned and monitored and that sampling utensils are appropriately cleaned or are single-use. (Apply)

C. **Material storage, identification, and rotation**

1. **Storage suitability** Ensure the storage environment is suitable, controlled, and monitored as required for the type of materials. (Apply)
2. **Labelling of stored materials** Confirm that the identification label for stored materials contains the required information. (Apply)
3. **Stock rotation** Define and use stock rotation requirements such as first-in/first-out (FIFO) and first-expired/first out (FEFO). (Apply)
4. **Retest dates vs. expiration dates** Describe the difference between retest dates and expiration dates. (Understand)
5. **Mix-up risk** Identify potential sources of mix-up and identify methods to minimize their risk (material segregation, labeling, special storage for rejects, control of material returns, lot-control methods, and special processes for materials with similar names). (Analyze)

D. **Shipping and distribution**

1. **Temperature-sensitive requirements** Identify special requirements for temperature-sensitive products including tertiary packaging, design, and monitoring devices. (Analyze)
2. **Special requirements** Determine specific product requirements and apply them to routine shipping processes. (Apply)
3. **Report requirements** Analyze shipping reports and transportation requirements in accordance with good distribution practices. (Analyze)
4. **Supply chain security** Identify and apply the various means to secure the supply chain including tamper-evident seals, shipping

manifests, verification of documentation, barcoding, radio frequency identification (RFID), and serialization. (Apply)

E. **Traceability and sourcing**

1. **Traceability requirements** Define and differentiate the requirements for traceability of incoming materials, intermediates, and finished drugs. (Apply)

2. **Biological agent requirements** Identify and apply the requirements related to biological agents such as bovine spongiform encephalopathy (BSE) and transmissible spongiform encephalopathy (TSE). (Apply)

3. **Pedigree and sourcing requirements** Identify and apply requirements for maintaining pedigree and sourcing details for active pharmaceutical ingredients (APIs), biological starting materials, excipients, intermediates, and finished products. Document the supply chain from raw materials through wholesale or retail to end user. (Apply)

F. **Salvaged/returned goods and destruction**

1. **Disposition** Review salvaged and returned goods and evaluate them for disposition. (Evaluate)

2. **Destruction facilities and processes**. Determine the destruction requirements for materials including suitable facilities and processes. (Apply)

VI. **Sterile and Nonsterile Manufacturing Systems (22 Questions)**

A. **Master batch and completed batch records**

1. **Required elements** Review batch records for required elements including proper issuance, sections on yields, critical manufacturing step verification, processing instructions, and hold times. (Apply)

2. **Record processing requirements** Confirm that batch records meet requirements for execution, review, and disposition decisions. (Apply)

B. **Production operations**

1. **Application factors** Describe and differentiate the requirements for manufacturing processes according to their application: human drugs, veterinary drugs, or biologics. (Apply)

2. **Process operations**
Understand and differentiate between continuous and batch manufacturing. (Apply)

3. **Utility requirements** Identify the facility and utility requirements that are appropriate for different production environments and product types including sterile vs. nonsterile manufacturing,

solid and semisolid dosage forms, liquids, creams, ointments, and combination products. (Analyze)

4. **Sanitization and protection** Identify various production operations that require gowning, sanitization, hygiene, and other product-protective steps. (Apply)

C. **In-process controls**

1. **In-process testing** Identify appropriate tests for each step in the manufacturing process and review results. (Analyze)
2. **Critical process parameters (CPPs)** Monitor critical process parameters (CPPs). (Analyze)
3. **Process capability** Understand the importance of conducting process capability studies, calculate Cp and Cpk, and monitor process capability. (Apply)
4. **Specification limits** Assess specification limits in relation to registration or compendial requirements. (Evaluate)

D. **Dispensing and weighing controls**

1. **Staging areas** Review product dispensing and after-dispensing staging areas to determine if they meet requirements. (Apply)
2. **Dispensing materials** Identify the requirements for using weighing equipment and handling utensils for dispensing raw materials or intermediates including proper cleaning, labeling, and environmental controls, based on the type of material and manufacturing process being used. (Analyze)

E. **Requirements for critical unit processes**

1. **Process parameters** Use required critical process parameters (CCPs) for unit processes such as sterilization or sterilizing filtration, aseptic filling, depyrogenation, lyophilization, other drying processes, tablet granulation and compression, terminal sterilization, and cream or ointment emulsification. (Apply)
2. **Validation studies** Explain and evaluate validation studies, specifically the methodologies and acceptance criteria required before implementing critical unit processes. Explain and evaluate validation studies such as requirements for aseptic processes including process simulations ("media fills") and temperature controls. (Evaluate)
3. **Unit operations**
   Assess unit processes and their validations for deviations requiring investigation. (Analyze)
4. **Operating procedures** Review qualification and validation results and confirm that validated and qualified parameters are reflected in operating procedures or batch records. (Analyze)

5. **Re-evaluation and revalidation** Determine appropriate criteria and frequency for re-evaluation and revalidation of unit processes. (Evaluate)

6. **Environmental monitoring requirements** Differentiate between environmental monitoring requirements for different manufacturing area classifications. (Apply)

7. **Environmental monitoring tools** Describe and use various monitoring tools to measure viable and nonviable particulates, pressure differentials, temperature, and humidity. (Apply)

F. **Contamination and cross-contamination**

1. **Sources** Identify potential sources for, and implement controls to minimize, contamination and cross-contamination. (Analyze)

2. **Risk mitigation** Describe and apply various techniques for mitigating the risk of contamination and cross-contamination including cleaning; facility, utility, and equipment design; material and personnel flow; qualified disinfectants; operator training; validation; and monitoring. (Apply)

G. **Reprocessed and reworked materials**

1. **Disposition process** Distinguish reprocessing from reworking and apply appropriate documentation, approval, and disposition methods for these materials. (Apply)

2. **Storage** Describe and apply requirements for segregation and secure storage of these materials. (Apply)

VII. **Filling, Packaging, and Labeling (18 Questions)**

A. **Filling operations and controls**

1. **Materials control** Develop and review procedures to ensure the identity, strength, and purity of specified materials (e.g., liquids, powders, ointments, tablets, capsules, and suspensions) and to prevent them from being altered. (Evaluate)

2. **Filling equipment control** Analyze the controls needed for various types of production equipment and processes and ensure that the appropriate controls are established to verify filling criteria. (Analyze)

3. **Contamination controls** Identify controls to prevent microbial and other contamination at all stages of filling. (Apply)

4. **Staged materials** Review staged materials and confirm that they are approved for use. (Apply)

5. **Status labeling** Identify and apply proper status labeling throughout the process. (Apply)

B. **Environmental monitoring**
Use various monitoring techniques (e.g., active air sampling, settling plates, swab sampling, nonviable particle counting, and contact plates for surfaces and people) to determine that appropriate environmental conditions are maintained during production operations. (Analyze)

C. **In-process and finished goods inspections**

1. **Inspections** Develop criteria for in-process and finished goods inspections of filled and packaged materials including seal tests, torque testing, and bottle rejection systems. (Create)
2. **Vision and detection systems** Ensure that vision and detection systems are qualified, calibrated, and challenged as required for the system. (Apply)
3. **Defect characterizations** Ensure that defect characterizations are identified for each product and can be detected by inspection or test. (Analyze)
4. **Equipment failure detection** Confirm by inspection or test that equipment failures can be detected. (Apply)

D. **Product inspection**

1. **Staff evaluation** Ensure that staff who perform manual and semi-automatic inspections are properly trained and that their inspections meet reproducibility requirements. (Analyze)
2. **Inspector requirements** Establish requirements for inspectors to have periodic eye examinations. Confirm and document that they take frequent breaks from inspection. (Apply)
3. **Automated inspection processes** Ensure that automated inspection processes are validated. (Apply)

E. **Packaging operations and controls**

1. **Content protection** Develop and apply procedures to prevent the environment or events from altering the identity, strength, purity, and quality of the package content. (Create)
2. **Qualification and maintenance of equipment** Ensure that equipment used in packaging operations is qualified and maintained. (Apply)
3. **Line clearance operations** Determine that line clearance is performed and documented. (Analyze)
4. **Quality check criteria** Identify and apply specified criteria when quality checks are performed. (Analyze)
5. **Cut-label procedures** Apply appropriate procedures for cut labels and splices. (Apply)
6. **Hand-applied label procedures** Ensure that hand-applied labels are 100% inspected. (Apply)

7. **Packaging controls** Distinguish between controls needed for different types of packaging processes. (Apply)

8. **Contamination controls** Identify controls to prevent microbial and other contamination at all stages of packaging. (Analyze)

9. **Tamper-evident packaging** Ensure that tamper-evident and child-proof packaging requirements are established for required products. (Apply)

F. **Labeling operations and controls 1. Label printing in packaging** Confirm and document that any printing done separately or during packaging is performed correctly. (Analyze)

2. **Quality of print used 3. Label changes** Determine whether regulatory notification and approval is required for proposed label changes. (Evaluate)

4. **Label reconciliation** Confirm that label reconciliation is performed and documented, and discrepancies are investigated. (Analyze)

5. **Unused labels** Confirm that procedures are established and used for controlled, unused batch-coded labels and labeling materials. (Analyze)

6. **Label production** Define terms related to offline printing, roll label splicing, gang printing, secure storage, and destruction. (Understand)

7. **Access control**
Ensure that any type of print information (engraved and embossed) on packaging materials is clear and resistant to fading, smudging, or erasure. (Apply)
Ensure that controls are established for the creation, storage, and issuance of labeling such as product labels, package inserts, and printed cartons. (Analyze)

G. **Filling and packaging records**

1. **Terms** Define terms related to these records including evidence of line clearance, printed material reconciliation, and yields. (Understand)

2. **Setup instructions** Ensure that packaging line setup instructions are appropriate for all components. (Apply)

VIII. **Product Development and Technology Transfer (13 Questions)**

A. **Quality by design concepts**

1. **Critical quality attributes (CQAs) and critical process parameters (CPPs)** Identify critical quality attributes (CQAs) for products and critical process parameters (CPPs) for processes. (Evaluate)

2. **Design space** Define the concept of design space as it is used throughout the product lifecycle. (Understand)

3. **Process analytical technology (PAT) tools** Identify process analytical technology(PAT) tools including multivariate data analysis, process analyzers, and process and endpoint controls. Describe their use in supporting the manufacture of quality products. (Understand)

B. **Phase-appropriate Good Manufacturing Practices (GMP) requirements**

1. **Product life cycle development** Apply phase appropriate Good Manufacturing Practices (GMPs) throughout the product life cycle. (Apply)
2. **Development phases** Identify recommendations and requirements in relation to phases of development including method qualification/validation, comparability protocols, and adoption of critical process parameters and specifications. (Understand)
3. **Combination products** Identify Good Manufacturing Practices (GMP) requirements and various studies required for combination drug-device or drug-delivery products. (Understand)
4. **Clinical trials material** Describe and apply requirements for production and packaging of clinical trials material and investigational medicinal products (IMPs). (Apply)

C. **Raw materials, packaging, and infrastructure for product development**
Select appropriate development studies for raw material selection and evaluate the results to determine their critical quality attributes. (Analyze)

D. **New product development studies and reports**
Analyze studies and reports including stability reports, material compatibility, method development, and development reports to support product development and submissions. (Evaluate)

E. **Scale-up and transfer activities**

1. **Development and validation principles** Identify and distinguish between development and validation studies. (Understand)
2. **Technology transfer types** Define different types of technology transfer including manufacturing site change and analytical laboratory site change. Analyze inter-site comparison of results. (Analyze)
3. **Successful technology transfer** Define various studies including ranging, capability, in-process control, hold times, and shipping to ensure successful transfer between development and commercial processes. (Evaluate)

LEVELS OF COGNITION BASED ON BLOOM'S TAXONOMY—REVISED (2001)

In addition to *content* specifics, the subtext detail also indicates the intended *complexity level* of the test questions for that topic. These levels are based on the revised "Levels of Cognition" (from Bloom's Taxonomy, 2001) and are presented below in rank order, from least complex to most complex.

Remember

Be able to remember or recognize terminology, definitions, facts, ideas, materials, patterns, sequences, methodologies, principles, etc. (Also commonly referred to as recognition, recall, or rote knowledge)

Understand

Be able to read and understand descriptions, communications, reports, tables, diagrams, directions, regulations, etc.

Apply

Be able to apply ideas, procedures, methods, formulas, principles, theories, etc., in job-related situations.

Analyze

Be able to break down information into its constituent parts and recognize the parts' relationship to one another and how they are organized; identify sublevel factors or salient data from a complex scenario.

Evaluate

Be able to make judgments regarding the value of proposed ideas, solutions, methodologies, etc., by using appropriate criteria or standards to estimate accuracy, effectiveness, economic benefits, etc.

Create

Be able to put parts or elements together in such a way as to show a pattern or structure not clearly there before; able to identify which data or information from a complex set is appropriate to examine further or from which supported conclusions can be drawn.

Glossary

A

Accreditation Body (AB)—An organization that provides accreditation services. Accreditation means that certification and inspection bodies have been assessed against recognized standards to demonstrate their competence, impartiality and capability.

accuracy—The degree to which the result of a measurement, calculation, or specification conforms to the correct value or a standard.

action level—Specification or limit (in fact, more correctly called an *action limit* rather than an action level) established in published regulation (or a guidance or guideline or standard that is regarded by competent authorities as de facto regulation) that, when exceeded, requires immediate intervention, including investigation of cause, immediate remediation, and/or corrective action.

active ingredient—Any component that is intended to furnish pharmacological activity or other direct effect in the diagnosis, cure, mitigation, treatment, or prevention of disease, or to affect the structure or any function of the body of humans or other animals. The term includes those components that may undergo chemical change in the manufacture of the drug product and be present in the drug product in a modified form intended to furnish the specified activity or effect.

actual yield—The quantity that is actually produced at any appropriate phase of manufacture, processing, or packing of a particular drug product.

adulterated (United States term)—Violative state of a drug or device as described in Section 501, "Adulterated Drugs and Devices," of the Food, Drug, and Cosmetic Act (FDCA). Any of the following are bases for adulteration: (i) if it is a drug or device containing a filthy, putrid, or decomposed substance (that is, an animal part or insect) or manufactured under unsanitary conditions, or if the drug or device may have been rendered injurious to health; (ii) if it is a drug and the methods used in, or the facilities or controls used for, its manufacture, processing, packing, or holding do not conform to current good manufacturing practice; (iii) if it is a drug recognized by the United States Pharmacopeia (USP) and its strength differs from, or its quality or purity falls below the standards set forth in the USP (unless exceptions are contained in the labeling); (iv) if it is a drug and any substance has been mixed or packed with it so as to reduce its quality or strength or substituted wholly or in part for the active pharmaceutical ingredient (API) (for example, oversulfated chondroitin sulfate for heparin).

affidavit—A written statement made under sworn oath, or affirmation, before an authorized officer.

aerosol generator—Instrument capable of producing particulate matter in an appropriate particle size (for example, 0.05 to 2 μm).

air exchange rate—Air changes per unit of time, which is calculated by dividing the volumetric flow rate of incoming air by the volume of the space/room.

alert level—Measurement magnitude set by an active pharmaceutical ingredient (API) or finished pharmaceutical manufacturer that gives early warning of a drift from normal conditions and that, when exceeded, should result in increased attention to the process.

annual review—An evaluation, conducted at least annually, that assesses the quality standards of each drug product to determine the need for changes in drug product specifications or manufacturing or control procedures.

antisepsis—Act or process of chemically reducing viable organisms on living tissue, including skin, oral cavities, and open wounds. (*Antiseptic* is the chemical agent.)

as-built—Occupancy state in which an installation is complete with all services connected and functioning but with no production equipment, materials, or personnel present.

aseptic processing area (APA)—Controlled environments in which the air supply, materials, equipment, and personnel are regulated to control microbial and particulate numbers to acceptable levels. APA consists of "critical (processing) area" and "direct support area."

aseptic technique—A technique requiring the use of various barriers to pre- vent the transfer of microorganisms and other contaminates from personnel and the environment to products being manufactured, inspected, labeled, or packaged.

assay—Qualitative or quantitative analysis of a drug to determine its components.

at rest—Occupancy state in which an installation is complete with equipment installed and operating in a defined manner but with zero personnel present.

audit—The on-site verification activity, such as inspection or examination, of a process or quality system, to ensure compliance to requirements. An audit can apply to an entire organization or might be specific to a function, process, or production step.

audit trails—An electronic record of the evidence behind any transaction in an electronic system that ensures all changes are apparent.

B

Back Room—An area where designated individuals gather and review information and documentation for regulatory inspections.

batch number—Any distinctive combination of letters, numbers, or symbols from which the complete history of the manufacture, processing, packing, holding, and distribution of a batch or lot of drug product or other material can be determined.

bioburden—The total quantity of recoverable microbes present on a defined surface or surface area, or within a nonsterile solid/powder or liquid drug, intermediate, or raw material.

biological product—A wide range of products such as vaccines, blood and blood components, allergenics, somatic cells, gene therapy, tissues, and recombinant therapeutic proteins. Biologics can be composed of sugars, proteins, or nucleic acids, or complex combinations of these substances, or may be living entities such as cells and tissues.

bulk production batch—A batch of product of a size described in the application for a marketing authorization, either ready for assembly into final containers or in individual containers ready for assembly into final packs. A bulk production batch may, for example, consist of a bulk quantity of liquid product, of solid dosage forms such as tablets or capsules, or of filled ampules.

C

calibration—The comparison of a measurement instrument or system of unverified accuracy to a measurement instrument or system of known accuracy to detect any variation from the required performance specification.

can—Indicates a possibility or a capability in regard to a standard.

Certification Body (CB)—An organization accredited to audit and issue certification confirming that an organization meets the requirements of a standard

certification of the finished product batch—Certification in a register or equivalent document by a European Union qualified person, as defined in Article 51 of Directive 2001/83/EC and Article 55 of Directive 2001/82/EC, before a batch is released for sale or distribution.

change control—A written procedure that describes the action to be taken if a change is proposed to facilities, materials, equipment, and/or processes used in the fabrication, packaging, and testing of drugs, or that may affect the operation of the quality or support system.

chiral—Not superimposable with its mirror image, as applied to molecules, conformations, and macroscopic objects, such as crystals. The term has been extended to samples of substances whose molecules are chiral, even if the macroscopic assembly of such molecules is racemic.

chromatogram—A graphical or other representation of detector response, effluent concentration, or other quantity used as a measure of effluent concentration, versus time, volume, or distance. Idealized chromatograms are represented as a sequence of Gaussian peaks on a baseline.

classification—Level (or process of specifying or determining the level) of airborne particulate cleanliness applicable to a cleanroom or clean zone, expressed in terms of an ISO Class N, which represents the maximum allowable concentrations (in particles per cubic meter of air) for considered sizes of particles (see ISO 14644-1).

classified area/cleanroom—Room or operating area designated by Grades A, B, C, or D as defined in *EU Guide to GMP*, or designated Class 4.8, 5, 6, 7, or 8 per ISO cleanroom standards; a classified area constructed (usually of durable walls as opposed to curtains) and used in a manner so as to minimize the introduction, generation, and retention of particles, and in which other relevant parameters, for example, temperature, humidity, and pressure, are controlled as necessary (see ISO 14644-1). A curtained area satisfying some of the definition of classified area, except its wall construction, is known as a *clean zone*, the distinction being necessary for purposes of the latter's moveable/flexible versus permanent nature and the degree of pre-use validation or qualification.

clean technique—A technique involving a reduction of the number of microorganisms and other contaminates to minimize the risk of transmission from the environment to the products being manufactured, inspected, labeled, or packaged.

cleaning—Physical removal of soil, organic debris, and particulate from surfaces.

cleaning validation—Documented evidence with a high degree of assurance that a cleaning process will consistently yield product contact surfaces that meet predetermined acceptance criteria and critical quality attributes (visual and residual levels).

cleaning validation master plan—Overview document that describes, at a high level, the entire site's cleaning validation strategy, structure, content, and actual plan or schedule. (The latter may be an attachment or a stand-alone referenced document.)

cleanroom—A room that is maintained virtually free of contaminants, such as dust or bacteria, used in laboratory work and in the production of precision parts for electronic or aerospace equipment.

combination products—Therapeutic and diagnostic products that combine drugs, devices, and/or biological products. Because combination products involve components that would normally be regulated under different types of regulatory authorities, and frequently by different FDA Centers, they raise challenging regulatory, policy, and review management challenges. Differences in regulatory pathways for each component can impact the regulatory processes for all aspects of product development and management, including preclinical testing, clinical investigation, marketing applications, manufacturing and quality control, adverse event reporting, promotion and advertising, and post- approval modifications.

competent authority—Country-specific agency or body designated to administer that country's laws/statutes governing the marketing of safe and effective drugs, medical devices, foods, dietary supplements, cosmetics, and so on; competent authority fulfils this role by enacting regulations, guidelines, guidances, and so on, in line with those statutes/laws. Examples include the

United States Food and Drug Administration (FDA), the United States Department of Agriculture (USDA), Health Canada, the United Kingdom's Medicines and Healthcare Products Regulatory Agency (MHRA) and Veterinary Medicines Directorate, Japan's Ministry of Health, Labor, and Welfare (MHLW), and Australia's Therapeutic Goods Administration (TGA).

component—Any ingredient intended for use in the manufacture of a drug product, including those that may not appear in such drug product.

confirmation—Signed statement that a process or test has been conducted in accordance with good manufacturing practices (GMP) and the relevant marketing authorization, as agreed in writing with the qualified person (QP) responsible for certifying the finished product batch before release. "Confirm" and "confirmed" have equivalent meanings.

conformity assessment—Any activity performed to determine, directly or indirectly, that a process, product, or service meets relevant technical standards and fulfills relevant requirements.

contact plate (touch plate)—A petri dish usually measuring 55 mm in diameter and containing convex-shaped agar that forms a dome above the dish to permit sampling of flat surfaces for microorganisms. Note: RODAC is the brand name (Replicate Organism Detection and Counting Plate) of a contact plate the trademark of which is currently owned by Becton Dickinson.

contamination—The undesired introduction of impurities of a chemical or microbiological nature, or of foreign matter, into or onto a raw material intermediate, active pharmaceutical ingredient (API), finished drug, primary surface, or equipment. A *contaminant* is any particulate, molecular, and/or biologic entity that can adversely affect a material or process (*biocontaminant* is a bacterial, mycoplasma, fungal, or protozoa contaminant).

continual improvement—Ongoing activities to evaluate and positively change products, processes, and the quality system to increase effectiveness.

control chart—A statistical tool including upper and lower control limits and the process average, which visually displays process performance.

control limits—The area three standard deviations on either side of the centerline, or mean, of data plotted on a control chart.

control number. See **batch number.**

control strategy—Planned set of controls, derived from current product and process understanding, that assures process performance and product quality. The controls can include parameters and attributes related to drug substance and drug product materials and components, facility and equipment operating conditions, in-process controls, finished product specifications, and the associated methods and frequency of monitoring and control.

controlled area—A nonclassified room or operating area designed to control/minimize the presence, proliferation, and/or ingress of particulates, and in which specific environmental conditions (for example, temperature, humidity, directional airflow, and viable and nonviable particulate limits) are defined and monitored to prevent contamination of exposed products.

correction—Repair, rework, or adjustment relating to the disposition of an existing discrepancy (also called *remedy, remediation*); usually the first step in a corrective action.

corrective action—Action taken to eliminate the causes of an existing discrepancy or other undesirable situation to prevent recurrence.

critical area—An area in which the sterilized drug product, containers, and closures are exposed to environmental conditions that must be designed to maintain product sterility.

critical process parameter (CPP)—Quantifiable equipment setting whose variability has an impact on a critical quality attribute and therefore should be monitored or controlled to ensure that the process produces the desired quality, purity, potency, and safety.

critical quality attribute (CQA)—Physical, chemical, biological, or microbiological property or characteristic that should be within an appropriate limit, range, or distribution to ensure the desired product quality.

customer—Person or organization (internal or external) that receives a product or service anywhere along the product's life cycle.

D

D-value—decimal reduction time is the time required at a given temperature to kill 90% of the exposed microorganisms.

decision maker(s)—Person(s) with the competence and authority to make appropriate and timely quality risk management decisions.

design of experiments (DOE) or **experimental design**—Statistical technique used for planning, conducting, analyzing, and interpreting sets of experiments, aimed at making sound decisions most efficiently.

design space—Multidimensional combination and interaction of input variables (for example, material attributes) and process parameters that have been demonstrated to provide assurance of quality. Note: Working within the design space is not considered as a change; however, movement out of the design space is considered to be a change and would normally initiate a regulatory post-approval change process.

detectability—Ability to discover or determine the existence, presence, or fact of a hazard.

discrepancy—Datum or result outside of the expected range; an unfulfilled requirement; may be called nonconformity, defect, deviation, out-of-specification, out-of-limit, or out-of-trend.

disinfection—Act or process of chemically or physically destroying or removing vegetative pathogens on inanimate objects (a *disinfectant* is a chemical or physical agent that assists disinfection).

don—Refers to the practice of putting on (donning) protective gear, clothing, and uniforms.

donning. See **don.**

doff—Refers to the practice of taking off (doffing) protective gear, clothing, and uniforms.

doffing. See **doff.**

drug—A substance recognized by an official pharmacopoeia or formulary. A substance intended for use in the diagnosis, cure, mitigation, treatment, or prevention of disease. A substance (other than food) intended to affect the structure or any function of the body. A substance intended for use as a component of a medicine but not a device or a component, part, or accessory of a device.

E

effectiveness—The state of having produced a decided-on or desired effect.

enantiomers—Compounds with the same molecular formula as the drug substance, which differ in the spatial arrangement of atoms within the molecule and are nonsuperimposable mirror images.

endotoxin—The lipopolysaccharide remnant of the outermost cell wall layer of a dead Gram-negative bacterium, which can be pyrogenic (fever-causing) to mammals, especially humans. Endotoxin is a potential contaminant of water, including pharmaceutical water. All pharmaceutical water for injection and pure steam have, among other specifications, an upper limit of not more than 0.25 endotoxin units/mL.

Establishment Inspection Report (EIR)—A formal written report by an FDA Inspector documenting all activities that occurred during the inspection.

excipient—An inert substance used as a diluent or vehicle for a drug.

F

failure mode analysis (FMA)—A procedure used to determine which malfunction symptoms appear immediately before or after a failure of a critical parameter in a system. After all possible causes are listed for each symptom, the product is designed to eliminate the problems.

failure mode and effects analysis (FMEA)—A systematized group of activities performed to recognize and evaluate the potential failure of a product or process and its effects, identify actions that could eliminate or reduce the occurrence of the potential failure, and document the process.

failure mode, effects, and criticality analysis (FMECA)—A procedure performed after a failure mode and effects analysis to classify each potential failure effect according to its severity and probability of occurrence.

FDA Form 482 Notice of Inspection—the form provided to top management or designee, formally notifying the organization of FDA's intention to conduct an inspection of the establishment.

FDA Form 483 Inspectional Observations—The form used to notify the inspected establishment's top management in writing, of significant objectionable conditions relating to products and/or processes, or other violations of the FD&C Act which were observed during the inspection.

FDA Form 484 Receipt for Samples—The form provided to top management or designee to document any physical product samples that an Inspector removed from the establishment.

finished product—Batch of product in its final pack for release to the market.

flora—Identified microbes (to species level) found at a specific time in an operating area(s), room(s), or the collection of rooms used in manufacturing, testing, and warehousing (as in the whole facility); this includes microbes isolated from primary surfaces and air, as well as product, intermediates, and raw materials tested and determined to be contaminated after a specific area or suite exposure.

Front Room—Area where designated individuals directly interface with an investigator during a regulatory inspection.

G

gage repeatability and reproducibility (GR&R)—The evaluation of a gauging instrument's accuracy by determining whether its measurements are repeatable (there is close agreement among a number of consecutive measurements of the output for the same value of the input under the same conditions) and reproducible (there is close agreement among repeated measurements of the output for the same value of input made under the same operating conditions over a period of time).

generic drug—A drug that is the same as a brand-name drug in dosage, safety, strength, how it is taken, quality, performance, and intended use.

gram-negative—Bacteria that lose the crystal violet stain (and take the color of the red counterstain) in Gram's method of staining. This is characteristic of bacteria that have a cell wall composed of a thin layer of a particular substance called peptidoglycan.

gram-positive—Bacteria that retain the color of the crystal violet stain in the Gram stain. This is characteristic of bacteria that have a cell wall composed of a thick layer of a particular substance called peptidologlycan.

H

harm—Damage to health, including the damage that can occur from loss of product quality or availability.

hazard—Potential source of harm (ISO/IEC Guide 51).

hazard analysis and critical control points (HACCP)—A quality management system for effectively and efficiently ensuring farm-to-table food safety in the United States. HACCP regulations for various sectors are established by the Department of Agriculture and the Food and Drug Administration.

housekeeping—General cleaning (including sweeping) and removal of accumulated process waste, dirty equipment, utensils, and other nonproduct materials resulting from normal manufacturing or warehousing activities.

I

importer—Holder of the marketing authorization required by Article 40.3 of Directive 2001/83/EC and Article 44.3 of Directive 2001/82/EC for importing medicinal products from countries outside the European Economic Community (EEC).

in vitro—Outside the living body and in an artificial environment.

in vivo—Within the living body.

inactive ingredient—Any component other than an active ingredient.

in-process control—One of various control strategies that measures a critical quality attribute (CQA) at a particular point in an active pharmaceutical ingredient (API) or drug manufacturing process.

installation (vis-à-vis cleanrooms)—The ensemble that includes a cleanroom or one or more clean zones, together with all associated structures, air treatment systems, services, and utilities.

intermediate—Material produced during manufacture that undergoes further change or purification.

investigation—generally focuses on certain aspects of a facility's operations or on a predetermined set of compliance concerns and do so in substantially more depth than an inspection.

L

laboratory—A test facility that can include chemical, metallurgical, dimensional, physical, electrical, and reliability testing or test validation.

laboratory scope—A record containing the specific tests, evaluations, and calibrations a laboratory has the ability and competency to perform, a list of the equipment it uses, and a list of the methods and standards to which it adheres for each of these.

Line clearance—the process that ensures equipment and work areas are free of products, materials, packaging, labeling, and procedures, work instructions, and forms from the previous job.

lot—A batch, or a specific identified portion of a batch, having uniform character and quality within specified limits; or, in the case of a drug product produced by a continuous process, it is a specific identified amount produced in a unit of time or quantity in a manner that assures its having uniform character and quality within specified limits.

lot number. See batch number.

lyophilization—Freeze-drying process in which water is removed from a product after it is frozen and placed under a vacuum, allowing the ice to change directly from solid to vapor without passing through a liquid phase.

M

Master cell bank (MCB) - An aliquot of a single pool of cells that generally has been prepared from the selected cell clone under defined conditions, dispensed into multiple containers, and stored under defined conditions. The MCB is used to derive all working cell banks. The testing performed on a new MCB (from a previous initial cell clone, MCB, or WCB) should be the same as for the MCB unless justified.

may—Indicates a permission/recommendation in regard to a standard.

medical device—An instrument, apparatus, implement, machine, contrivance, implant, in vitro reagent, or other similar or related article, including a component part or accessory that is:

- Recognized in the official National Formulary, or the United States Pharmacopoeia, or any supplement to them,
- Intended for use in the diagnosis of disease or other conditions, or in the cure, mitigation, treatment, or prevention of disease, in humans or other animals, or
- Intended to affect the structure or any function of the body of man or other animals, and which does not achieve its primary intended purposes through chemical action within or on the body of humans or other animals and that is not dependent on being metabolized for the achievement of any of its primary intended purposes.

microbiological contamination—Presence of one or more various bacteria, yeasts, mold, protozoa, or their toxins or by-products (for example, endotoxins or exotoxins) that could adversely affect the product or a patient's health and safety.

microorganisms (or microbes)—Unicellular living creatures/organisms that include the bacteria, molds, yeasts, protozoa, mycoplasma, and archaea. Because viruses and prions do not satisfy the modern biology text definition of life, they are not organisms and, hence, not microorganisms.

Mid Room—This serves as the staging area for SME (Subject Matter Experts) preparation during regulatory inspections.

most responsible person—The designated individual who owns the most accountability for the operations of the facility and to ensure the legalities of the facility is compliant.

must—Indicates a requirement in regard to a standard.

mutual recognition agreement (MRA)—Appropriate arrangement between the European Economic Community (EEC) and an exporting third country mentioned in Article 51(2) of Directive 2001/83/EC and Article 55(2) of Directive 2001/82/EC.

mycobacterial—Of or relating to *Mycobacterium tuberculosis.*

N

nonconforming record (NCR)—A permanent record—made in writing—for accounting and preserving the knowledge of a nonconforming condition for the purposes of documenting facts or events.

nonconformity—Deficiency in a characteristic, product specification, process parameter, record, or procedure that renders the quality of a product unacceptable, indeterminate, or not according to specified requirements.

Notified Body (NB)—An organization in the European Union that has been designated by a Member State to assess the conformity of certain products, before being placed on the E.U. market, with the applicable essential technical requirements.

O

occupancy state—Status of a classified area with regard to the presence/absence of equipment and personnel and operation of the heating, ventilation, and air conditioning (HVAC).

official article—Term used by the United States Pharmacopeia (USP) to refer to any of the following materials: a drug substance, drug product, dietary ingredient, dietary substance, or excipient.

official compendium—Publication known as the United States Pharmacopeia (USP) according to Section 501(b) of the Federal Food, Drug, and Cosmetic Act (FDCA) (as amended) of the United States.

operational—Occupancy state in which an installation is functioning in a defined manner, with a specified number of personnel present and working in a defined manner.

out-of-control process—A process in which the statistical measure being evaluated is not in a state of statistical control. In other words, the variations between the observed sampling results cannot be attributed to a constant system of chance causes.

out-of-specification (OOS)—All in-process laboratory tests that are outside of established specifications.

out of tolerance (OOT)—During calibration, an item that is not within specification limits.

P

particle/particulate matter—A minute piece of matter with defined physical boundaries; a solid or liquid object that, for purposes of classification of cleanrooms, falls within the size range of 0.1 to 5 microns (µm). A *viable* particle is one that consists of, or supports, one or more live microorganisms; a *nonviable* particle is one that consists of zero culturable microorganisms.

particle size—Diameter of a sphere that produces a response, by a given particle-sizing instrument, that is equivalent to the response produced by the particle being measured.

pathogenic—Regarding a microbe that is harmful to humans or animals.

percentage of theoretical yield—The ratio of the actual yield (at any appropriate phase of manufacture, processing, or packing of a particular drug product) to the theoretical yield (at the same phase), stated as a percentage.

pest control—System for preventing, evaluating, and eliminating infestation by rodents, insects, birds, and other vermin.

pharmacovigilance—the detection, assessment, understanding and prevention of adverse effects, particularly long-term and short-term side effect of medicines.

pilot scale—Manufacturing of drug product by a procedure fully representative of and simulating that used for full manufacturing scale.

polymorphism—The occurrence of different crystalline forms of the same drug substance. This may include solvation or hydration products (also known as *pseudopolymorphs*) and amorphous forms.

precision—The number of significant digits to which a value has been reliably measured.

preventive action—Action taken to eliminate the cause of a potential discrepancy or other potential undesirable situation to make such an occurrence less probable.

primary surface—The first solid surface contacted in moving away from an exposed product. In a room, this includes the walls, floors, ceilings, and doors, the room side of air diffusers but not the duct side, and floor drain covers, but not the piping system. A primary surface is not a product contact surface.

prion—Infectious agent composed of protein, which can transmit its (misfolded) configuration to native, similar proteins in an infected animal's central nervous system, causing transmissible spongiform encephalopathy.

procedure—The steps in a process and how these steps are to be performed in order for the process to fulfill a customer's requirements; usually documented.

process capability—A statistical measure of the inherent process variability of a given characteristic. The most widely accepted formula for process capability is six sigma.

process capability index—The value of the tolerance specified for the characteristic, divided by the process capability. The several types of process capability indexes include the widely used Cpk and Cp.

process challenge device (PCD)—An item designed to constitute a defined resistance to a sterilization process and used to assess performance of the process parametric release - declaration that product is sterile, based on records demonstrating that the process parameters were delivered within specified tolerances

process control—The method for keeping a process within boundaries; the act of minimizing the variation of a process.

process development studies—Experiments that help rule in or rule out the choice and sequence of specific unit operations and unit processes and their associated detailed choices of equipment models and critical process parameters for the manufacturing of an API or a finished drug.

product/service—Intended results of activities or processes; products/services can be tangible or intangible.

product life cycle—All phases in the life of the product, from the initial development through marketing and the product's discontinuation.

Q

qualified person (QP)—The person defined in Article 48 of Directive 2001/83/EC and Article 52 of Directive 2001/82/EC.

qualitative analysis—Testing performed to establish the composition of natural or synthetic substances.

quality—A measure of a product's or service's ability to satisfy the customer's stated or implied needs; also, the degree to which a set of inherent properties of a product, system, or process fulfills requirements (see ICH Q6A definition specifically for quality of drug substance and drug products).

quality assurance (QA)—Proactive and retrospective activities that provide confidence that active pharmaceutical ingredient or drug product requirements are fulfilled.

quality by design (QbD)—A systematic approach to development that begins with predefined objectives and emphasizes product and process understanding and process control based on sound science and quality risk management.

quality control (QC)—Steps taken during the generation of a product or service to ensure that it meets requirements and that the product or service is reproducible.

quality management—Accountability for the successful implementation of the quality system.

quality objectives—Specific, measurable activities or processes designed to meet the organization's intentions and directions as defined in the quality policy.

quality plan—Documented result of quality planning that is disseminated to all relevant levels of the organization.

quality planning—Management activity that sets quality objectives and defines the operational and/or quality system processes and the resources needed to fulfill the objectives.

quality policy—Statement of intentions and direction issued by the highest level of the organization, related to satisfying customer needs. It is similar to a strategic direction, which communicates quality expectations that the organization is striving to achieve.

quality system—Formalized business practices that define management responsibilities for organizational structure, processes, procedures, and resources needed to fulfill product/service requirements, customer satisfaction, and continual improvement.

quality unit (QU)—Group organized within an organization to promote quality in general practice.

quantitative analysis—Analytical techniques used to identify the amount or concentration of an analyte and quantify any compound or substance in a sample.

R

reagent—A substance, other than a starting material or solvent, that is used in the manufacture of a new drug substance.

reconciliation—An accounting for the labels that were issued to the production order, consumed, returned to inventory, and/or destroyed.

reference sample—A sample of a batch of starting material, packaging material, or finished product stored for the purpose of being analyzed should the need arise during the shelf life of the batch concerned.

reprocessing—Subjecting all or part of a batch or lot of an in-process drug, a bulk process intermediate (final biological bulk intermediate), or a bulk drug of a single batch/lot to a previous step in the validated manufacturing process due to failure to meet predetermined specifications. Reprocessing procedures are foreseen as occasionally necessary, and are validated and preapproved by the quality control department or as part of the marketing authorization. Similarly, introducing an intermediate or active pharmaceutical ingredient, including one that does not conform to standards or specifications, back into the process and repeating a crystallization step or other appropriate chemical or physical manipulation steps (for example, distillation, filtration, chromatography, milling) that are part of the established manufacturing process. Continuation of a process step after an in-process control test has shown that the step is incomplete is considered to be part of the normal process, and not reprocessing.

representative sample—A sample that consists of a number of units that are drawn based on rational criteria, such as random sampling, intended to assure that the sample accurately portrays the material being sampled.

requirements—Explicit or implicit needs or expectations of the patients or their surrogates (for example, healthcare professionals, regulators, and legislators). In this document, "requirements" refers not only to statutory, legislative, or regulatory requirements, but also to such needs and expectations.

reserve sample—A sample representative of each lot in each shipment of each active ingredient, or a sample representative of each lot of drug product.

retain sample. See **reserve sample.**

retention sample—A sample of a fully packaged unit from a batch of finished product stored for identification purposes.

reworking—Subjecting an in-process drug, a bulk process intermediate (final biological bulk intermediate), or final product of a single batch or lot to an alternate manufacturing process due to a failure to meet predetermined specifications. Reworking is an unexpected occurrence and is not preapproved as part of the marketing authorization.

risk—Combination of the probability of occurrence of harm and the severity of that harm.

risk acceptance—Decision to accept risk (ISO Guide 73).

risk analysis—Estimation of the risk associated with the identified hazards.

risk assessment—Systematic process for organizing information to support a risk decision that is made within a risk management process (the process consists of the identification of hazards and the analysis and evaluation of risks associated with exposure to those hazards).

risk communication—The sharing of information about risk and risk management between the decision maker and other stakeholders.

risk control—Actions implementing risk management decisions (ISO Guide 73).

risk evaluation—Comparison of the estimated risk to given risk criteria using a quantitative or qualitative scale to determine the significance of the risk.

risk identification—Systematic use of information to identify potential sources of harm (hazards) referring to the risk question or problem description.

risk management—Systematic application of quality management policies, procedures, and practices to the tasks of assessing, controlling, communicating, and reviewing risk.

risk reduction—Actions taken to lessen the probability of occurrence of harm and the severity of that harm.

risk review—Review or monitoring of output/results of the risk management process considering (if appropriate) new knowledge and experience about the risk.

rounding—Used to shorten numbers by either increasing or decreasing a number to the next digit.

run chart—Plot of process output over time without superimposed control limits.

S

sample—In acceptance sampling, one or more units of product (or a quantity of material) drawn from a lot for purposes of inspection to reach a decision regarding acceptance of the lot.

sample item portion (SIP)—Defined portion of a health care product that is tested.

sanitization—The act or process, physical or chemical, of reducing viable organisms on a surface to a defined acceptance level (a *sanitizer* is a physical or chemical agent that assists sanitization); *sterilization* is a subset of sanitization. Surface sanitization concerns product contact surfaces as well as durable surfaces not far removed from those that contact product (for example, outside surfaces of equipment, table tops, floors, walls, ceilings, exterior of drains, air registers).

senior management—Top management officials in a firm who have the authority and responsibility to mobilize resources.

settle plate—Suitable container (for example, Petri dish) of appropriate size, containing an appropriate sterile culture medium, that is left open for a defined period to collect viable particles depositing from the air.

severity—A measure of the possible consequences of a hazard.

shall—Indicates a requirement in regard to a standard.

should—Indicates a recommendation in regard to a standard.

significant digit—Any of the digits of a number, beginning with the digit farthest to the left, that is not zero, and ending with the last digit farthest to the right that is either not zero or that is a zero but is considered to be exact.

small molecule drug—A low molecular weight organic compound that may help regulate a biological process. Most drugs are small molecules.

sporicide—A chemical agent that destroys bacterial and fungal spores when used in sufficient concentration for a specified contact time. It is expected to kill all vegetative organisms. Health Canada prohibits a supplier from labeling a non-sporicidal disinfectant with claims against the vegetative cells of spore-forming bacteria whose spores may be the primary means of spread of healthcare-associated infections, which could mislead users into assuming that the disinfectant has sporicidal effectiveness.

stakeholder—An individual or organization having an ownership or interest in the delivery, results, and metrics of the quality system framework or business process improvements; also, any individual, group, or organization that can affect, be affected by, or perceive itself to be affected by a risk. Decision makers might also be stakeholders. The primary stakeholders are the patient, healthcare professional, regulatory authority, and industry.

statistical process control chart—Plot of process output over time with a superimposed central tendency line and upper and lower control limit lines.

sterilization—The act or process, physical or chemical, of destruction or elimination of all viable organisms (including bacterial and fungal spores, viruses, protozoa) in the inanimate environment; never being considered absolute, it is characterized by a probability of the presence of one or more viable organisms, expressed as 10^{-n}. A *sterilant* is an agent that can assist sterilization; sterilants are liquid or vapor-phase agents. All sterilants are sporicides, but not all sporicides are sterilants.

stratified sampling—The process of collecting a representative sample by selecting units deliberately from various identified locations.

Subject Matter Experts (SMEs)—specific individuals who will be called upon to answer regulatory inspector questions.

T

theoretical yield—The quantity that would be produced at any appropriate phase of manufacture, processing, or packing of a particular drug product, based on the quantity of components to be used, in the absence of any loss or error in actual production.

total organic (or **oxidizable**) **carbon (TOC)**—An indirect measure of organic molecules, measured as carbon, present in pharmaceutical (that is, compendial) waters.

trend—A variable/attribute/characteristic's tendency, over time, to increase, decrease, or remain unchanged.

U

unclassified area—A room or area not designated by grades but that needs to be designed and maintained such that its environment does not adversely impact the quality, purity, and integrity of the products. Unclassified areas may or may not be controlled areas.

unidirectional airflow—Controlled airflow through the entire cross-section of a clean zone with a steady velocity and approximately parallel streamlines.

unit operation—Distinct physical changes or unit actions (for example, micronizing, mixing, drying, filtration).

unit process—Unit operations involving chemical changes.

Untitled Letter—A letter issued by FDA as an initial correspondence with regulated industry that cites violations that do not meet the threshold of regulatory significance for a Warning Letter.

V

validation—Confirmation, through the provision of objective evidence, that the requirements for a specific intended use or application have been fulfilled.

validation life cycle—Process, equipment, and facility assurance that depends on the following stages (separated in time) being executed properly:

- *Stage 1—Process, equipment, and/or facility design.* The commercial manufacturing process is defined during this stage based on knowledge gained through development and scale-up activities.
- *Stage 2—Process, equipment, and/or facility qualification.* During this stage, the process design is evaluated (according to preapproved protocol) to determine whether the process is capable of reproducible commercial manufacturing.
- *Stage 3—Process, equipment, and/or facility continued verification/ monitoring.* Ongoing assurance is gained during routine production that the process remains in a state of control.

- *Stage 4—Process, equipment, and/or facility retirement.* During this stage, any electronically stored data are transferred to replacement system(s) or simply preserved according to record retention policy. Also, to achieve "bookending," a final proving of proper calibration since the preceding set of calibrations is necessary to demonstrate that all instruments remained in a state of calibration prior to retirement.

verification—Confirmation, through the provision of objective evidence, that specified requirements have been fulfilled.

W

Warning Letter—A letter issued by the FDA if a company fails to adequately address all of the issues in a form FDA 483, or if the compliance issues identified during an inspection are more serious or widespread. Warning Letters typically are issued for violations of regulatory significance.

Working Cell Bank (WCB)—The WCB is prepared from aliquots of a homogeneous suspension of cells obtained from culturing the MCB under defined culture conditions.

working shift—Scheduled period of work or production, usually no greater than 12 hours in length, during which operations are conducted by a single, defined group of workers.

References and Suggested Reading

PUBLICATIONS

Anderson, P. F., and B. L. Wortman. 2004. *The Quality Auditor Primer.* 6th ed. West Terre Haute, IN: Quality Council of Indiana.

Arter, D. R. 2002. *Quality Audits for Improved Performance.* 3rd ed. Milwaukee: ASQ Quality Press.

Austin, P. R. 2000. *Encyclopedia of Cleanrooms, Bio-Cleanrooms, and Aseptic Areas.* 3rd ed. Acorn Industries.

Baker, G. 2010. "Only You Can Prevent Cross-Contamination: A Guide to Designing Cleaning Procedures for CMOs." *Contract Pharma.* Accessed 3/31/14. www.contractpharma.com/issues/2010-05/view_features/only-you-can-prevent-cross-contamination.

Bossert, J. L., editor. 2004. *The Supplier Management Handbook.* 6th ed. Milwaukee: ASQ Quality Press.

Dean, D. A., E. R. Evans, and I. H. Hall, eds. 2000. *Pharmaceutical Packaging Technology.* London and New York: Taylor & Francis.

Durivage, M. A., editor. 2014. *The Certified Pharmaceutical GMP Professional Handbook.* Milwaukee: ASQ Quality Press.

Durivage, M. A. 2014. *Practical Engineering, Process, and Reliability Statistics.* Milwaukee: ASQ Quality Press.

Durivage, M. A. 2015. *Practical Attribute and Variable Measurement Systems Analysis (MSA).* Milwaukee: ASQ Quality Press.

Durivage, M. A. 2016. *Practical Design of Experiments (DOE).* Milwaukee: ASQ Quality Press.

Ellis, J. R., and P. M. Bentz. 2007. *Modules for Basic Nursing Skills.* 7th ed. Philadelphia: Lippincott Williams & Wilkins.

Fraise, A. P., P. A. Lambert, and J.-Y. Maillard, eds. 2004. *Principles and Practice of Disinfection, Preservation and Sterilization.* Oxford, UK and Malden, MA: Blackwell.

Mehta, B. 2013. *Implementing ISO/IEC 17025:2005: A Practical Guide.* Milwaukee: ASQ Quality Press.

Pawar, H., N. Banerjee, S. Pawar, and P. Pawar. 2011. "Current Perspectives on Cleaning Validation in Pharmaceutical Industry: A Scientific and Risk Based Approach." *International Journal of Pharmaceutical and Phytopharmacological Research* 1 (August): 8–16.

Perry, A. G., and P. A. Potter. 2006. *Clinical Nursing Skills and Techniques.* 6th ed. St. Louis, MO: Elsevier Mosby.

PIC/S. 2007. P006-03 "Recommendations on Validation Master Plan Installation and Operational Qualification Non-Sterile Process Validation Cleaning Validation." September 25.

Pronovost D. 2000. *Internal Quality Auditing.* Milwaukee: ASQ Quality Press.

Rodríguez-Pérez, J. 2014. *The FDA and Worldwide Current Good Manufacturing Practices and Quality System Requirements Guidebook for Finished Pharmaceuticals.* Milwaukee: ASQ Quality Press

Russell, J.P., ed. 2013. *The ASQ Auditing Handbook: Principles, Implementation, and Use.* 4th ed. Milwaukee: ASQ Quality Press.
Russell, J.P., and T. L. Regel. 2000. *After the Quality Audit: Closing the Loop on the Audit Process.* 2nd ed. Milwaukee: ASQ Quality Press.
Shaw, A. B. 2007. "Drug Master Files." DIA webinar. March 19. Revised November 7, 2008.
Sinclair, U. 1906. *The Jungle.* New York: New American Library.
Whyte, W. 2010. *Cleanroom Technology: Fundamentals of Design, Testing and Operation.* 2nd ed. United Kingdom: John Wiley & Sons.
United States Geological Survey. 2002. *Pharmaceuticals, Hormones, and Other Organic Wastewater Contaminants in U. S. Streams, 1999–2000: A National Reconnaissance.* Lincoln, NE: University of Nebraska.

AUSTRALIAN GUIDELINES

TGA *Guidelines for Sterility Testing of Therapeutic Goods*
TGA Amended EU (EMEA/410/01) *Guideline Note for Guidance on Minimizing the Risk of Transmitting Animal Spongiform Encephalopathy Agents via Human and Veterinary Medicinal Products*

EUROPEAN AND OTHER NATIONAL GUIDELINES

Eudralex Volume 4, Parts I and II and Annexes
Final Version of Annex 15 to the *EU Guide to Good Manufacturing Practice Qualification and Validation.* July 2001.
Health Products Food Branch (Health Canada) Guide-0069, *Guidelines for Temperature Control of Drug Products During Storage and Transportation.*
Irish Medicines Board. *Guide to Control and Monitoring of Storage and Transportation Temperature Conditions for Medicinal Products and Active Substances.* March 2006.

EUROPEAN DIRECTIVES

Directive 2001/82/EC on the Community Code Relating to Veterinary Medicinal Products
Directive 2007/92/25/EEC Article 10 and Rules and Guidance for Pharmaceutical Distributors
Directive 2003/94/EC Laying Down the Principles and Guidelines of Good Manufacturing Practice in Respect of Medicinal Products for Human and Investigational Medicinal Products for Human Use as Amended
Directive 2001/83/EC on the Community Code Relating to Human Medicinal Products

EUROPEAN PHARMACOPEIA

2.2.44 Total Organic Carbon in Water for Pharmaceutical Use
2.2.46 Chromatographic Separation Techniques
2.6.7 Mycoplasmas

FDA GUIDANCES

Bioanalytical Method Validation

Biological Indicator (BI) Premarket Notification [510(k)] Submissions
Biological Product Deviation Reporting for Licensed Manufacturers of Biological Products Other than Blood and Blood Components

Changes to an Approved NDA or ANDA
Characterization and Qualification of Cell Substrates and Other Biological Starting Materials Used in the Production of Viral Vaccines for the Prevention and Treatment of Infectious Diseases
Compliance Program Guidance Manual: Active Pharmaceutical Ingredient (API) Process Inspection
Compliance Program Guidance Manual for FDA Staff: Drug Manufacturing Inspections Program
Container and Closure Integrity Testing in Lieu of Sterility Testing as a Component of the Stability Protocol for Sterile Products
Container Closure Systems for Packaging Human Drugs and Biologics
Current Good Manufacturing Practice for Combination Products for the Submission Documentation for Sterilization Process Validation in Applications for Human and Veterinary Drug Products
General Principles of Process Validation General Principles of Software Validation
Guidance for Industry Standards for Securing the Drug Supply Chain—Standardized Numerical Identification for Prescription Drug Packages
Inspections of Oral Solid Dosage Forms Pre/Post Approval Issues for Development and Validation Inspections of Sterile Drug Substance Manufacturers
Investigating Out of Specification Test Results
Investigational New Drug Applications (INDs) for Phase 1 Studies of Drugs
Medical Devices; Current Good Manufacturing Practice (CGMP) Final Rule; Quality System Regulation
Points to Consider (PTC) in the Characterization of Cell Lines Used to Produce Biologicals Preapproval Inspections
Product Recalls, Including Removals and Corrections
Quality Systems Approach to Pharmaceutical CGMP Regulations
Questions and Answers on Current Good Manufacturing Practices, Good Guidance Practices, Level 2 Guidance Records and Reports
Sterile Drug Process Inspections Program
Sterile Drug Products Produced by Aseptic Processing
SUPAC: Immediate Release Solid Oral Dosage Forms
SUPAC: Manufacturing Equipment Addendum
SUPAC: Modified Release Solid Oral Dosage Forms
SUPAC: Nonsterile Semisolid Dosage Forms
Validation of the Limulus Amebocyte Lysate Test

FDA GUIDE TO INSPECTIONS

Biotechnology

Biotechnology Inspection Guide (11/91)

Computer Issues

Computerized Systems in Drug Establishments (2/83)

Devices

Quality Systems
Bioresearch Monitoring Inspections in Vitro Diagnostics Devices Medical Device Manufacturers

Drugs

High Purity Water System (7/93)
Lyophilization of Parenterals (7/93)
Microbiological Pharmaceutical Quality Control Labs (7/93)
Pharmaceutical Quality Control Labs (7/93)
Validation of Cleaning Processes (7/93)
Dosage Form Drug Manufacturers cGMPs (10/93)
Oral Solid Dosage Forms Pre/Post Approval Issues (1/94)
Sterile Drug Substance Manufacturers (7/94)
Topical Drug Products (7/94)
Oral Solutions and Suspensions (8/94)

Miscellaneous

Foreign Medical Device Manufacturers (9/95)
Foreign Pharmaceutical Manufacturers (5/96)

HEALTH CANADA GUIDELINES

Guidelines for Temperature Control of Drug Products During Storage and Transportation Good Manufacturing Practices Guidelines

INTERNATIONAL CONFERENCE ON HARMONIZATION GUIDELINESS

Quality Guidelines

Q1A–Q1F *Stability*
Q2 *Analytical Validation*
Q3A–Q3D *Impurities*
Q4–Q4B *Pharmacopoeias*
Q5A–Q5E *Quality of Biotechnological Products*
Q6A–Q6B *Specifications*
Q7 *Good Manufacturing Practice*
Q8 *Pharmaceutical Development*
Q9 *Quality Risk Management*
Q10 *Pharmaceutical Quality System*
Q11 *Development and Manufacture of Drug Substances*
Q12 *Lifecycle Management*

Safety Guidelines

S1AS1C *Carcinogenicity Studies*
S2 *Genotoxicity Studies*
S3A–S3B *Toxicokinetics and Pharmacokinetics*
S4 *Toxicity Testing*
S5 *Reproductive Toxicology*
S6 *Biotechnological Products*
S7A–S7B *Pharmacology Studies*

S8 *Immunotoxicology Studies*
S9 *Nonclinical Evaluation for Anticancer Pharmaceuticals*
S10 *Photosafety Evaluation*
S11 *Nonclinical Safety Testing*

Efficacy Guidelines

E1 *Clinical Safety for Drugs Used in Long-Term Treatment*
E2A–E2F *Pharmacovigilance*
E3 *Clinical Study Reports*
E4 *Dose-Response Studies*
E5 *Ethnic Factors*
E6 *Good Clinical Practice*
E7 *Clinical Trials in Geriatric Population*
E8 *General Considerations for Clinical Trials*
E9 *Statistical Principles for Clinical Trials*
E10 *Choice of Control Group in Clinical Trials*
E11 *Clinical Trials in Pediatric Population*
E12 *Clinical Evaluation by Therapeutic Category*
E14 *Clinical Evaluation*
E15 *Definitions in Pharmacogenetics/Pharmacogenomics*
E16 *Qualification of Genomic Biomarkers*
E17 *Multi-Regional Clinical Trials*
E18 *Genomic Sampling Methodologies*

Multidisciplinary Guidelines

M1 *MedDRA Terminology*
M2 *Electronic Standards*
M3 *Nonclinical Safety Studies*
M4 *Common Technical Document*
M5 *Data Elements and Standards for Drug Dictionaries*
M6 *Gene Therapy*
M7 *Genotoxic Impurities*
M8 *Electronic Common Technical Document (eCTD)*

INTERNATIONAL SOCIETY FOR PHARMACEUTICAL ENGINEERING GUIDELINES

ISPE GAMP V *The Good Automated Manufacturing Practice (GAMP) Guide for Validation of Automated Systems in Pharmaceutical Manufacture*

INTERNATIONAL ORGANIZATION FOR STANDARDIZATION (ISO) DOCUMENTS

ISO 9001:2015 *Quality management systems—Requirements*
ISO 11137-1:2006 *Sterilization of health care products—Radiation—Part 1: Requirements for development, validation and routine control of a sterilization process for medical devices*
ISO 11137-2:2013 *Sterilization of health care products—Radiation—Part 2: Establishing the sterilization dose*

ISO/DIS 11137-3 *Sterilization of health care products—Radiation—Part 3: Guidance on dosimetric aspects of development, validation and routine control*
ISO 13485:2003 *Medical devices—Quality management systems—Requirements for regulatory purposes*
ISO 13485:2016 *Medical devices—Quality management systems—Requirements for regulatory purposes*
ISO 14644-1:2015 *Cleanrooms and associated controlled environments—Part 1: Classification of air cleanliness by particle concentration*
ISO 14644-2:2015 *Cleanrooms and associated controlled environments—Part 2: Monitoring to provide evidence of cleanroom performance related to air cleanliness by particle concentration*
ISO 14644-3:2005 *Cleanrooms and associated controlled environments—Part 3: Test methods*
ISO 14644-4:2001 *Cleanrooms and associated controlled environments—Part 4: Design, construction and start-up*
ISO 14644-5:2004 *Cleanrooms and associated controlled environments—Part 5: Operations*
ISO 14644-7:2004 *Cleanrooms and associated controlled environments—Part 7: Separative devices (clean air hoods, gloveboxes, isolators and mini-environments)*
ISO 14644-8:2013 *Cleanrooms and associated controlled environments—Part 8: Classification of air cleanliness by chemical concentration (ACC)*
ISO 14644-9:2012 *Cleanrooms and associated controlled environments—Part 9: Classification of surface cleanliness by particle concentration*
ISO 14644-10:2013 *Cleanrooms and associated controlled environments—Part 10: Classification of surface cleanliness by chemical concentration*
ISO 14698-1:2003 *Cleanrooms and associated controlled environments—Biocontamination control—Part 1: General principles and methods)*
ISO 14698-2:2003 *Cleanrooms and associated controlled environments—Biocontamination control—Part 2: Evaluation and interpretation of biocontamination data*
ISO 14969:2004 *Medical devices—Quality management systems—Guidance on the application of ISO 13485:2003*
ISO 17025 *General requirements for the competence of testing and calibration laboratories*
ISO 19011:2011 *Guidelines for auditing management systems*
ISO 22000:2005 *Food safety management systems—Requirements for any organization in the food chain*
ISO 22301:2012 *Societal security—Business continuity management systems—requirements*

JAPAN

Pharmaceutical Administration, Regulation, and Drug Development in Japan

PARENTERAL DRUG ASSOCIATION TECHNICAL REPORTS

PDA Technical Report No. 1 *Validation of Steam Sterilization Cycles*
PDA Technical Report No. 17 *Current Practices in the Validation of Aseptic Processing*

PHARMACEUTICAL INSPECTION CONVENTION/ CO-OPERATION SCHEME (PIC/S)

PE 008-2 *Explanatory Notes for Industry on the Preparation of a Site Master File*
PE 010-3 *PIC/S Guide to Good Practices for the Preparation of Medicinal Products in Healthcare Establishments*
PI 006-3 *Recommendations on Validation Master Plan Installation and Operational Qualification Non-Sterile Process Validation Cleaning Validation*

PI 007-4 *Recommendation on the Validation of Aseptic Processes*
PI 009-3 *Aide-Memoire on Inspection of Utilities*
PI 010-3 *Procedure for Handling Rapid Alerts and Recalls Arising from Quality Defects* PI 011-3 *Good Practices for Computerized Systems in Regulated "GXP" Environments* PI 012-3 *Recommendation on Sterility Testing*
PI 014-3 *Recommendation on Isolators Used for Aseptic Processing and Sterility Testing*

UNITED STATES CODE OF FEDERAL REGULATIONS 9 CFR

Part 101 Definitions
Part 102 Licenses for biological products
Part 103 Experimental production, distribution, and evaluation of biological products prior to licensing
Part 104 Permits for biological products
Part 105 Suspension, revocation, or termination of biological licenses or permits
Part 106 Exemption for biological products used in department programs or under department control or supervision
Part 107 Exemptions from preparation pursuant to an unsuspended and unrevoked license
Part 108 Facility requirements for licensed establishments
Part 109 Sterilization and pasteurization at licensed establishments
Part 112 Packaging and labeling
Part 113 Standard requirements
Part 114 Production requirements for biological products
Part 115 Inspections
Part 116 Records and reports
Part 117 Animals at licensed establishments
Part 118 Detention; seizure and condemnation
Part 121 Possession, use, and transfer of select agents and toxins
Part 122 Organisms and vectors

UNITED STATES CODE OF FEDERAL REGULATIONS 21 CFR

Part 1 General enforcement regulations
Part 2 General administrative rulings and decisions
Part 3 Product jurisdiction
Part 4 Regulation of combination products
Part 7 Enforcement policy
Part 10 Administrative practices and procedures
Part 11 Electronic records; electronic signatures
Part 26 Mutual recognition of pharmaceutical good manufacturing practice reports, medical device quality system audit reports, and certain medical device product evaluation reports: United States and the European Community
Part 50 Protection of human subjects
Part 56 Institutional review boards
Part 58 Good laboratory practice for nonclinical laboratory studies
Part 81 General specifications and general restrictions for provisional color additives for use in foods, drugs, and cosmetics
Part 111 Current good manufacturing practice in manufacturing, packaging, labeling, or holding operations for dietary supplements

Part 120 Hazard Analysis and Critical Control Point (HACCP) systems Part 190 Dietary supplements
Part 200 General
Part 201 Labeling
Part 202 Prescription drug advertising
Part 203 Prescription drug marketing
Part 205 Guidelines for state licensing of wholesale prescription drug distributors
Part 206 Imprinting of solid oral dosage form drug products for human use
Part 207 Registration of producers of drugs and listing of drugs in commercial distribution
Part 208 Medication guides for prescription products
Part 209 Requirements for authorized dispensers and pharmacies to distribute a side effects statement
Part 210 Current good manufacturing practice in manufacturing, processing, packing, or holding of drugs: general
Part 211 Current good manufacturing practice for finished pharmaceuticals
Part 212 Current good manufacturing practice for positron emission tomography drugs
Part 310 New drugs
Part 312 Investigational new drug application
Part 314 Applications for FDA approval to market a new drug
Part 315 Diagnostic radiopharmaceuticals
Part 316 Orphan drugs
Part 320 Bioavailability and bioequivalence requirements
Part 330 Over-the-counter (OTC) human drugs which are generally recognized as safe and effective and not misbranded
Part 361 Prescription drugs for human use generally recognized as safe and effective and not misbranded: drugs used in research
Part 369 Interpretative statements re: warnings on drugs and devices for over-the-counter sale
Part 600 Biological products: General
Part 601 Licensing
Part 606 Current good manufacturing practice for blood and blood components
Part 607 Establishment registration and product listing for manufacturers of human blood and blood products
Part 610 General biological products standards
Part 630 Requirements for blood and blood components intended for transfusion or for further manufacturing use
Part 801 Labeling
Part 803 Medical device reporting
Part 806 Medical devices; reports of corrections and removals
Part 807 Establishment registration and device listing for manufacturers and initial importers of devices
Part 808 Exemptions from Federal preemption of State and local medical device requirements
Part 809 In vitro diagnostic products for human use
Part 810 Medical device recall authority
Part 812 Investigational device exemptions
Part 814 Premarket approval of medical devices
Part 820 Quality system regulation
Part 821 Medical device tracking requirements
Part 822 Postmarket surveillance
Part 830 Unique device identification

Part 860 Medical device classification procedures
Part 1270 Human tissue intended for transplantation
Part 1271 Human cells, tissues, and cellular and tissue-based products
Part 1301 Registration of manufacturers, distributors, and dispensers of controlled substances
Part 1302 Labeling and packaging requirements for controlled substances Part 1308 Schedules of controlled substances
Part 1308.03 Administrative controlled substances code number
Part 1308.11 List of Schedule I drugs
Part 1308.12 List of Schedule II drugs
Part 1308.13 List of Schedule III drugs
Part 1308.14 List of Schedule IV drugs
Part 1308.15 List of Schedule V drugs

UNITED STATES CODE OF FEDERAL REGULATIONS 40 CFR

Part 141 National Primary Drinking Water Regulations

UNITED STATES PHARMACOPEIA

US Pharmacopeia General chapters related to "Pharmaceutical analytical methods and practices"
US Pharmacopeia General Notices and Requirements—"Preservation, Packaging, Storage, and Labeling"
US Pharmacopeia <797> Pharmaceutical Compounding—Sterile Preparations
US Pharmacopeia <1010> Analytical Data—Interpretation and Treatment
US Pharmacopeia <1072> Disinfectants and Antiseptics
US Pharmacopeia <1079> Good Storage and Shipping Practices
US Pharmacopeia <1118> Monitoring Devices—Time, Temperature, and Humidity
US Pharmacopeia <1150> Pharmaceutical Stability
US Pharmacopeia <1231> Water for Pharmaceutical Purposes

WORLD HEALTH ORGANIZATION

WHO "Expert Committee on Specifications for Pharmaceutical Preparations: Thirty- Seventh Report"
WHO "Good Distribution Practices (GDP) for Pharmaceutical Products"
WHO "Guidelines for Sampling of Pharmaceutical Products and Related Materials"
WHO "Quality Assurance of Pharmaceuticals: A Compendium of Guidelines and Related Materials—Volume 1"
WHO "Quality Assurance of Pharmaceuticals: A Compendium of Guidelines and Related Materials—Volume 2: Good Manufacturing Practices and Inspection"
WHO Good Manufacturing Practices: "Starting Materials"

Index

A

B

C

D

G

H

I

J

K

L

Q

T

U

V

W

Y

www.ingramcontent.com/pod-product-compliance
Ingram Content Group UK Ltd.
Pitfield, Milton Keynes, MK11 3LW, UK
UKHW050950290926
15470UKWH00005B/1737

9 781636 941493